The ARRL Advanced Class License Manual

Edited By: Larry D. Wolfgang, WR1B

Production Staff:

Dan Wolfgang

David Pingree, N1NAS

Steffie Nelson, KA1IFB

Jodi Morin, KA1JPA

Sue Fagan

Michelle Bloom, WB1ENT

Published By:
The American Radio Relay League, Inc.
Newington, CT 06111-1494

TABLE OF CONTENTS

FOREWORD

For more than 60 years, the ARRL has produced the most widely used, most complete license preparation materials available to prospective radio amateurs. Since 1985, after the FCC began issuing question pools for each license exam, the League's series of License Manuals has included individual study guides for each license class. We continue to revise and improve our study materials to meet the changing needs of the amateur community. The ARRL now offers video study materials and computerized exam review and drill software in addition to printed license manuals.

This fourth edition of *The ARRL Advanced Class License Manual* represents the next generation of improvements to our study guides. This completely revised and expanded book is designed to simplify the process of studying for your Advanced class license exam.

Earning the Advanced class license is a significant expression of your commitment to Amateur Radio. This step up the license-class ladder earns you some small frequency segments reserved for the exclusive use of Amateur Extra and Advanced class licensees. Your knowledge of electronics technology will grow considerably, however, as will your understanding and familiarity with some of the more "exotic" communications modes used by amateurs. You will learn about Thevenin's Theorem, resonant circuits and how mixer circuits work, for example. You will learn about slow-scan television and fax communications, as well as sporadic-E and auroral propagation modes.

Here you will find all of the material you need to pass the Element 4A written exam! There is study material to explain every question in the Advanced class question pool. The text in Chapter 1 covers all the FCC Rules questions in this question pool. Appropriate sections of the current FCC Rules, Part 97, are quoted where the actual rules text will best help you understand the material. Chapters 2 through 9 explain the electronics theory and amateur operating practices that you'll need to know to pass your exam.

Chapter 10 is the complete Advanced class question pool as released by the Volunteer-Examiner Coordinators' Question Pool Committee in December 1994 for use beginning July 1, 1995. It has been laid out so you can drill yourself on the actual examination questions, to check your understanding of the study material. The answer key is included on the same page as each question, for your convenience. If you do well answering the questions from the pool, you can be confident when you go to take the actual exam.

We aren't satisfied just to help you pass your license exam, though. We want you to be an active ham, enjoying lots of on-the-air operating time. The ARRL provides additional technical and operating material in the many books and aids that make up our "Radio Amateur's Library." Contact the Publication Sales Office at ARRL Headquarters to request the latest publications catalog or to place an order.

This fourth edition of *The ARRL Advanced Class License Manual* is not just the product of the many ARRL staff members who have helped bring the book to you. Readers of the earlier editions sent comments and suggestions; you can, too. First, use this book to prepare for your exam. Then, write your suggestions (or any corrections you think need to be made) on the Feedback Form at the back of the book, and send the form to us. Your comments and suggestions are important to us. Thanks, and good luck!

David Sumner, K1ZZ
Executive Vice President, ARRL
Newington, Connecticut

HOW TO USE THIS BOOK

To earn an Advanced class Amateur Radio license, you will have to know some electronics theory and the rules and regulations governing the Amateur Service, as contained in Part 97 of the FCC Rules. You'll also have to be able to send and receive the international Morse code at a rate of 13 wpm. This book provides a brief description of the Amateur Service, and the Advanced class license in particular. Applicants for the Advanced class license must pass a 50-question written exam drawn from the Element 4A question pool, as released by the Volunteer-Examiner Coordinators' Question Pool Committee (QPC), and the Element 1B (13 wpm) Morse code exam.

Please note that Novice licensees wishing to upgrade directly to General class must pass both the Element 3A — Technician — and the Element 3B — General — exams. (If you don't already have an Amateur Radio license, you will also have to pass the Element 2 — Novice — exam.) Chapter 10 of this book contains the complete Element 4A question pool and multiple-choice answers released by the VEC QPC in December 1994 for use starting July 1, 1995.

At the beginning of each chapter, you will find a list of **Key Words** that appear in that chapter, along with a simple definition for each word or phrase. As you read the text, you will find these words printed in **boldface type** the first time they appear. You may want to refer to the **Key Words** list at the beginning of the chapter when you come to a **boldface** word, to review the definition. After you have studied the material in that chapter, you may also want to review those definitions. Most of the key words are terms that will appear on exam questions, so a quick review of the definitions just before you go to take the test may also be helpful. Appendix C is a glossary of all the key words used in the book. They are arranged in alphabetical order for your convenience as you review before the exam.

The Question Pool

As you study the material, you will be instructed to turn to sections of the questions in Chapter 10. Be sure to use these questions to review your understanding of the material at the suggested times. This breaks the material into bite-sized pieces, and makes it easier for you to learn. Do not try to memorize all of the questions and answers. With over 500 questions, that will be nearly impossible! Instead, by using the questions for review, you will be familiar with them when you take the test, but you will also understand the electronics theory or Rules point behind the questions.

Most people learn more when they are actively involved in the learning process. Study the text, rather than passively reading it. Use the questions to review your progress rather than just reading the question and looking at the correct answer letter. Fold the answer-key column under at the dashed line down the page before you begin answering questions on that page. Look at the answer key only to verify your answer. This will help you check your progress and ensure you know the answer to each question. If you missed one, go back to the supporting text and review that material. Page numbers are included with each answer letter. These indicate where to turn in the book to find the explanatory text for that question. You may have to read more than one page for the complete explanation. Paper clips make excellent place markers to help you find your spot in the text and question pool.

Other ARRL Study Materials

If you need to learn Morse code, ARRL offers a set of cassette tapes to teach Morse code: *Your Introduction to Morse Code*. The two 90-minute cassettes in this package introduce each of the required characters and drill you on that character. Then the character is used in words and text before proceeding to the next character. After all characters have been introduced, there is plenty of practice at 5 wpm to help you prepare for the Element 1A (5 wpm) exam. ARRL also offers several cassette tape sets to help you increase your speed: *Increasing Your Code Speed*; *5 to 10 WPM*, *10 to 15 WPM* and *15 to 22 WPM*. Each set includes two 90-minute cassettes with a variety of practice at gradually increasing speeds over the range for that set. The Morse code on these tapes is sent using 18-wpm characters, with extra space between characters to slow the overall code speed, for code up to 18 wpm. This same technique is used on ARRL/VEC Morse code exams for 5 and 13 wpm.

For those who prefer a computer program to learn and practice Morse code, ARRL offers the excellent GGTE *Morse Tutor* program for IBM PC and compatible computers.

Even with the tapes or computer program, you'll want to tune in the code-practice sessions transmitted by W1AW, the ARRL Headquarters station. When you are able to copy the code at 13 wpm, you may find it helpful to listen to code at 15 wpm at the beginning of your practice session, and then decrease the speed to 13 wpm. The W1AW fast-code-practice sessions start at 35 wpm and decrease to 30, 25 and 20 wpm before getting to 15 and 13-wpm practice. You can tune in to the practice session a little late and still get the practice you need. You might also want to try copying the W1AW CW bulletins, which are sent at 18 wpm. (There is a W1AW code-practice schedule on page 7 in the Introduction.) For more information about W1AW or how to order any ARRL publication, software or set of code tapes, write to ARRL Headquarters, 225 Main St, Newington, CT 06111-1494.

When to Use This Book

Amateur-license-exam questions are written by a Question Pool Committee (QPC). The QPC consists of representatives from the various Volunteer Examiner Coordinators (VECs). The QPC has established the following schedule for implementing new question pools:

Pool	Effective Date
Novice (Element 2)	July 1, 1997*
Technician (Element 3A)	July 1, 1997*
General (Element 3B)	July 1, 1998*
Advanced (Element 4A)	July 1, 1999*
Amateur Extra (Element 4B)	July 1, 1996

*Tentative date

When the schedules for each new question pool is confirmed, the information will be published in *QST*. New editions of ARRL License Manuals that cover the new question pools will be available before the effective dates of the new pools.

Code Examination Credit for Handicapped Persons

Effective February 14, 1991, FCC adopted new Rules pertaining to Morse code examinations for handicapped persons. Applicants must have a current FCC amateur license that required a code test. If the license has expired, it must be within the grace period for renewal. Valid licenses include:

- Novice
- Technician Plus
- Technician and a Certificate of Successful Completion of Examination (CSCE) for passing Element 1A or 1B
- Technician issued before February 14, 1991
- General
- Advanced

Effective March 1, 1994, all applicants must use a new Form 610 (dated November 1993 or later). The back of the new form includes the "PHYSICIAN'S CERTIFICATION OF DISABILITY" section, which must be completed and signed by a person licensed to practice in a place where the Amateur Service is regulated by the FCC as a doctor of medicine (MD) or doctor of osteopathy (DO). After the physician signs the Form 610, the applicant must also sign in the same section, granting the physician permission to disclose medical information relevant to the application. The back of the Form 610 also includes a "NOTICE TO PHYSICIAN CERTIFYING TO A DISABILITY," which describes the Morse code exam procedures, including exam accommodations that can be made for handicapped persons.

To obtain a new Form 610, send a business-sized, self-addressed, stamped envelope to: Federal Communications Commission, Forms Distribution Center, 2803 52nd Ave, Hyattsville, MD 20781 (specify "Form 610" on the envelope), or: Form 610, ARRL, 225 Main St, Newington, CT 06111-1494.

INTRODUCTION

THE ADVANCED CLASS LICENSE

An Advanced class Amateur Radio license is a worthy goal. Whether you now hold a Novice, Technician, Technician Plus or General class license, or are as yet unlicensed, you will find the extra operating privileges available to an Advanced class licensee to be worth the time spent learning about your hobby. After passing the FCC Element 4A exam (and written Elements 2, 3A and 3B along with the Morse code Element 1B or 1C if you haven't already passed those exam elements), you will be able to operate on every frequency band assigned to the Amateur Radio Service. Segments of the 80, 40, 20 and 15-meter phone bands are reserved exclusively for Advanced and Amateur Extra class operators. So if you find the General-class portions of the bands getting too crowded, just move down to these less-used segments.

To earn those additional privileges, you'll have to demonstrate that you know Advanced-class electronics theory, operating practices, and FCC rules and regulations. This book was carefully designed to teach you everything you need to know to pass the Element 4A exam, and earn the Advanced class Amateur Radio license.

Chapter 1 of this book, **Commission's Rules**, covers those sections of the FCC Part 97 Rules that you will be tested on for your Advanced class license. We recommend that you also obtain a copy of *The FCC Rule Book*, published by ARRL. *The FCC Rule Book* includes a complete copy of Part 97, along with detailed explanations for all the rules governing Amateur Radio.

Once you make the commitment to study and learn what it takes to pass the exam, you will be able to accomplish your goal. It often takes more than one attempt to pass the Advanced class license exam, but many amateurs do pass on their first try. The key is that you must make the commitment and be willing to study.

If you haven't passed the 13-wpm Morse code exam (Element 1B), there are many good Morse-code training techniques. These include the ARRL code tapes, W1AW code practice and even some computer programs such as the *Morse Tutor* program for the IBM PC and compatible computers, which is available from ARRL. Of course the most enjoyable way to increase your code speed is through on-the-air operating!

IF YOU'RE A NEWCOMER TO AMATEUR RADIO

Earning an Amateur Radio license, at whatever level, is a special achievement. The half a million or so people in the US who call themselves Amateur Radio operators, or hams, are part of a global fraternity. Radio amateurs serve the public as a voluntary, noncommercial, communication service. This is especially true during natural disasters or other emergencies. Hams have made many important contributions to the field of electronics and communications: this tradition continues today. Amateur Radio experimentation is yet another reason many people become part of this self-disciplined group of trained operators, technicians and electronics experts — an asset to any country. Hams pursue their hobby purely for personal enrichment in technical and operating skills, without consideration of any type of payment except the personal satisfaction they feel from a job well done!

Figure 1 — A guest operator at ARRL's Hiram Percy Maxim Memorial Station, W1AW, enjoys a chat (QSO) with another ham.

Radio signals do not know territorial boundaries, so hams have a unique ability to enhance international goodwill. Hams become ambassadors of their country every time they put their stations on the air.

Amateur Radio has been around since before World War I, and hams have always been at the forefront of technology. Today, hams relay signals through their own satellites in the OSCAR (Orbiting Satellite Carrying Amateur Radio) series, bounce signals off the moon, relay messages automatically through computerized radio networks and use any number of other "exotic" communications techniques. Amateurs talk from hand-held transceivers through mountaintop repeater stations that can relay their signals to transceivers in other hams' cars or homes. Hams send their own pictures by television, talk with other hams around the world by voice or, keeping alive a distinctive traditional skill, tap out messages in Morse code. When emergencies arise, radio amateurs are on the spot to relay information to and from disaster-stricken areas that have lost normal lines of communication.

The US government, through the Federal Communications Commission (FCC), grants all US Amateur Radio licenses. This licensing procedure ensures operating skill and electronics know-how. Without this skill, radio operators might unknowingly cause interference to other services using the radio spectrum because of improperly adjusted equipment or neglected regulations.

Who Can Be a Ham?

The FCC doesn't care how old you are or whether you're a US citizen. If you pass the examination, the Commission will issue you an amateur license. Any person (except the agent of a foreign government) may take the exam and, if successful, receive an amateur license. It's important to understand that if a citizen of a foreign country receives an amateur license in this manner, he or she is a US Amateur Radio operator. (This should not be confused with a reciprocal permit for alien amateur licensee, which allows visitors from certain countries who hold valid amateur licenses in their homelands to operate their own stations in the US without having to take an FCC exam.)

Licensing Structure

By examining Table 1, you'll see that there are six amateur license classes. Each class has its own requirements and privileges. The FCC requires proof of your ability to operate an amateur station properly. The required knowledge is in line with the

Figure 2 — Many active Amateur Radio operators enjoy collecting colorful QSL cards, many of them from countries around the world.

Table 1
Amateur Operator Licenses†

Class	Code Test	Written Examination	Privileges
Novice	5 wpm (Element 1A)	Novice theory and regulations (Element 2)	Telegraphy on 3675-3725, 7100-7150 and 21,100-21,200 kHz with 200 W PEP output maximum; telegraphy, RTTY and data on 28.100-28,300 kHz and telegraphy and SSB voice on 28,300-28,500 kHz with 200 W PEP max; all amateur modes authorized on 222-225 MHz, 25 W PEP max; all amateur modes authorized on 1270-1295 MHz, 5 W PEP max.
Technician	None	Novice theory and regulations; Technician-level theory and regulations. (Elements 2, 3A)*	All amateur privileges above 50.0 MHz.
Technician Plus	5 wpm (Element 1A)	Novice theory and regulations; Technician-level theory and regulations. (Elements 2, 3A)*	All Novice HF privileges in addition to all Technician privileges.
General	13 wpm (Element 1B)	Novice theory and regulations; Technician and General theory and regulations. (Elements 2, 3A and 3B)	All amateur privileges except those reserved for Advanced and Amateur Extra class; see Table 2.
Advanced	13 wpm (Element 1B)	All lower exam elements, plus Advanced theory. (Elements 2, 3A, 3B and 4A)	All amateur privileges except those reserved for Amateur Extra class; see Table 2.
Amateur Extra	20 wpm (Element 1C)	All lower exam elements plus Extra-class theory (Elements 2, 3A, 3B, 4A and 4B)	All amateur privileges.

†A licensed radio amateur will be required to pass only those elements that are not included in the examination for the amateur license currently held.
*If you hold a valid Technician class license issued before March 21, 1987, you also have credit for Element 3B. You must be able to prove your Technician license was issued before March 21, 1987 to claim this credit.

Table 2
Amateur
Operating
Privileges

US AMATEUR BANDS

December 20, 1994

160 METERS

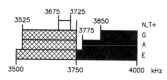

E,A,G

1800   1900   2000   kHz

Amateur stations operating at 1900–2000 kHz must not cause harmful interference to the radiolocation service and are afforded no protection from radiolocation operations.

80 METERS

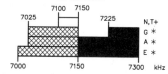

3675   3725
3525   3850
        3775
                N,T+
                G
                A
                E

3500   3750   4000   kHz

5167.5 kHz (SSB only): Alaska emergency use only.

40 METERS

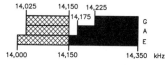

7100   7150
7025   7225
                N,T+
                G *
                A *
                E *

7000   7150   7300   kHz

* Phone operation is allowed on 7075–7100 kHz in Puerto Rico, US Virgin islands and areas of the Caribbean south of 20 degrees north latitude; and in Hawaii and areas near ITU Region 3, including Alaska.

30 METERS

E,A,G

10,100   10,150   kHz

Maximum power on 30 meters is 200 watts PEP output. Amateurs must avoid interference to the fixed service outside the US.

20 METERS

14,025   14,150   14,225
          14,175
                        G
                        A
                        E

14,000   14,150   14,350   kHz

17 METERS

E,A,G

18,068   18,110   18,168   kHz

15 METERS

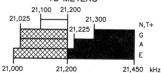

21,100   21,200
21,025   21,300
          21,225
                N,T+
                G
                A
                E

21,000   21,200   21,450   kHz

12 METERS

E,A,G

24,890   24,930   24,990   kHz

10 METERS

28,100   28,500
                N,T+
                E,A,G

28,000   28,300   29,700   kHz

Novices and Technicians are limited to 200 watts PEP output on 10 meters.

6 METERS

50.1
                E,A,G,T+,T
50.0   54.0   MHz

2 METERS

144.1
                E,A,G,T+,T
144.0   148.0   MHz

1.25 METERS

                E,A,G,T+,T,N
222.0   225.0   MHz

Novices are limited to 25 watts PEP output from 222 to 225 MHz.

70 CENTIMETERS **

                E,A,G,T+,T
420.0   450.0   MHz

33 CENTIMETERS **

                E,A,G,T+,T
902.0   928.0   MHz

23 CENTIMETERS **

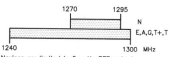

1270   1295
                N
                E,A,G,T+,T
1240   1300   MHz

Novices are limited to 5 watts PEP output from 1270 to 1295 MHz.

US AMATEUR POWER LIMITS

At all times, transmitter power should be kept down to that necessary to carry out the desired communications. Power is rated in watts PEP output. Unless otherwise stated, the maximum power output is 1500 W. Power for all license classes is limited to 200 W in the 10,100–10,150 kHz band and in all Novice subbands below 28,100 kHz. Novices and Technicians are restricted to 200 W in the 28,100–28,500 kHz subbands. In addition, Novices are restricted to 25 W in the 222–225 MHz band and 5 W in the 1270–1295 MHz subband.

Operators with Technician class licenses and above may operate on all bands above 50 MHz. For more detailed information see The FCC Rule Book.

--- KEY ---

= CW, RTTY and data

= CW, RTTY, data, MCW, test, phone and image

= CW, phone and image

= CW and SSB

= CW, RTTY, data, phone, and image

= CW only

E = EXTRA CLASS
A = ADVANCED
G = GENERAL
T+ = TECHNICIAN PLUS
T = TECHNICIAN
N = NOVICE

** Geographical and power restrictions apply to these bands. See The FCC Rule Book for more information about your area.

Above 23 Centimeters:

All licensees except Novices are authorized all modes on the following frequencies:
2300–2310 MHz
2390–2450 MHz
3300–3500 MHz
5650–5925 MHz
10.0–10.5 GHz
24.0–24.25 GHz
47.0–47.2 GHz
75.5–81.0 GHz
119.98–120.02 GHz
142–149 GHz
241–250 GHz
All above 300 GHz

For band plans and sharing arrangements, see The ARRL Operating Manual or The FCC Rule Book.

privileges of the license you hold. Higher license classes require more knowledge — and offer greater operating privileges. So as you upgrade your license class, you must pass more challenging written examinations. The specific operating privileges for Advanced class licensees are listed in Table 2.

In addition, you must demonstrate an ability to receive international Morse code at 5 wpm for Novice, 13 wpm for General and Advanced, and 20 wpm for Amateur Extra. It's important to stress that although you may intend to use voice rather than code, this doesn't excuse you from the code test. By international treaty, knowing the international Morse code is a basic requirement for operating on any amateur band below 30 MHz.

The FCC allows the Volunteer Examiners to use a variety of procedures to accommodate applicants with various disabilities. In addition, a handicapped individual may be able to receive credit for passing the 13 or 20 wpm Morse code exams if they have passed the 5 wpm exam, and a physician completes the section on the back of the Form 610, certifying that the individual has a disability that prevents them from passing a higher-speed code exam. For more information, see the "Code Examination Credit for Handicapped Persons" box before this **Introduction**. If you believe you qualify for special examination procedures or for the code examination credit, contact the ARRL/VEC Office, 225 Main Street, Newington, CT 06111-1494 (203-666-1541) or the VEC that will coordinate the exam session you plan to attend. Ask for more information about the special exam procedures their Volunteer Examiners will use to administer the code exam to you.

Learning Morse code is a matter of practice. Instructions on learning the code, how to handle a telegraph key, and so on, can be found in *Now You're Talking!*, published by ARRL. In addition, *Your Introduction to Morse Code*, ARRL's package to teach Morse code, includes two cassettes for beginners and a booklet that will help you learn the code. *Increasing Your Code Speed* is a series of additional cassettes for code practice at speeds of 5 to 10, 10 to 15 and 15 to 22 wpm, available from the American Radio Relay League, Newington, CT 06111.

Code Practice

Besides listening to code tapes, some on-the-air operating experience will be a great help in building your code speed. When you are in the middle of a contact via Amateur Radio, and have to copy the code the other station is sending to continue the

Figure 3 — As you collect QSL cards, you may want to work toward the ARRL's prestigious 5-Band Worked All States award, earned by collecting QSL cards from each of the 50 United States on five different amateur bands.

Table 3

W1AW schedule

Pacific	Mtn	Cent	East	Sun	Mon	Tue	Wed	Thu	Fri	Sat
6 am	7 am	8 am	9 am			Fast Code	Slow Code	Fast Code	Slow Code	
7 am	8 am	9 am	10 am	Code Bulletin						
8 am	9 am	10 am	11 am	Teleprinter Bulletin						
9 am	10 am	11 am	noon							
10 am	11 am	noon	1 pm	**Visiting Operator Time**						
11 am	noon	1 pm	2 pm							
noon	1 pm	2 pm	3 pm							
1 pm	2 pm	3 pm	4 pm	Slow Code	Fast Code	Slow Code	Fast Code	Slow Code	Fast Code	Slow Code
2 pm	3 pm	4 pm	5 pm	Code Bulletin						
3 pm	4 pm	5 pm	6 pm	Teleprinter Bulletin						
4 pm	5 pm	6 pm	7 pm	Fast Code	Slow Code	Fast Code	Slow Code	Fast Code	Slow Code	Fast Code
5 pm	6 pm	7 pm	8 pm	Code Bulletin						
6 pm	7 pm	8 pm	9 pm	Teleprinter Bulletin						
6⁴⁵ pm	7⁴⁵ pm	8⁴⁵ pm	9⁴⁵ pm	Voice Bulletin						
7 pm	8 pm	9 pm	10 pm	Slow Code	Fast Code	Slow Code	Fast Code	Slow Code	Fast Code	Slow Code
8 pm	9 pm	10 pm	11 pm	Code Bulletin						
9 pm	10 pm	11 pm	Mdnte	Teleprinter Bulletin						
9⁴⁵ pm	10⁴⁵ pm	11⁴⁵ pm	12⁴⁵ am	Voice Bulletin						

• MORSE CODE TRANSMISSIONS:

Frequencies are 1.818, 3.5815, 7.0475, 14.0475, 18.0975, 21.0675, 28.0675 and 147.555 MHz.
Slow Code = practice sent at 5, 7½, 10, 13 and 15 WPM.
Fast Code = practice sent at 35, 30, 25, 20, 15, 13 and 10 WPM.
Code practice text is from the pages of *QST*. The source is given at the beginning of each practice session and alternate speeds within each session. For example, "Text is from July 1992 *QST*, pages 9 and 81," indicates that the plain text is from the article on page 9 and mixed number/letter groups are from page 81.
Code bulletins are sent at 18 WPM.

• TELEPRINTER TRANSMISSIONS:

Frequencies are 3.625, 7.095, 14.095, 18.1025, 21.095, 28.095 and 147.555 MHz.
Bulletins are sent at 45.45-baud Baudot and 100-baud AMTOR, FEC Mode B. 110-baud ASCII will be sent only as time allows.
On Tuesdays and Saturdays at 6:30 PM Eastern time, Keplerian elements for many amateur satellites are sent on the regular teleprinter frequencies.

• VOICE TRANSMISSIONS:

Frequencies are 1.855, 3.99, 7.29, 14.29, 18.16, 21.39, 28.59 and 147.555 MHz.

• MISCELLANEA:

On Fridays, UTC, a DX bulletin replaces the regular bulletins.
W1AW is open to visitors during normal operating hours: from 1 PM until 1 AM on Mondays, 9 AM until 1 AM Tuesday through Friday, and from 3:30 PM to 1 AM on Saturdays and Sundays. FCC licensed amateurs may operate the station from 1-4 PM Monday through Friday. Be sure to bring your current FCC amateur license or a photocopy. Special arrangements must be made for weekend operation; please call or write at least a week in advance.
In a communications emergency, monitor W1AW for special bulletins as follows: voice on the hour, teleprinter at 15 minutes past the hour, and CW on the half hour.
Headquarters and W1AW are closed on New Year's Day, President's Day, Good Friday, Memorial Day, Independence Day, Labor Day, Thanksgiving and the following Friday, and Christmas Day. On the first Thursday of September, Headquarters and W1AW will be closed during the afternoon.

conversation, your copying ability will improve quickly!

ARRL's Hiram Percy Maxim Memorial Station, W1AW, transmits code practice and information bulletins of interest to all amateurs. These code-practice sessions and Morse code bulletins provide an excellent opportunity for code practice. Table 3 is an abbreviated W1AW operating schedule. When we change from Standard Time to Daylight Saving Time, the same local times are used.

Station Call Signs

Many years ago, by international agreement, the nations of the world decided to allocate certain call-sign prefixes to each country. This means that if you hear a radio station call sign beginning with W or K, for example, you know the station is licensed by the United States. A call sign beginning with the letter G is licensed by Great Britain, and a call sign beginning with VE is from Canada. Table 4 is the International Telecommunication Union (ITU) International Call Sign Allocation Table, which will help you determine what country a certain call sign is from. *The ARRL DXCC Countries List* is an operating aid no ham who is active on the HF

Table 4
Allocation of International Call Signs

Call Sign Series	Allocated to	Call Sign Series	Allocated to
AAA-ALZ	United States of America	CPA-CPZ	Bolivia
		CQA-CUZ	Portugal
AMA-AOZ	Spain	CVA-CXZ	Uruguay
APA-ASZ	Pakistan	CYA-CZZ	Canada
ATA-AWZ	India	C2A-C2Z	Nauru
AXA-AXZ	Australia	C3A-C3Z	Andorra
AYA-AZZ	Argentina	C4A-C4Z	Cyprus
A2A-A2Z	Botswana	C5A-C5Z	Gambia
A3A-A3Z	Tonga	C6A-C6Z	Bahamas
A4A-A4Z	Oman	* C7A-C7Z	World Meteorological Organization
A5A-A5Z	Bhutan		
A6A-A6Z	United Arab Emirates	C8A-C9Z	Mozambique
A7A-A7Z	Qatar	DAA-DRZ	Germany
A8A-A8Z	Liberia	DSA-DTZ	South Korea
A9A-A9Z	Bahrain	DUA-DZZ	Philippines
BAA-BZZ	China	D2A-D3Z	Angola
CAA-CEZ	Chile	D4A-D4Z	Cape Verde
CFA-CKZ	Canada	D5A-D5Z	Liberia
CLA-CMZ	Cuba	D6A-D6Z	Comoros
CNA-CNZ	Morocco	D7A-D9Z	South Korea
COA-COZ	Cuba	EAA-EHZ	Spain

Table 4 (continued)
Allocation of International Call Signs

Call Sign Series	Allocated to	Call Sign Series	Allocated to
EIA-EJZ	Ireland	IAA-IZZ	Italy
† EKA-EKZ	Armenia	JAA-JSZ	Japan
ELA-ELZ	Liberia	JTA-JVZ	Mongolia
† EMA-EOZ	Ukraine	JWA-JXZ	Norway
EPA-EQZ	Iran	JYA-JYZ	Jordan
† ERA-ERZ	Moldova	JZA-JZZ	Indonesia
† ESA-ESZ	Estonia	J2A-J2Z	Djibouti
ETA-ETZ	Ethiopia	J3A-J3Z	Grenada
EUA-EWZ	Belarus	J4A-J4Z	Greece
EXA-EXZ	Russian Federation	J5A-J5Z	Guinea-Bissau
† EYA-EYZ	Tajikistan	J6A-J6Z	Saint Lucia
† EZA-EZZ	Turkmenistan	J7A-J7Z	Dominica
† E2A-E2Z	Thailand	† J8A-J8Z	St. Vincent and the Grenadines
FAA-FZZ	France		
GAA-GZZ	United Kingdom of Great Britain and Northern Ireland	KAA-KZZ	United States of America
		LAA-LNZ	Norway
HAA-HAZ	Hungary	LOA-LWZ	Argentina
HBA-HBZ	Switzerland	LXA-LXZ	Luxembourg
HCA-HDZ	Ecuador	† LYA-LYZ	Lithuania
HEA-HEZ	Switzerland	LZA-LZZ	Bulgaria
HFA-HFZ	Poland	L2A-L9Z	Argentina
HGA-HGZ	Hungary	MAA-MZZ	United Kingdom of Great Britain and Northern Ireland
HHA-HHZ	Haiti		
HIA-HIZ	Dominican Republic		
HJA-HKZ	Colombia	NAA-NZZ	United States of America
HLA-HLZ	South Korea		
HMA-HMZ	North Korea	OAA-OCZ	Peru
HNA-HNZ	Iraq	ODA-ODZ	Lebanon
HOA-HPZ	Panama	OEA-OEZ	Austria
HQA-HRZ	Honduras	OFA-OJZ	Finland
HSA-HSZ	Thailand	OKA-OLZ	Czech Republic
HTA-HTZ	Nicaragua	OMA-OMZ	Slovak Republic
HUA-HUZ	El Salvador	ONA-OTZ	Belgium
HVA-HVZ	Vatican City	OUA-OZZ	Denmark
HWA-HYZ	France	PAA-PIZ	Netherlands
HZA-HZZ	Saudi Arabia	PJA-PJZ	Netherlands Antilles
H2A-H2Z	Cyprus	PKA-POZ	Indonesia
H3A-H3Z	Panama	PPA-PYZ	Brazil
H4A-H4Z	Solomon Islands	PZA-PZZ	Suriname
H6A-H7Z	Nicaragua	P2A-P2Z	Papua New Guinea
H8A-H9Z	Panama	P3A-P3Z	Cyprus

Table 4 (continued)
Allocation of International Call Signs

Call Sign Series	Allocated to	Call Sign Series	Allocated to
† P4A-P4Z	Aruba	† T9A-T9Z	Bosnia and Herzegovina
P5A-P9Z	North Korea	UAA-UIZ	Russian Federation
RAA-RZZ	Russian Federation	† UJA-UMZ	Uzbekistan
SAA-SMZ	Sweden	UNA-UQZ	Russian Federation
SNA-SRZ	Poland	URA-UTZ	Ukraine
• SSA-SSM	Egypt	† UUA-UZZ	Ukraine
• SSN-SSZ	Sudan	VAA-VGZ	Canada
STA-STZ	Sudan	VHA-VNZ	Australia
SUA-SUZ	Egypt	VOA-VOZ	Canada
SVA-SZZ	Greece	VPA-VSZ	United Kingdom of Great Britain and Northern Ireland
S2A-S3Z	Bangladesh		
† S5A-S5Z	Slovenia		
S6A-S6Z	Singapore	VTA-VWZ	India
S7A-S7Z	Seychelles	VXA-VYZ	Canada
S9A-S9Z	Sao Tome and Principe	VZA-VZZ	Australia
		† V2A-V2Z	Antigua and Barbuda
TAA-TCZ	Turkey	† V3A-V3Z	Belize
TDA-TDZ	Guatemala	† V4A-V4Z	Saint Kitts and Nevis
TEA-TEZ	Costa Rica	† V5A-V5Z	Namibia
TFA-TFZ	Iceland	† V6A-V6Z	Micronesia
TGA-TGZ	Guatemala	† V7A-V7Z	Marshall Islands
THA-THZ	France	† V8A-V8Z	Brunei
TIA-TIZ	Costa Rica	WAA-WZZ	United States of America
TJA-TJZ	Cameroon		
TKA-TKZ	France	XAA-XIZ	Mexico
TLA-TLZ	Central Africa	XJA-XOZ	Canada
TMA-TMZ	France	XPA-XPZ	Denmark
TNA-TNZ	Congo	XQA-XRZ	Chile
TOA-TQZ	France	XSA-XSZ	China
TRA-TRZ	Gabon	XTA-XTZ	Burkina Faso
TSA-TSZ	Tunisia	XUA-XUZ	Cambodia
TTA-TTZ	Chad	XVA-XVZ	Viet Nam
TUA-TUZ	Ivory Coast	XWA-XWZ	Laos
TVA-TXZ	France	XXA-XXZ	Portugal
TYA-TYZ	Benin	XYA-XZZ	Myanmar
TZA-TZZ	Mali	YAA-YAZ	Afghanistan
T2A-T2Z	Tuvalu	YBA-YHZ	Indonesia
T3A-T3Z	Kiribati	YIA-YIZ	Iraq
T4A-T4Z	Cuba	YJA-YJZ	Vanuatu
T5A-T5Z	Somalia	YKA-YKZ	Syria
T6A-T6Z	Afghanistan	† YLA-YLZ	Latvia
† T7A-T7Z	San Marino		

Table 4 (continued)
Allocation of International Call Signs

Call Sign Series	Allocated to	Call Sign Series	Allocated to
YMA-YMZ	Turkey	4DA-4IZ	Philippines
YNA-YNZ	Nicaragua	† 4JA-4KZ	Azerbaijan
YOA-YRZ	Romania	† 4LA-4LZ	Georgia
YSA-YSZ	El Salvador	4MA-4MZ	Venezuela
YTA-YUZ	Yugoslavia	4NA-4OZ	Yugoslavia
YVA-YYZ	Venezuela	4PA-4SZ	Sri Lanka
YZA-YZZ	Yugoslavia	4TA-4TZ	Peru
Y2A-Y9Z	Germany	* 4UA-4UZ	United Nations
ZAA-ZAZ	Albania	4VA-4VZ	Haiti
ZBA-ZJZ	United Kingdom of Great Britain and Northern Ireland	4XA-4XZ	Israel
		* 4YA-4YZ	International Civil Aviation Organization
ZKA-ZMZ	New Zealand	4ZA-4ZZ	Israel
ZNA-ZOZ	United Kingdom of Great Britain and Northern Ireland	5AA-5AZ	Libya
		5BA-5BZ	Cyprus
		5CA-5GZ	Morocco
ZPA-ZPZ	Paraguay	5HA-5IZ	Tanzania
ZQA-ZQZ	United Kingdom of Great Britain and Northern Ireland	5JA-5KZ	Colombia
		5LA-5MZ	Liberia
		5NA-5OZ	Nigeria
ZRA-ZUZ	South Africa	5PA-5QZ	Denmark
ZVA-ZZZ	Brazil	5RA-5SZ	Madagascar
† Z2A-Z2Z	Zimbabwe	5TA-5TZ	Mauritania
† Z3A-Z3Z	Macedonia (Former Yugoslav Republic)	5UA-5UZ	Niger
		5VA-5VZ	Togo
2AA-2ZZ	United Kingdom of Great Britain and Northern Ireland	5WA-5WZ	Western Samoa
		5XA-5XZ	Uganda
		5YA-5ZZ	Kenya
3AA-3AZ	Monaco	6AA-6BZ	Egypt
3BA-3BZ	Mauritius	6CA-6CZ	Syria
3CA-3CZ	Equatorial Guinea	6DA-6JZ	Mexico
• 3DA-3DM	Swaziland	6KA-6NZ	South Korea
• 3DN-3DZ	Fiji	6OA-6OZ	Somalia
3EA-3FZ	Panama	6PA-6SZ	Pakistan
3GA-3GZ	Chile	6TA-6UZ	Sudan
3HA-3UZ	China	6VA-6WZ	Senegal
3VA-3VZ	Tunesia	6XA-6XZ	Madagascar
3WA-3WZ	Viet Nam	6YA-6YZ	Jamaica
3XA-3XZ	Guinea	6ZA-6ZZ	Liberia
3YA-3YZ	Norway	7AA-7IZ	Indonesia
3ZA-3ZZ	Poland	7JA-7NZ	Japan
4AA-4CZ	Mexico	7OA-7OZ	Yemen

Table 4 (continued)
Allocation of International Call Signs

Call Sign Series	Allocated to	Call Sign Series	Allocated to
7PA-7PZ	Lesotho	9BA-9DZ	Iran
7QA-7QZ	Malawi	9EA-9FZ	Ethiopia
7RA-7RZ	Algeria	9GA-9GZ	Ghana
7SA-7SZ	Sweden	9HA-9HZ	Malta
7TA-7YZ	Algeria	9IA-9JZ	Zambia
7ZA-7ZZ	Saudi Arabia	9KA-9KZ	Kuwait
8AA-8IZ	Indonesia	9LA-9LZ	Sierra Leone
8JA-8NZ	Japan	9MA-9MZ	Malaysia
8OA-8OZ	Botswana	9NA-9NZ	Nepal
8PA-8PZ	Barbados	9OA-9TZ	Zaire
8QA-8QZ	Maldives	9UA-9UZ	Burundi
8RA-8RZ	Guyana	9VA-9VZ	Singapore
8SA-8SZ	Sweden	9WA-9WZ	Malaysia
8TA-8YZ	India	9XA-9XZ	Rwanda
8ZA-8ZZ	Saudi Arabia	9YZ-9ZZ	Trinidad and Tobago
† 9AA-9AZ	Croatia		

• Half-series
* Series allocated to an international organization
† Provisional allocation in accordance with RR2088

bands should be without. That booklet, available from ARRL, includes the common call-sign prefixes used by amateurs in virtually every location in the world. It also includes a check-off list to help you keep track of the countries you contact as you work towards collecting QSL cards from 100 or more countries to earn the prestigious DX Century Club award. (DX is ham lingo for distance, generally taken to mean any country outside the one you are operating from on the HF bands.)

The ITU radio regulations outline the basic principles used in forming amateur call signs. According to these regulations, an amateur call sign must be made up of one or two characters (the first one may be a numeral) as a prefix, followed by a numeral, and then a suffix of not more than three letters. The prefixes W, K, N and A are used in the United States. The continental US is divided into 10 Amateur Radio call districts (sometimes called areas), numbered 0 through 9. Figure 4 is a map showing the US call districts.

All US Amateur Radio call signs assigned by the FCC after March 1978 can be categorized into one of five groups, each corresponding to a class, or classes, of license. Call signs are issued systematically by the FCC. For further information on the

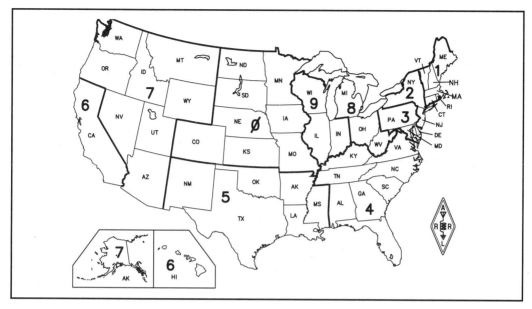

Figure 4 — There are 10 US call areas. Hawaii is part of the sixth call area, and Alaska is part of the seventh.

FCC's call-sign assignment system, and a table listing the blocks of call signs for each license class, see the ARRL publication, *The FCC Rule Book*. If you already have an amateur call sign, you may keep the same one when you change license class, if you wish. You must indicate that you want to receive a new call sign when you fill out an FCC Form 610 to apply for the exam or change your address.

Effective March 24, 1995, the FCC has established a vanity call sign system. Under this system the FCC will issue a call sign selected from a list of preferred available call signs. While there is no fee for an Amateur Radio license, there is a fee for the selection of a vanity call sign. At press time, the fee was $7 per year, paid upon application for a vanity call sign and at license renewal after that. This fee is $70 for a 10-year Amateur Radio license. Complete details about the vanity call sign system are available from the Regulatory Information Branch at ARRL Headquarters, 225 Main St, Newington, CT 06111-1494.

EARNING A LICENSE

Applying for an Exam: FCC Form 610

Before you can take an FCC exam, you'll have to fill out a Form 610. This form is used as an application for a new license, an upgraded license, a renewal of a license or a modification to a license.

Licenses are normally good for ten years. Your application for a license renewal must be submitted to the FCC no more than 90 days before the license expires. (We recommend you submit the application between 90 and 60 days before your license expires.) If the FCC receives your renewal application before the license expires, you may continue to operate until your new license arrives, even if it is past the expiration date. If you forget to apply before your license expires, you may still be able to renew without taking another exam. There is a two-year grace period, during which you may apply for renewal of your expired license. Use a Form 610 to apply for reinstatement (and your old call sign). If you apply for reinstatement of your expired license under this two-year grace period, you may not operate your station until your new license is issued.

The FCC has a set of detailed instructions for the Form 610, which are included with the form. To obtain a new Form 610, send a business-sized, self-addressed, stamped envelope to: Federal Communications Commission, Forms Distribution Center, 2803 52nd Ave, Hyattsville, MD 20781 (specify "Form 610" on the envelope), or: Form 610, ARRL, 225 Main Street, Newington, CT 06111-1494. Many clubs and Amateur Radio equipment dealers may also have a supply of these forms. If you obtain one from a source other than the FCC or ARRL, be sure it is the latest version. (As of this writing, the latest Form 610 is dated November 1993 in the lower-right corner.)

Figure 5 shows a sample 610 Form. All previous versions are obsolete and no longer acceptable. The back of the new form is used only for a physician's certification that the applicant has a handicap that makes it impossible for him or her to pass a 13 wpm or 20 wpm Morse code exam. You can ignore the back of the form unless you feel you qualify for such a code exemption. You do need to attach a copy of your current license (if you have one) to the back of the form, however.

Volunteer Examiner Program

All US amateur exams are administered by Volunteer Examiners who are certified by a Volunteer Examiner Coordinator (VEC). *The FCC Rule Book* contains details about the Volunteer Examiner program.

To qualify for an Advanced class license, you must pass Element 1B (the Morse code exam) and Elements 2, 3A, 3B and 4A. If you already hold a valid license,

Figure 5 — This sample FCC Form 610 shows the sections you should complete before going for your license exam. The form shown here is the current version, dated November 1993 in the lower right corner. All previous versions of the Form 610 are obsolete, and no longer acceptable. Attach a photocopy of your current license to the back of the form. The back of the form also includes a section to be completed by your physician, if you are claiming credit for the 13 wpm or 20 wpm Morse code exam because of a handicap that prevents you from passing such an exam.

FEDERAL COMMUNICATIONS COMMISSION
GETTYSBURG, PENNSYLVANIA

Approved by OMB
3060-0003
Expires 08/31/96
See instructions for
information regarding
public burden estimate.

APPLICATION FORM 610 FOR
AMATEUR OPERATOR/PRIMARY STATION LICENSE

SECTION 1 - TO BE COMPLETED BY APPLICANT (See instructions)

1. Print or type last name	Suffix	First name	Middle initial	2. Date of birth
Kleinman		Joel	P	0 5 - 2 3 - 4 8

month day year

3. Mailing address (Number and street)	City	State code	ZIP code
55 Corrigan Ave	Meriden	CT	06451-2874

4. I HEREBY APPLY FOR (make an X in the appropriate box(es)):

4A. ☐ **EXAMINATION** for a new license

4B. ☒ **EXAMINATION** for upgrade of my operator license class

4C. ☐ **CHANGE** my name on my license to my new name in Item 1. My former name was:

_____
(Last name)   (Suffix)   (First name)   (MI)

4D. ☐ **CHANGE** my mailing address on my license to my new address in Item 3

4E. ☐ **CHANGE** my station call sign systematically (See instructions)
Applicant's Initials _____

4F. ☐ **RENEWAL** of my license

5. Unless you are requesting a new license, attach the original or a photocopy of your license to the back of this Form 610 and complete Items 5A and 5B.	5A. Call sign shown on license	5B. Operator class shown on license
	N1BKE	General

6. Would an FCC grant of your request be an action that may have a significant environmental effect?   ☒ NO   ☐ YES (Attach required statement)

7. If you have filed another Form 610 that we have not acted upon, complete Items 7A and 7B.	7A. Purpose of other form	7B. Date filed
		_ _ - _ _ - _ _

month day year

WILLFUL FALSE STATEMENTS MADE ON THIS FORM ARE PUNISHABLE BY FINE AND/OR IMPRISONMENT, (U.S. CODE, TITLE 18, SECTION 1001), AND/OR REVOCATION OF ANY STATION LICENSE OR CONSTRUCTION PERMIT (U.S. CODE, TITLE 47, SECTION 312(A)(1)) AND/OR FORFEITURE (U.S. CODE, TITLE 47, SECTION 503).

I CERTIFY THAT ALL STATEMENTS AND ATTACHMENTS ARE TRUE, COMPLETE, AND CORRECT TO THE BEST OF MY KNOWLEDGE AND BELIEF AND ARE MADE IN GOOD FAITH; THAT I AM NOT A REPRESENTATIVE OF A FOREIGN GOVERNMENT; THAT I WAIVE ANY CLAIM TO THE USE OF ANY PARTICULAR FREQUENCY REGARDLESS OF PRIOR USE BY LICENSE OR OTHERWISE; AND THAT THE STATION TO BE LICENSED WILL BE INACCESSIBLE TO UNAUTHORIZED PERSONS.

8. Signature of applicant (Do not print, type, or stamp.) (Must match name in Item 1.)	9. Date signed
✗ Joel P. Kleinman	0 4 - 1 5 - 9 5

month day year

SECTION 2 - TO BE COMPLETED BY ALL ADMINISTERING VE's

A. Applicant is qualified for operator license class:

☐ NOVICE   (Elements 1(A), 1(B), or 1(C) and 2)
☐ TECHNICIAN   (Elements 2 and 3(A))
☐ TECHNICIAN PLUS   (Elements 1(A), 1(B), or 1(C), 2 and 3(A))
☐ GENERAL   (Elements 1(B) or 1(C), 2, 3(A) and 3(B))
☐ ADVANCED   (Elements 1(B) or 1(C), 2, 3(A), 3(B) and 4(A))
☐ AMATEUR EXTRA   (Elements 1(C), 2, 3(A), 3(B), 4(A) and 4(B))

B. VEC receipt date:

C. Name of Volunteer-Examiner Coordinator (VEC):

D. Date of VEC coordinated examination session:	E. Examination session location:

I CERTIFY THAT I HAVE COMPLIED WITH THE ADMINISTERING VE REQUIREMENTS IN PART 97 OF THE COMMISSION'S RULES AND WITH THE INSTRUCTIONS PROVIDED BY THE COORDINATING VEC AND THE FCC

1st VE's name (Print First, MI, Last, Suffix)	VE's station call sign	VE's signature (must match name)	Date signed
2nd VE's name (Print First, MI, Last, Suffix)	VE's station call sign	VE's signature (must match name)	Date signed
3rd VE's name (Print First, MI, Last, Suffix)	VE's station call sign	VE's signature (must match name)	Date signed

FCC Form 610
November 1993

then you have credit for passing at least some of those elements, and will not have to retake them when you go for your Advanced class exam. See Table 1 for details.

The Element 4A exam consists of 50 questions taken from a pool of more than 500. The question pools for all amateur exams are maintained by a Question Pool Committee selected by the Volunteer Examiner Coordinators. The FCC allows Volunteer Examiners to select the questions for an amateur exam, but they must use the questions exactly as they are released by the VEC Question Pool Committee. If you attend a test session coordinated by the ARRL/VEC, your test will be designed by the ARRL/VEC or by a computer program designed by the VEC, and the questions and answers will be exactly as they are printed in Chapter 10 of this book.

The FCC specifies the number of questions from each subelement that must appear on the exam. The Advanced Class Syllabus printed at the end of this chapter (Table 5) lists the number of questions to be taken from each subelement. The syllabus is further broken into sections, and the VEC Question Pool Committee recommends one question be taken from each of these sections. The beginning of Chapter 10 includes a summary of how many questions are taken from each subelement. The number of questions also appears at the beginning of each subelement in the question pool printed in Chapter 10.

Finding an Exam Opportunity

To determine where and when an exam will be given, contact the ARRL/VEC office, or watch for announcements in the Hamfest Calendar and Coming Conventions columns in *QST*. Many local clubs sponsor exams, so they are another good source of information on exam opportunities. ARRL officials, such as Directors, Vice Directors and Section Managers receive notices about test sessions in their area. See page 8 in the latest issue of *QST* for names and addresses.

To register for an exam, send a completed Form 610 to the Volunteer Examiner team responsible for the exam session if preregistration is required. Otherwise, bring the form to the session. Registration deadlines, and the time and location of the exams, are mentioned prominently in publicity releases about upcoming sessions.

Taking the Exam

By the time examination day rolls around, you should have already prepared yourself. This means getting your schedule, supplies and mental attitude ready. Plan your schedule so you'll get to the examination site with plenty of time to spare. There's no harm in being early. In fact, you might have time to discuss hamming with another applicant, which is a great way to calm pretest nerves. Try not to discuss the material that will be on the examination, as this may make you even more nervous. By this time, it's too late to study anyway!

What supplies will you need? First, be sure you bring your current original Amateur Radio license, if you have one. Bring a photocopy of your license, too, as well as the original and a photocopy of any Certificates of Successful Completion of Examination (CSCE) that you plan to use for exam credit. Bring along several sharpened number 2 pencils and two pens (blue or black ink). Be sure to have a good eraser. A pocket calculator will also come in handy. You may use a programmable calculator if that is the kind you have, but take it into your exam "empty" (cleared of all programs and constants in memory). Don't program equations ahead of time, because you may be asked to demonstrate that there is nothing in the calculator memory.

The Volunteer Examiner Team is required to check two forms of identification before you enter the test room. This includes your *original* Amateur Radio license, if you have one. A photo ID of some type is best for the second form of ID, but is not required by the FCC. Other acceptable forms of identification include a driver's license, a piece of mail addressed to you, a birth certificate or some other such document.

The following description of the testing procedure applies to exams coordinated by the ARRL/VEC, although many other VECs use a similar procedure.

Code Tests

The code tests are usually given before the written exams. The 20 wpm exam normally is given first, then the 13 wpm exam and finally the 5 wpm test. There is no harm in trying the 20 wpm exam, even if you have already passed the 13 wpm exam and are testing for the Advanced license.

Before you take the code test, you'll be handed a piece of paper to copy the code as it is sent. The test will begin with about a minute of practice copy. Then comes the actual test: at least five minutes of Morse code. You are responsible for knowing the 26 letters of the alphabet, the numerals 0 through 9, the period, comma, question mark, and the procedural signals $\overline{AR}$ (+), $\overline{SK}$, $\overline{BT}$ (= or double dash) and $\overline{DN}$ (/ or fraction bar, sometimes called the "slant bar"). You may copy the entire text word for word, or just take notes on the content. At the end of the transmission, the examiner will hand you 10 questions about the text. Depending on the test format, answer the multiple-choice questions or simply fill in the blanks with your answers. (You must spell each answer exactly as it was sent.) If you get at least seven correct, you pass! Alternatively, the exam team has the option to look at your copy sheet if you fail the 10-question exam. If you have one minute of solid copy (65 characters for the 13 wpm exam), the examiners can certify that you passed the test on that basis. The format of the test transmission is generally similar to one side of a normal on-the-air amateur conversation.

A sending test may not be required. The Commission has decided that if applicants can demonstrate receiving ability, they most likely can also send at that speed. But be prepared for a sending test, just in case! Subpart 97.503(a) of the FCC Rules says, "A telegraphy examination must be sufficient to prove that the examinee has the ability to send correctly by hand and to receive correctly by ear texts in the international Morse code at not less than the prescribed speed..."

Written Tests

After the code tests are administered, you'll take the written examination. The examiner will give each applicant a test booklet, an answer sheet and scratch paper.

After that, you're on your own. The first thing to do is read the instructions. Be sure to sign your name every place it's called for. Do all of this at the beginning to get it out of the way.

Next, check the examination to see that all pages and questions are there. If not, report this to the examiner immediately. When filling in your answer sheet, make sure your answers are marked next to the numbers that correspond to each question.

Go through the entire exam, and answer the easy questions first. Next, go back to the beginning and try the harder questions. Leave the really tough questions for last. Guessing can only help, as there is no additional penalty for answering incorrectly.

If you have to guess, do it intelligently: At first glance, you may find that you can eliminate one or more "distracters." Of the remaining responses, more than one may seem correct; only one is the best answer, however. To the applicant who is fully prepared, incorrect distracters to each question are obvious. Nothing beats preparation!

After you've finished, check the examination thoroughly. You may have read a question wrong or goofed in your arithmetic. Don't be overconfident. There's no rush, so take your time. Think, and check your answer sheet. When you feel you've done your best and can do no more, return the test booklet, answer sheet and scratch pad to the examiner.

The Volunteer Examiner Team will grade the exam while you wait. The passing mark is 74%. (That means no more than 13 incorrect answers on the 50-question Element 4A exam.) You will receive a Certificate of Successful Completion of Examination (CSCE) showing all exam elements that you pass at that exam session. If you are already licensed, and you pass the exam elements required to earn a higher license class, the CSCE authorizes you to operate with your new privileges. When you use these new privileges, you must sign your call sign, followed by the slant mark ("/"; on voice, say "stroke" or "slant") and the letters "AA." You only have to follow this special identification procedure until your new license arrives, however.

If you pass only some of the exam elements required for a higher class license you will still receive a CSCE. That certificate shows what exam elements you passed, and is valid for 365 days. Use it as proof that you passed those exam elements so you won't have to take them over again next time you try for the upgrade.

AND NOW, LET'S BEGIN

The complete Advanced class question pool (Element 4A) is printed in Chapter 10. Chapters 1 through 9 explain the material covered in subelements A1 through A9 of the question pool.

This book provides the background in FCC Rules, operating procedures, radio-wave propagation, Amateur Radio practice, electrical principles, circuit components, practical circuits, signals and emissions, and antennas and feed lines that you will need to pass the Element 4A Advanced class written exam.

Table 5 shows the study guide or syllabus for the Element 4A exam. This study guide was released by the Volunteer-Examiner Coordinators' Question Pool Committee in 1994. The syllabus lists the topics to be covered by the Advanced class exam, and so forms the basic outline for the remainder of this book. Use the syllabus to guide your study, and to ensure that you have studied the material for all of the topics listed.

The question numbers used in the question pool refer to this syllabus. Each question number begins with a syllabus-point number (A1C, A2A, A3B). The question numbers end with a two-digit number. For example, question A1C03 is the third question in the series about syllabus point A1C. Question A3B09 is the ninth question about the A3B syllabus point.

The Question Pool Committee designed the syllabus and question pool so there are the same number of points in each subelement as there are exam questions from that subelement. For example, the FCC specifies that six questions on the Advanced exam must come from the "Commission's Rules" subelement. There are six groups of questions for this subelement, A1A through A1F. Two exam questions must be from the "Radio-Wave Propagation" subelement, so there are two groups for that point. These are numbered A3A and A3B. While not a requirement of the FCC Rules, the Question Pool Committee recommends that one question be taken from each group to make the best possible Advanced class license exam.

Good luck with your studies!

Table 5

Advanced Class (Element 4A) Syllabus

(Required for Advanced and Amateur Extra licenses.)

SUBELEMENT A1 — COMMISSION RULES
[6 Exam Questions — 6 Groups]

A1A   Advanced control operator frequency privileges; station identification; emissions standards.

A1B   Definition and operation of remote control and automatic control; control link.

A1C   Type acceptance of external RF power amplifiers and external RF power amplifier kits

A1D   Definition and operation of spread spectrum; auxiliary station operation.

A1E   "Line A"; National Radio Quiet Zone; business communications; restricted operation; antenna structure limitations.

A1F   Volunteer examinations: when examination is required; exam credit; examination grading; Volunteer Examiner requirements; Volunteer Examiner conduct.

SUBELEMENT A2 — OPERATING PROCEDURES
[1 Exam Question — 1 Group]

A2A   Facsimile communications; slow-scan TV transmissions; spread-spectrum transmissions; HF digital communications (ie, PacTOR, CLOVER, HF packet); automatic HF forwarding.

SUBELEMENT A3 — RADIO-WAVE PROPAGATION
[2 Exam Questions — 2 Groups]

A3A   Sporadic-E; auroral propagation; ground-wave propagation (distances and coverage, and frequency versus distance in each of these topics).

A3B   Selective fading; radio-path horizon; take-off angle over flat or sloping terrain; earth effects on propagation.

SUBELEMENT A4 — AMATEUR RADIO PRACTICE
[4 Exam Questions — 4 Groups]

A4A   Frequency measurement devices (ie, frequency counter, oscilloscope Lissajous figures, dip meter); component mounting techniques (ie, surface, dead bug {raised}, circuit board).

A4B   Meter performance limitations; oscilloscope performance limitations; frequency counter performance limitations.

A4C   Receiver performance characteristics (ie, phase noise, desensitization, capture effect, intercept point, noise floor, dynamic range {blocking and IMD}, image rejection, MDS, signal-to-noise-ratio).

A4D   Intermodulation and cross-modulation interference

SUBELEMENT A5 — ELECTRICAL PRINCIPLES
[10 Exam Questions — 10 Groups]

A5A   Characteristics of resonant circuits.

A5B   Series resonance (capacitor and inductor to resonate at a specific frequency).

A5C   Parallel resonance (capacitor and inductor to resonate at a specific frequency).

A5D   Skin effect; electrostatic and electromagnetic fields.

A5E   Half-power bandwidth.

A5F   Circuit Q.

A5G   Phase angle between voltage and current.

A5H   Reactive power; power factor.

A5I   Effective radiated power; system gains and losses.

A5J   Replacement of voltage source and resistive voltage divider with equivalent voltage source and one resistor (Thevenin's Theorem).

SUBELEMENT A6 — CIRCUIT COMPONENTS
[6 Exam Questions — 6 Groups]

A6A   Semiconductor material: Germanium, Silicon, P-type, N-type.

A6B   Diodes: Zener, tunnel, varactor, hot-carrier, junction, point contact, PIN and light emitting.

A6C   Toroids: Permeability, core material, selecting, winding.

A6D   Transistor types: NPN, PNP, junction, unijunction, power.

A6E   Silicon controlled rectifier (SCR); triac; neon lamp.

A6F   Quartz crystal (frequency determining properties as used in oscillators and filters); monolithic amplifiers (MMICs).

SUBELEMENT A7 — PRACTICAL CIRCUITS
[10 Exam Questions — 10 Groups]

A7A   Amplifier circuits: Class A, Class AB, Class B, Class C, amplifier operating efficiency (ie, DC input versus PEP), transmitter final amplifiers.

A7B   Amplifier circuits: tube, bipolar transistor, FET.

A7C   Impedance-matching networks: Pi, L, Pi-L.

A7D   Filter circuits: constant K, M-derived, band-stop, notch, crystal lattice, pi-section, T-section, L-section, Butterworth, Chebyshev, elliptical.

A7E   Voltage-regulator circuits: discrete, integrated and switched mode.

A7F   Oscillators: types, applications, stability.

A7G   Modulators: reactance, phase, balanced.

A7H   Detectors; filter applications (audio, IF, digital signal processing {DSP}).

A7I   Mixer stages; frequency synthesizers.

A7J   Amplifier applications: AF, IF, RF.

Table 5 (continued)

Advanced Class (Element 4A) Syllabus

(Required for Advanced and Amateur Extra licenses.)

SUBELEMENT A8 — SIGNALS AND EMISSIONS
[6 Exam Questions — 6 Groups]

A8A   FCC emission designators versus emission types.

A8B   Modulation symbols and transmission characteristics.

A8C   Modulation methods; modulation index; deviation ratio.

A8D   Electromagnetic radiation; wave polarization; signal-to-noise (S/N) ratio.

A8E   AC waveforms: sine wave, square wave, sawtooth wave.

A8F   AC measurements: peak, peak-to-peak and root-mean-square (RMS) value, peak-envelope-power (PEP) relative to average.

SUBELEMENT A9 — ANTENNAS AND FEED LINES
[5 Exam Questions — 5 Groups]

A9A   Basic antenna parameters: radiation resistance and reactance (including wire dipole, folded dipole), gain, beamwidth, efficiency.

A9B   Free-space antenna patterns: E and H plane patterns (ie, azimuth and elevation in free-space); gain as a function of pattern; antenna design (computer modeling of antennas).

A9C   Antenna patterns: elevation above real ground, ground effects as related to polarization, take-off angles as a function of height above ground.

A9D   Losses in real antennas and matching: resistivity losses, losses in resonating elements (loading coils, matching networks, etc. {ie, mobile, trap}); SWR bandwidth; efficiency.

A9E   Feed lines: coax versus open-wire; velocity factor; electrical length; transformation characteristics of line terminated in impedance not equal to characteristic impedance.

CHAPTER 1 KEYWORDS KEYWORDS KEYWORDS

Automatic control — The operation of an amateur station without a control operator present at the control point. In §97.3 (a) (6), the FCC defines automatic control as "The use of devices and procedures for control of a station when it is transmitting so that compliance with the FCC Rules is achieved without the control operator being present at a control point."

Auxiliary station — An amateur station, other than in a message forwarding system, that is transmitting communications point-to-point within a system of cooperating amateur stations. {§97.3 (a) (7)}

Control link — A device used by a control operator to manipulate the station adjustment controls from a location other than the station location. A control link provides the means of control between a control point and a remotely controlled station.

Emissions — Any signals produced by a transmitter that reach the antenna connector to be radiated.

International Telecommunication Union (ITU) — The international organization with responsibility for dividing the range of communications frequencies between the various services for the entire world.

Line A — A line roughly parallel to, and south of, the US-Canadian border. {See §97.3 (a) (27).}

National Radio Quiet Zone — The area in Maryland, Virginia and West Virginia bounded by 39° 15' N on the north, 78° 30' W on the east, 37° 30' N on the south and 80° 30' W on the west {§97.3 (a) (30)}.

Remote control — The operation of an Amateur Radio station using a **control link** to manipulate the station operating adjustments from somewhere other than the station location.

Repeater operation — Operation of an amateur station that simultaneously retransmits the transmission of another amateur station on a different channel or channels.

Spread spectrum — A signal-transmission technique in which the transmitted carrier is spread out over a wide frequency band. The FCC refers to this as a form of bandwidth-expansion modulation.

Spurious emissions — Any **emission** that is not part of the desired signal. The FCC defines this term as "an emission, on frequencies outside the necessary bandwidth of a transmission, the level of which may be reduced without affecting the information being transmitted."

Telecommand — The FCC defines telecommand as a one-way transmission to initiate, modify, or terminate functions of a device at a distance. {§97.3 (a) (41)} A telecommand station is used to remotely control an amateur station through a radio link.

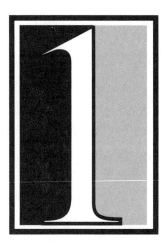

COMMISSION'S RULES

Part 97 of the Federal Communications Commission's Rules govern the Amateur Radio Service in the United States. Each Amateur Radio license exam includes questions covering sections of these rules. The Advanced-class license exam includes six rules questions. In this chapter of *The ARRL Advanced Class License Manual*, you will learn about the specific rules covered on the Advanced-class (Element 4A) license exam.

This book does not contain a complete listing of the Part 97 Rules. The FCC modifies Part 97 on an irregular basis, and it would be impossible to keep an up-to-date set of rules in a book designed to be revised every four years, to coincide with the release of a new Advanced-Class Question Pool. We do recommend, however, that every Amateur Radio operator have an up-to-date copy of the rules in their station for reference. *The FCC Rule Book*, published by ARRL, contains the complete text of Part 97, along with detailed explanations of all the regulations. *The FCC Rule Book* is updated as necessary to keep it current with the latest rule changes.

The six rules questions on your Advanced-class license exam will come from six exam-question groups. Each group forms a subelement section on the syllabus or study guide, and the question pool. The six groups for the Commission's Rules subelement are:

A1A   Advanced control operator frequency privileges; station identification; emissions standards;
A1B   Definition and operation of remote control and automatic control; control link;
A1C   Type acceptance of external RF power amplifiers and external RF power amplifier kits;
A1D   Definition and operation of spread spectrum; auxiliary station operation;
A1E   "Line A;" National Radio Quiet Zone; business communications; restricted operation; antenna structure limitations;
A1F   Volunteer examinations: when examination is required; exam credit; examination grading; Volunteer Examiner requirements; Volunteer Examiner conduct.

ADVANCED CONTROL OPERATOR FREQUENCY PRIVILEGES

As an Advanced-class Amateur Radio licensee, you will have all Amateur Radio frequency privileges above 50 MHz, just as a Technician, Technician Plus and General class licensee has. These frequency privileges are listed in Section 97.301(a) of the FCC Rules. (Section is often represented by the § character when referencing a specific portion of the FCC Rules.) You will also have extensive privileges on the high-frequency (HF) bands, as listed in §97.301(c). Table 1-1 shows this section of the Rules, along with the appropriate notes from §97.303.

The **International Telecommunication Union (ITU)** is the organization responsible for international regulation of the radio spectrum. The ITU regulations state that communications between amateurs in foreign countries must be limited to technical topics or "remarks of a personal nature of relative unimportance" — you can talk about your new rig or what you had for dinner, but don't get into a serious political discussion with a foreign ham.

Under the ITU, the world has been separated into three areas, or Regions. See Figure 1-1. Amateurs in different Regions may not be able to use all the same frequencies or operating modes as US operators. The United States, including Alaska and Hawaii, is in ITU Region 2. Region 3 comprises most of the islands in the Pacific Ocean, including American Samoa, the Northern Mariana Islands, Guam and Wake Island. (Many Pacific islands are under FCC regulation.) Also included in Region 3 are China, India and parts of the Middle East. Region 1 comprises Europe, Africa, Russia, the former Soviet Republics and the remainder of the Middle East.

If you will be operating in an FCC-regulated area in ITU Regions 1 or 3, be sure to look in §97.301 (a) and (c) for your Advanced class frequency privileges in those Regions. In addition to the land areas described, your operation is under FCC regulation if you are operating from onboard a US-registered ship sailing in international waters. In such cases, you must follow the frequency privileges

Table 1-1

Advanced Class HF Bands and Sharing Requirements

§97.301 Authorized frequency bands.

The following transmitting frequency bands are available to an amateur station located within 50 km of the Earth's surface, within the specified ITU Region, and outside any area where the amateur service is regulated by any authority other than the FCC.

(c) For a station having a control operator who has been granted an operator license of Advanced Class:

Wavelength Band	ITU Region 1	ITU Region 2	ITU Region 3	Sharing requirements See §97.303, Paragraph:
MF	*kHz*	*kHz*	*kHz*	
160 m	1810 - 1850	1800 - 2000	1800 - 2000	(a), (b), (c)
HF	*MHz*	*MHz*	*MHz*	
80 m	3.525 - 3.750	3.525 - 3.750	3.525 - 3.750	(a)
75 m	3.775 - 3.800	3.775 - 4.000	3.775 - 3.900	(a)
40 m	7.025 - 7.100	7.025 - 7.300	7.025 - 7.100	(a)
30 m	10.10 - 10.15	10.10 - 10.15	10.10 - 10.15	(d)
20 m	14.025 - 14.150	14.025 - 14.150	14.025 - 14.150	
20 m	14.175 - 14.350	14.175 - 14.350	14.175 - 14.350	
17 m	18.068 - 18.168	18.068 - 18.168	18.068 - 18.168	
15 m	21.025 - 21.200	21.025 - 21.200	21.025 - 21.200	
15 m	21.225 - 21.450	21.225 - 21.450	21.225 - 21.450	
12 m	24.89 - 24.99	24.89 - 24.99	24.89 - 24.99	
10 m	28.0 - 29.7	28.0 - 29.7	28.0 - 29.7	

§97.303 Frequency sharing requirements.

The following is a summary of the frequency sharing requirements that apply to amateur station transmissions on the frequency bands specified in §97.301 of this Part. (For each ITU Region, each frequency band allocated to the amateur service is designated as either a secondary service or a primary service. A station in a secondary service must not cause harmful interference to, and must accept interference from, stations in a primary service. See §§2.105 and 2.106 of the FCC Rules, *United States Table of Frequency Allocations* for complete requirements.)

(a) Where, in adjacent ITU Regions or Subregions, a band of frequencies is allocated to different services of the same category, the basic principle is the equality of right to operate. The stations of each service in one region must operate so as not to cause harmful interference to services in the other Regions or Subregions. (See ITU *Radio Regulations*, No. 346 (Geneva, 1979).)

(b) No amateur station transmitting in the 1900 - 2000 kHz segment, the 70 cm band, the 33 cm band, the 13 cm band, the 9 cm band, the 5 cm band, the 3 cm band, the 24.05 - 24.25 GHz segment, the 76 - 81 GHz segment, the 144 - 149 GHz segment and the 241 - 248 GHz segment shall cause harmful interference to, nor is protected from interference due to the operation of, the Government radiolocation service.

(c) No amateur station transmitting in the 1900 - 2000 kHz segment, the 3 cm band, the 76 - 81 GHz segment, the 144 - 149 GHz segment and the 241 - 248 GHz segment shall cause harmful interference to, nor is protected from interference due to the operation of, stations in the non-Government radiolocation service.

(d) No amateur station transmitting in the 30 meter band shall cause harmful interference to stations authorized by other nations in the fixed service. The licensee of the amateur station must make all necessary adjustments, including termination of transmissions, if harmful interference is caused.

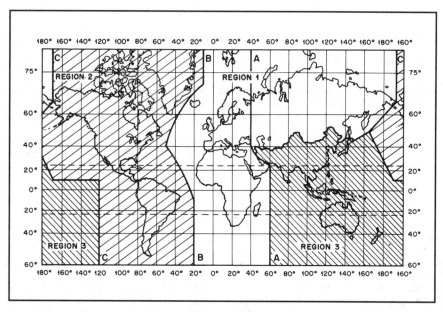

Figure 1-1 — This map shows the world divided into three ITU Regions.

authorized for use in the ITU region in which you are located.

Table 1-2 summarizes the Advanced-class HF privileges for ITU Region 2, and serves to help you memorize the frequencies for your exam. This table also shows the emission types authorized for use on certain portions of these bands. An emission is any type of signal transmitted from an Amateur Radio station. Several of the Advanced-Class Question-Pool questions refer to frequency privileges in ITU Region 2.

Study the frequency limits for each band shown in Table 1-2, and learn the segments designated for CW, RTTY and data as well as the segments designated for CW, phone and image.

Although you can operate CW on any frequency you are authorized to operate on, it is common practice to refer to the lower-frequency section of each band as the *CW Band* and the higher-frequency segment as the *Phone Band*. §97.3(c)(1, 2, 3, 5 and 7) list the definitions of these emission types.

[After you have learned the HF frequency privileges described in Table 1-2, turn to Chapter 10 and study questions A1A01 through A1A04. Review this section of the chapter if you have difficulty with any of these questions. As you study the questions, turn the edge of the question-pool page under to hide the answer key.]

Table 1-2
Advanced Class High-Frequency Operating Privileges

Band	Operating Privileges	Frequency Privileges
160 m	CW, RTTY, Data, Phone, Image	1800 kHz to 2000 kHz
80 m	CW, RTTY, Data	3525 kHz to 3750 kHz
	CW, Phone, Image	3775 kHz to 4000 kHz
40 m	CW, RTTY, Data	7025 kHz to 7150 kHz
	CW, Phone, Image	7150 kHz to 7300 kHz
30 m	CW, RTTY, Data	10,100 kHz to 10,150 kHz
20 m	CW, RTTY, Data	14,025 kHz to 14,150 kHz
	CW, Phone, Image	14,175 kHz to 14,350 kHz
17 m	CW, RTTY, Data	18,068 kHz to 18,110 kHz
	CW, Phone, Image	18,110 kHz to 18,168 kHz
15 m	CW, RTTY, Data	21,025 kHz to 21,200 kHz
	CW, Phone, Image	21,225 kHz to 21,450 kHz
12 m	CW, RTTY, Data	24,890 kHz to 24,930 kHz
	CW, Phone, Image	24,930 kHz to 24,990 kHz
10 m	CW, RTTY, Data	28,000 kHz to 28,300 kHz
	CW, Phone, Image	28,300 kHz to 29,700 kHz

STATION IDENTIFICATION

At the end of your Advanced class license exam you will receive a Certificate of Successful Completion of Examination (CSCE) showing those exam elements you passed. (See Figure 1-2.) This certificate is proof that you passed those elements, and is valid for 365 days. So if you don't pass all the exam elements to earn your new license, you can retake only the ones you didn't pass to earn the license. You can also use this certificate to show that you passed certain elements if you want to go back to another exam session before your license actually arrives from the FCC. If you want to upgrade as fast as possible, you might go to another exam session the following month (or sooner) and take the Amateur Extra exam, for example.

If you pass all the exam elements for the Advanced class license, and have a Novice, Technician, Technician Plus or General license, you may begin using your new Advanced class license privileges on the air immediately. Your CSCE serves

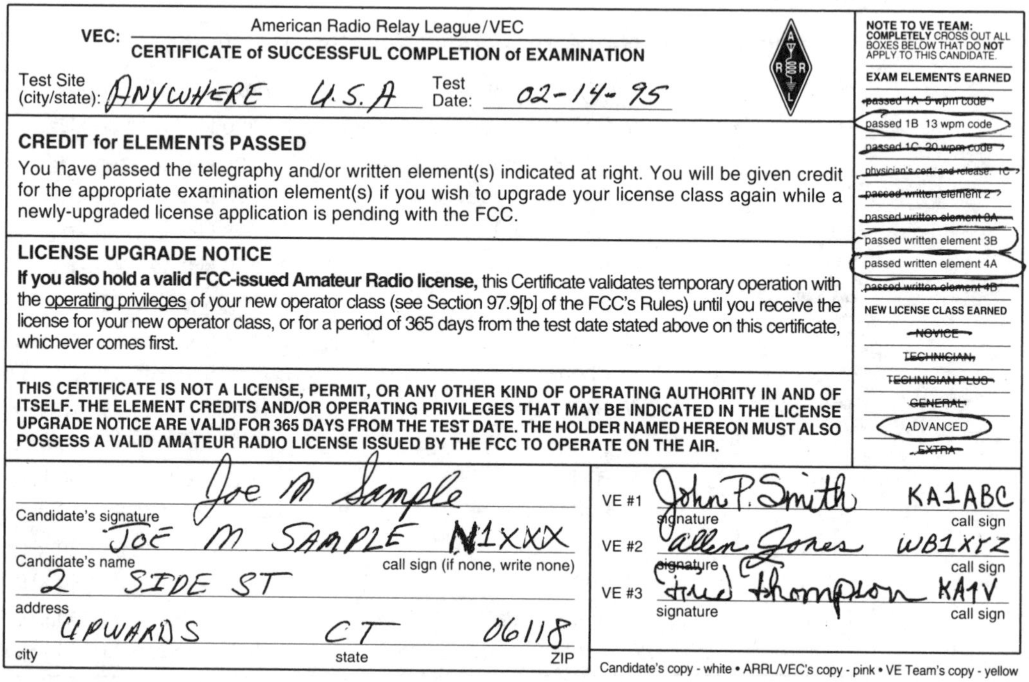

VEC: American Radio Relay League/VEC
CERTIFICATE of SUCCESSFUL COMPLETION of EXAMINATION

Test Site (city/state): *ANYWHERE U.S.A* Test Date: *02-14-95*

NOTE TO VE TEAM: COMPLETELY CROSS OUT ALL BOXES BELOW THAT DO NOT APPLY TO THIS CANDIDATE.

EXAM ELEMENTS EARNED
~~passed 1A 5 wpm code~~
passed 1B 13 wpm code
~~passed 1C 20 wpm code~~
~~physician's cert. and release 1C~~
~~passed written element 2~~
~~passed written element 3A~~
passed written element 3B
passed written element 4A
~~passed written element 4B~~

NEW LICENSE CLASS EARNED
~~NOVICE~~
~~TECHNICIAN~~
~~TECHNICIAN PLUS~~
~~GENERAL~~
ADVANCED
~~EXTRA~~

CREDIT for ELEMENTS PASSED

You have passed the telegraphy and/or written element(s) indicated at right. You will be given credit for the appropriate examination element(s) if you wish to upgrade your license class again while a newly-upgraded license application is pending with the FCC.

LICENSE UPGRADE NOTICE

If you also hold a valid FCC-issued Amateur Radio license, this Certificate validates temporary operation with the operating privileges of your new operator class (see Section 97.9[b] of the FCC's Rules) until you receive the license for your new operator class, or for a period of 365 days from the test date stated above on this certificate, whichever comes first.

THIS CERTIFICATE IS NOT A LICENSE, PERMIT, OR ANY OTHER KIND OF OPERATING AUTHORITY IN AND OF ITSELF. THE ELEMENT CREDITS AND/OR OPERATING PRIVILEGES THAT MAY BE INDICATED IN THE LICENSE UPGRADE NOTICE ARE VALID FOR 365 DAYS FROM THE TEST DATE. THE HOLDER NAMED HEREON MUST ALSO POSSESS A VALID AMATEUR RADIO LICENSE ISSUED BY THE FCC TO OPERATE ON THE AIR.

Joe M Sample
Candidate's signature

JOE M SAMPLE N1XXX
Candidate's name   call sign (if none, write none)

2 SIDE ST
address

UPWARDS CT 06118
city   state   ZIP

VE #1 *John P. Smith* KA1ABC
signature   call sign

VE #2 *Allen Jones* WB1XYZ
signature   call sign

VE #3 *Fred Thompson* KA1V
signature   call sign

Candidate's copy - white • ARRL/VEC's copy - pink • VE Team's copy - yellow

Figure 1-2 — This is an example of the CSCE you will get after you pass the Element 4A exam to upgrade from a General class license if your test session is coordinated by the ARRL/VEC. The certificates issued by other VECs will differ from this, but will serve the same purpose.

as a temporary "endorsement" to your license, until the FCC actually grants your new license.

You must follow a special identification procedure when you use these new privileges with the CSCE, however. In §97.119(e) we read that "When the control operator who is exercising the rights and privileges authorized by §97.9(b) of this Part, an indicator must be included after the call sign as follows: ... (3) For a control operator who has requested a license modification from Novice, Technician, or General class operator to Advanced class: AA."

Section 97.9(b) is the Rule point that allows you to use your new privileges after an upgrade, as long as you have a CSCE to prove your new privileges, until the FCC takes final action on your application, or until 365 days following the passing of the examination, whichever comes first.

There is a bit more detail about how to use this special identifier in §97.119(c): "An indicator may be included with the call sign. It must be separated from the call sign by the slant mark or by any suitable word that denotes the slant mark..." So if you are a Technician Plus licensee with a CSCE for Advanced privileges transmitting on 14.185 MHz, for example, you would give your station identification with your call sign, followed by the slant mark, and the identifier "AA." If you were operating phone on 14.220 MHz, it would sound like "This is N1ODI slant AA." You may also hear stations use words like "stroke," "interim," or "temporary" instead of "slant."

Remember that you only have to give this temporary identifier when you operate using your new privileges. If you previously had a Novice license, and are operating in the Novice segment of the 10-meter band, you need not use the "slant AA." Or if you had a Technician license and are operating on the 2-meter band, you don't have to say "temporary AA."

As you expand your operating horizons you will find yourself in some other "special identification" situations. For example, several operators often get together as a "multioperator" entry in Amateur Radio contests. One operator usually serves as control operator for these events. (The control operator may or may not be the licensee of the station you are using.) Most of the time the group will select an Amateur Extra class licensee as the control operator, and will use that operator's call sign for the contest.

Suppose you are invited to participate in such a contest group, as an Advanced class licensee. What are the Rules regarding station identification and operating frequencies? First, the station licensee may designate you to be the control operator while you are at the controls. In that case, you use the Amateur Extra class station licensee's call sign for identification, but you must stay within the Advanced class frequency privileges. Under a second scenario, the Amateur Extra class control operator may simply allow you to operate as a third party. In that case, the control operator remains "within reach" of the controls, and monitors your operating. You may still work all the station controls, give the station identification, and so on. You use the Amateur Extra class call sign and may operate on Amateur Extra class frequencies, because you have an Amateur Extra class control operator present.

Perhaps the most confusing point about this second scenario is that many people believe the control operator must remain "at the controls" pushing buttons, turning the radio dial and throwing switches as required. This simply is not true. The control operator may allow a third party (even an unlicensed third party) to "work the controls," as long as he or she is monitoring the operation and is in a position to stop the transmission if anything should go wrong.

Another station identification question that often comes up involves the operation of a station by a control operator who holds a higher license class than the station licensee. For example, suppose you, as an Advanced class licensee, were invited to operate a Technician Plus class operator's station. How should you identify?

Here again, there are two possible answers. The first, and easiest, is that the Technician Plus operator may choose to "loan" you the station equipment. In that case you can operate on Advanced class frequencies and identify the operation with your call sign. The Technician Plus operator may prefer that you use their call sign. In that case, you may still operate on Advanced class frequencies, but when you give the Technician Plus call sign, you must follow it with the slant mark or some word that means the same thing, and then give your Advanced class call sign.

[Turn to Chapter 10 now and check your understanding of these station identification procedures by studying questions A1A05 through A1A07. Review this section if you are uncertain of the answers to any of these questions.]

EMISSION STANDARDS

The signals produced by your transmitter are called **emissions**. These emissions must meet certain technical standards established by the FCC in the Part 97 Rules. These emission standards are given in §97.307 of the Rules. Paragraphs a, b and c establish some general guidelines:

§97.307 Emission standards.

(a) No amateur station transmission shall occupy more bandwidth than necessary for the information rate and emission type being transmitted, in accordance with good amateur practice.

(b) Emissions resulting from modulation must be confined to the band or segment available to the control operator. Emissions outside the necessary bandwidth must not cause splatter or keyclick interference to operations on adjacent frequencies.

(c) All spurious emissions from a station transmitter must be reduced to the greatest extent practicable. If any spurious emission, including chassis or power line radiation, causes harmful interference to the reception of another radio station, the licensee of the interfering amateur station is required to take steps to eliminate the interference, in accordance with good engineering practice.

In addition, paragraphs d and e list some more specific requirements:

(d) The mean power of any spurious emission from a station transmitter or external RF power amplifier transmitting on a frequency below 30 MHz must not exceed 50 mW and must be at least 40 dB below the mean power of the fundamental emission. For a transmitter of mean power less than 5 W, the attenuation must be at least 30 dB. A transmitter built before April 15, 1977, or first marketed before January 1, 1978, is exempt from this requirement.

(e) The mean power of any spurious emission from a station transmitter or external RF power amplifier transmitting on a frequency between 30-225 MHz must be at least 60 dB below the mean power of the fundamental. For a transmitter having a mean power of 25 W or less, the mean power of any spurious emission supplied to the antenna transmission line must not exceed 25 µW and must be at least 40 dB below the mean power of the fundamental emission, but need not be reduced below the power of 10 µW. A transmitter built before April 15, 1977, or first marketed before January 1, 1978, is exempt from this requirement.

Notice that there are two frequency ranges specified in this Rule point, below 30 MHz and between 30 and 225 MHz. Also notice that both frequency ranges have **spurious emissions** limits that vary with transmitter power. Let's take a look at what those numbers really mean.

Suppose you have an HF transmitter that produces a mean power of 500 watts. What is the maximum mean power permitted for any spurious emission from this transmitter? There are two limits listed in §97.307 (d) for transmitters with power above 5 W. Which will apply here? Let's run through a calculation to find out. First we'll assume that the spurious emissions are 50 mW, and calculate how many decibels below the mean power of the fundamental emission this is.

$$dB = 10 \log \left(\frac{P_1}{P_2} \right)$$ (Equation 1-1)

where:
  P_1 = mean power of the fundamental.
  P_2 = mean power of spurious emission.

$$dB = 10 \log \left(\frac{P_1}{P_2} \right) = 10 \log \left(\frac{500 \text{ W}}{50 \times 10^{-3} \text{ W}} \right) = 10 \log \left(1 \times 10^4 \right) = 10 \times 4 = 40 \text{ dB}$$

In this case the spurious emissions meet both the 50 mW maximum and they are at least 40 dB below the mean power of the fundamental signal.

Suppose your transmitter (or amplifier) produces 1500 W. A spurious emission that is 40 dB down from the fundamental will have a mean power of 150 mW! That does not meet the maximum 50 mW spec, so we'll need more filtering to suppress the spurious signals some more. In fact, we'll need almost 45 dB of suppression in this case, to keep the unwanted signals below the 50 mW level. (You can calculate all these numbers using Equation 1-1. To find the mean power of the spurious signals for 40 dB of suppression, you'll have to solve that equation for P_2.)

For a 100 W transmitter, with 40-dB suppression of the spurious signals, they will have no more than 10 mW of mean power. If the transmitter power is below 5 W, we only need 30 dB of suppression. In that case the maximum strength of the spurious signals will be 5 mW.

For the frequency range between 30 and 225 MHz, spurious emissions from high-power signals must be at least 60 dB below the mean power of the fundamental. If the transmitter power is 25 W or less, spurious signals must be no stronger than 25 µW, and at least 40 dB below the mean power of the fundamental signal.

[Turn to Chapter 10 and study questions A1A08 through A1A11 to check your understanding of these spurious emission limits. Review this section if you have trouble with any of the questions.]

REMOTELY AND AUTOMATICALLY CONTROLLED STATIONS

When you studied for your Novice or Technician class written exams, you learned that every Amateur Radio station must have a control operator whenever it is transmitting. There are times, however, when that control operator need not be located physically at the radio itself.

A station operating under **remote control** uses a **control link** to manipulate the station operating adjustments. A remotely controlled station is operated indirectly through a control link. The control link may be through a wire line, such as a telephone-line link, or through a radio link. Remote control through a radio link is called **telecommand**. The FCC Rules specify the requirements for telecommand of an amateur station.

---

§97.213 Telecommand of an amateur station.

An amateur station on or within 50 km of the Earth's surface may be under telecommand where:

(a) There is a radio or wireline control link between the control point and the station sufficient for the control operator to perform his/her duties. If radio, the control link must use an auxiliary station. A control link using a fiber optic cable or another telecommunication service is considered wireline.

(b) Provisions are incorporated to limit transmission by the station to a period of no more than 3 minutes in the event of malfunction in the control link.

(c) The station is protected against making, willfully or negligently, unauthorized transmissions.

(d) A photocopy of the station license and a label with the name, address, and telephone number of the station licensee and at least one designated control operator is posted in a conspicuous place at the station location.

Notice that these rules specify that the transmitter must include provisions to turn it off after no more than 3 minutes if the control link fails. This is to prevent the transmitter from becoming locked in the transmit mode with no way to turn it off.

If a radio control link is used, you must also follow the rules for auxiliary stations. An **auxiliary station** is part of a system of cooperating amateur stations that send point-to-point communications. One common use for an auxiliary station is for remote control of a station in repeater operation. Auxiliary stations only communicate with other stations in the system of cooperating amateur stations. The most significant point of these rules, for remote control, is that the auxiliary station can only transmit on the 222-MHz (1.25-meter) band and higher frequencies (shorter wavelengths). There are some frequencies you can't use even in these bands.

§97.201 Auxiliary station.

(a) Any amateur station licensed to a holder of a Technician, General, Advanced or Amateur Extra Class operator license may be an auxiliary station. A holder of a Technician, General, Advanced or Amateur Extra Class operator license may be the control operator of an auxiliary station, subject to the privileges of the class of operator license held.

(b) An auxiliary station may transmit only on the 1.25 m and shorter wavelength frequency bands, except the 222.00-222.15 MHz, 431-433 MHz and 435-438 MHz segments.

(c) Where an auxiliary station causes harmful interference to another auxiliary station, the licensees are equally and fully responsible for resolving the interference unless one station's operation is recommended by a frequency coordinator and the other station's is not. In that case, the licensee of the non-coordinated auxiliary station has primary responsibility to resolve the interference.

(d) An auxiliary station may be automatically controlled.

(e) An auxiliary station may transmit one-way communications.

Auxiliary stations are often identified in Morse code using an automatic keying device. In that case the identification Rules of §97.119 (b) (1) apply: "By a CW emission. When keyed by an automatic device used only for identification, the speed must not exceed 20 words per minute." Of course the same identification rules apply to auxiliary stations as to other amateur stations. They must be identified at least once every ten minutes during and at the end of any operating activity. If an auxiliary station is using a digital code for at least part of its communication, then that digital code may be used for identification.

Automatic control refers to amateur station operation in which the control operator is not present at the control point. The station *does* have a control operator, as do all amateur stations when they are transmitting. A station under remote control *may not* also be under automatic control at the same time. You can use remote control for the operation and control of a model craft, but such stations may not operate under automatic control.

Probably the most common type of amateur operation involving remote and automatic control is **repeater operation**. There are several other types of amateur stations that may be automatically controlled. These include beacon stations and auxiliary stations.

Portions of several amateur bands are designated for repeater operation. This doesn't mean no other modes can be used on those frequencies, but it does mean that repeaters can only operate on a frequency in one of these designated bands. Repeaters can operate only on the 10-meter and shorter wavelength (higher frequency) bands. There are segments of several of these bands where repeaters may not operate, however. These prohibited frequencies are listed in §97.205 (b).

Table 1-3

Repeater Station Frequencies on Several Popular Bands

Wavelength Band	Frequencies
10 m	29.5 - 29.7 MHz
6 m	51.0 - 54.0 MHz
2 m	144.5 - 145.5 MHz and 146.0 - 148.0 MHz
1.25 m	222.15 - 225.00 MHz
70 cm	420 - 431 MHz, 433 - 435 MHz and 438 - 450 MHz
23 cm	1240 - 1300 MHz

Table 1-3 lists the specific frequencies in several popular bands that *can* be used for repeater operation.

[Before you go on to the next section, turn to Chapter 10 and study questions A1B01 through A1B15. Also study questions A1D04 through A1D11. Review this section as needed.]

RF POWER AMPLIFIERS

RF power amplifiers capable of operating on frequencies below 144 MHz may require FCC type acceptance. Sections 97.315 and 97.317 describe the conditions under which type acceptance is required, and set out the standards to be met for type acceptance. See Table 1-4.

Many of these rules apply to manufacturers of amplifiers or kits, but several points are important for individual amateurs. Amateurs may build their own amplifiers or modify amplifiers for use in their own station without concern for the type acceptance rules. If you build or modify an amplifier for use by another amateur operator, then §97.315(a) applies. This rule says you cannot build or modify more than one amplifier of any particular model during any calendar year without obtaining a grant of type acceptance from the FCC. An unlicensed person may not build or modify any amplifier capable of operating below 144 MHz without a grant of FCC type acceptance.

To receive a grant of type acceptance, an amplifier must satisfy the spurious emission standards specified in §97.307(d) or (e) when operated at full power output. The amplifier must also meet the spurious emission standards when it is being driven with at least 50 W mean RF input power (unless a higher drive level is specified.) In addition, the amplifier must meet the spurious emission standards when it is placed in the "standby" or "off" position but is still connected to the transmitter. The amplifier must not be capable of reaching its designed output power when driven with less than 50 watts. The amplifier must also not be capable of operation on any frequency between 24 and 35 MHz. (This is to prevent the amplifier from being used illegally on the

Table 1-4

RF Amplifier Type Acceptance Standards

§97.315 Type acceptance of external RF power amplifiers.

(a) No more than 1 unit of 1 model of an external RF power amplifier capable of operation below 144 MHz may be constructed or modified during any calendar year by an amateur operator for use at a station without a grant of type acceptance. No amplifier capable of operation below 144 MHz may be constructed or modified by a non-amateur operator without a grant of type acceptance from the FCC.

(b) Any external RF power amplifier or external RF power amplifier kit (see §2.815 of the FCC Rules), manufactured, imported or modified for use in a station or attached at any station must be type accepted for use in the amateur service in accordance with Subpart J of Part 2 of the FCC Rules. This requirement does not apply if one or more of the following conditions are met:

(1) The amplifier is not capable of operation on frequencies below 144 MHz. For the purpose of this part, an amplifier will be deemed to be incapable of operation below 144 MHz if it is not capable of being easily modified to increase its amplification characteristics below 120 MHz and either:

(i) The mean output power of the amplifier decreases, as frequency decreases from 144 MHz, to a point where 0 dB or less gain is exhibited at 120 MHz; or

(ii) The amplifier is not capable of amplifying signals below 120 MHz even for brief periods without sustaining permanent damage to its amplification circuitry.

(2) The amplifier was manufactured before April 28, 1978, and has been issued a marketing waiver by the FCC, or the amplifier was purchased before April 28, 1978, by an amateur operator for use at that amateur operator's station.

(3) The amplifier was:

(i) Constructed by the licensee, not from an external RF power amplifier kit, for use at the licensee's station; or

(ii) Modified by the licensee for use at the licensee's station.

(4) The amplifier is sold by an amateur operator to another amateur operator or to a dealer.

(5) The amplifier is purchased in used condition by an equipment dealer from an amateur operator and the amplifier is further sold to another amateur operator for use at that operator's station.

(c) A list of type accepted equipment may be inspected at FCC headquarters in Washington, DC or at any FCC field location. Any external RF power amplifier appearing on this list as type accepted for use in the amateur service may be marketed for use in the amateur service.

§97.317 Standards for type acceptance of external RF power amplifiers.

(a) To receive a grant of type acceptance, the amplifier must satisfy the spurious emission standards of §97.307(d) or (e) of this Part, as applicable, when the amplifier is:

(1) Operated at its full output power;

(continued on next page)

(2) Placed in the "standby" or "off" positions, but still connected to the transmitter; and

(3) Driven with at least 50 W mean RF input power (unless higher drive level is specified).

(b) To receive a grant of type acceptance, the amplifier must not be capable of operation on any frequency or frequencies between 24 MHz and 35 MHz. The amplifier will be deemed incapable of such operation if it:

(1) Exhibits no more than 6 dB gain between 24 MHz and 26 MHz and between 28 MHz and 35 MHz. (This gain will be determined by the ratio of the input RF driving signal (mean power measurement) to the mean RF output power of the amplifier); and

(2) Exhibits no amplification (0 dB gain) between 26 MHz and 28 MHz.

(c) Type acceptance may be denied when denial would prevent the use of these amplifiers in services other than the amateur service. The following features will result in dismissal or denial of an application for the type acceptance:

(1) Any accessible wiring which, when altered, would permit operation of the amplifier in a manner contrary to the FCC Rules;

(2) Circuit boards or similar circuitry to facilitate the addition of components to change the amplifier's operating characteristics in a manner contrary to the FCC Rules;

(3) Instructions for operation or modification of the amplifier in a manner contrary to the FCC Rules;

(4) Any internal or external controls or adjustments to facilitate operation of the amplifier in a manner contrary to the FCC Rules;

(5) Any internal RF sensing circuitry or any external switch, the purpose of which is to place the amplifier in the transmit mode;

(6) The incorporation of more gain in the amplifier than is necessary to operate in the amateur service; for purposes of this paragraph, the amplifier must:

(i) Not be capable of achieving designed output power when driven with less than 40 W mean RF input power;

(ii) Not be capable of amplifying the input RF driving signal by more than 15 dB, unless the amplifier has a designed transmitter power of less than 1.5 kW (in such a case, gain must be reduced by the same number of dB as the transmitter power relationship to 1.5 kW; This gain limitation is determined by the ratio of the input RF driving signal to the RF output power of the amplifier where both signals are expressed in peak envelope power or mean power);

(iii) Not exhibit more gain than permitted by paragraph (c)(6)(ii) of this Section when driven by an RF input signal of less than 50 W mean power; and

(iv) Be capable of sustained operation at its designed power level.

(7) Any attenuation in the input of the amplifier which, when removed or modified, would permit the amplifier to function at its designed transmitter power when driven by an RF frequency input signal of less than 50 W mean power; or

(8) Any other features designed to facilitate operation in a telecommunication service other than the Amateur Radio Services, such as the Citizens Band (CB) Radio Service.

Citizen's Band frequencies.) The amplifier may be capable of operation on all amateur bands with frequencies below 24 MHz, however.

A manufacturer must obtain a separate grant of type acceptance for each amplifier model. The manufacturer must obtain another grant of type acceptance for future amplifier models as they are developed.

An amplifier may not be built with accessible wiring that can be modified to permit the amplifier to operate in a manner contrary to FCC Rules, and must not include instructions for operation or modification of the amplifier in a way that would violate FCC Rules. The amplifier may not have any features designed to be used in a telecommunication service other than the Amateur Service. The amplifier must not produce 3 dB or more gain for input signals between 26 MHz and 28 MHz. Failure to meet any of these conditions would disqualify the amplifier from being granted FCC type acceptance.

There is one condition by which an equipment dealer may sell an amplifier that has not been granted FCC type acceptance. A dealer may purchase a used amplifier from an amateur operator and sell it to another licensed amateur for use at that operator's station.

[Now turn to Chapter 10 and study questions A1C01 through A1C11. If any of these questions give you difficulty, review this section.]

SPREAD SPECTRUM COMMUNICATION

Spread spectrum communication is a signal-transmission technique in which the transmitted carrier is spread out over a wide bandwidth. The FCC refers to this as *bandwidth expansion* modulation {§97.3 (c) (8)}.The technical details for spread spectrum are covered on the Amateur Extra question pool, so we won't go into those details here. If you are interested in learning more about this fascinating mode you can find the details in *The ARRL Handbook*. ARRL's *Spread Spectrum Sourcebook* also contains a wealth of information.

The FCC permits amateurs to experiment with two forms of spread spectrum communication. These are called *frequency hopping* spread spectrum and *direct sequence* spread spectrum. With frequency hopping SS, the center frequency of the carrier is altered many times per second according to a *pseudo*random list of channels. (Pseudo means it appears to be random, but actually follows a predetermined pattern known by the transmitting and receiving stations. With direct sequence SS, a very fast binary bit stream is used to shift the phase of the RF carrier. Again, the bit stream code appears to be random, but is known by the transmitting and receiving stations.

The FCC limits the maximum transmitter power for spread spectrum communications to 100 watts. The idea behind this communications mode is to spread a little power over a wide bandwidth to minimize interference, not concentrate a lot of power in a narrow bandwidth.

[Check your understanding of spread spectrum by turning to Chapter 10 and studying questions A1D01 through A1D03. Review this section if necessary.]

SHARED RESOURCES

When most hams talk about Amateur Radio frequency bands, they refer to the bands as though Amateur Radio operators are the exclusive occupants of those bands. That simply is not true, however, especially with regard to large portions of our UHF and microwave bands. We share many of these bands, or portions of them, with other users, often government stations of one sort or another. Sometimes amateurs are the primary users, and other times we are secondary. When Amateur Radio has a primary allocation, the other users are supposed to consider amateur activity and needs, and to modify their operations to avoid interference with the amateur stations. When Amateur Radio is the secondary user on a band, we must ensure that our amateur operations do not interfere with the other services. Amateurs must also accept any interference they receive from the primary users in this case.

Table 1-1 shows the Advanced class HF privileges, and also lists the sharing requirements for these bands. That table is taken from §97.301 (c) and §97.303 (a, b, c and d).

Table 1-5, taken from §97.301 (a) and §97.303 (f through q), shows the VHF and higher frequency bands and the sharing requirements for those bands. The UHF and microwave bands are more often shared with other users, partly because of their shorter distance propagation characteristics. A user in one geographic area is not likely to interfere with another user in a different geographic area even though they are operating on the same frequency.

The 420 to 430-MHz band segment is allocated to the fixed and mobile services in the international allocations table on a primary basis, worldwide. Canada has allocated this band segment to its fixed and mobile services, so US amateurs along the Canadian border are not permitted to transmit on these frequencies. An imaginary line, called **Line A**, runs roughly parallel to, and south of, the US-Canadian border. US stations north of this line may not transmit on the 420 to 430-MHz band. Part 97 gives an exact definition of Line A.

---

§97.3 (27) *Line A.* Begins at Aberdeen, WA, running by great circle arc to the intersection of 48° N, 120° W, thence along parallel 48° N, to the intersection of 95° W, thence by great circle arc through the southernmost point of Duluth, MN, thence by great circle arc to 45° N, 85° W, thence southward along meridian 85° W, to its intersection with parallel 41° N, thence along parallel 41° N, to its intersection with meridian 82° W, thence by great circle arc through the southernmost point of Bangor, ME, thence by great circle arc through the southernmost point of Searsport, ME, at which point it terminates.

Table 1-5

UHF and Higher Frequency Bands and Sharing Requirements

§97.301 Authorized frequency bands.

The following transmitting frequency bands are available to an amateur station located within 50 km of the Earth's surface, within the specified ITU Region, and outside any area where the amateur service is regulated by any authority other than the FCC.

(a) For a station having a control operator who has been granted an operator license of Technician, Technician Plus, General, Advanced, or Amateur Extra Class:

Wavelength Band	ITU Region 1	ITU Region 2	ITU Region 3	Sharing requirements See §97.303, Paragraph:
VHF	MHz	MHz	MHz	
6 m	—	50 - 54	50 - 54	(a)
2 m	144 - 146	144 - 148	144 - 148	(a)
1.25 m	—	222 - 225	—	(a)
UHF	MHz	MHz	MHz	
70 cm	430 - 440	420 - 450	420 - 450	(a), (b), (f)
33 cm	—	902 - 928	—	(a), (b), (g)
23 cm	1240 - 1300	1240 - 1300	1240 - 1300	(h), (i)
13 cm	2300 - 2310	2300 - 2310	2300 - 2310	(a), (b), (j)
13 cm	2390 - 2450	2390 - 2450	2390 - 2450	(a), (b), (j)
SHF	GHz	GHz	GHz	
9 cm	—	3.3 - 3.5	3.3 - 3.5	(a), (b), (k), (l)
5 cm	5.650 - 5.850	5.650 - 5.925	5.650 - 5.850	(a), (b), (m)
3 cm	10.00 - 10.50	10.00 - 10.50	10.00 - 10.50	(b), (c), (i), (n)
1.2 cm	24.00 - 24.25	24.00 - 24.25	24.00 - 24.25	(a), (b), (h), (o)
EHF	GHz	GHz	GHz	
6 mm	47.0 - 47.2	47.0 - 47.2	47.0 - 47.2	
4 mm	75.5 - 81.0	75.5 - 81.0	75.5 - 81.0	(b), (c), (h)
2.5 mm	119.98 - 120.02	119.98 - 120.02	119.98 - 120.02	(k), (p)
2 mm	142 - 149	142 - 149	142 - 149	(b), (c), (h), (k)
1 mm	241 - 250	241 - 250	241 - 250	(b), (c), (h), (q)
—	above 300	above 300	above 300	(k)

§97.303 Frequency sharing requirements.

(f) In the 70 cm band:

(1) No amateur station shall transmit from north of Line A in the 420 - 430 MHz segment.

(2) The 420 - 430 MHz segment is allocated to the amateur service in the United States on a secondary basis, and is allocated in the fixed and mobile (except aeronautical mobile) services in the International Table of Allocations on a primary basis. No amateur station transmitting in this

(Continued)

Commission's Rules   1-17

band shall cause harmful interference to, nor is protected from interference due to the operation of, stations authorized by other nations in the fixed and mobile (except aeronautical mobile) services.

(3) The 430 - 440 MHz segment is allocated to the amateur service on a secondary basis in ITU Regions 2 and 3. No amateur station transmitting in this band in ITU Regions 2 and 3 shall cause harmful interference to, nor is protected from interference due to the operation of, stations authorized by other nations in the radiolocation service. In ITU Region 1, the 430 - 440 MHz segment is allocated to the amateur service on a co-primary basis with the radiolocation service. As between these two services in this band in ITU Region 1, the basic principle that applies is the equality of right to operate. Amateur stations authorized by the United States and radiolocation stations authorized by other nations in ITU Region 1 shall operate so as not to cause harmful interference to each other.

(4) No amateur station transmitting in the 449.75 - 450.25 MHz segment shall cause interference to, nor is protected from interference due to the operation of stations in, the space operation service and the space research service or Government or non-Government stations for space telecommand.

(g) In the 33 cm band:

(1) No amateur station shall transmit from within the States of Colorado and Wyoming, bounded on the south by latitude 39° N, on the north by latitude 42° N, on the east by longitude 105° W, and on the west by longitude 108° W.[1] This band is allocated on a secondary basis to the amateur service subject to not causing harmful interference to, and not receiving protection from any interference due to the operation of, industrial, scientific and medical devices, automatic vehicle monitoring systems or Government stations authorized in this band.

(2) No amateur station shall transmit from those portions of the States of Texas and New Mexico bounded on the south by latitude 31° 41' N, on the north by latitude 34° 30' N, on the east by longitude 104° 11' W, and on the west by longitude 107° 30' W.

(h) No amateur station transmitting in the 23 cm band, the 3 cm band, the 24.05 - 24.25 GHz segment, the 76 - 81 GHz segment, the 144 - 149 GHz segment and the 241 - 248 GHz segment shall cause harmful interference to, nor is protected from interference due to the operation of, stations authorized by other nations in the radiolocation service.

(i) In the 1240 - 1260 MHz segment, no amateur station shall cause harmful interference to, nor is protected from interference due to the operation of, stations in the radionavigation-satellite service, the aeronautical radionavigation service, or the radiolocation service.

(j) In the 13 cm band:

(1) The amateur service is allocated on a secondary basis in all ITU Regions. In ITU Region 1, no amateur station shall cause harmful interference to, and is not protected from interference due to the operation of, stations authorized by other nations in the fixed service. In ITU Regions 2 and 3, no station shall cause harmful interference to, and is not protected from interference due to the operation of, stations authorized by other nations in the fixed, mobile and radiolocation services.

(2) In the United States, the 2300 - 2310 MHz segment is allocated to the amateur service on a co-secondary basis with the Government fixed and mobile services. In this segment, the fixed and mobile services must not cause harmful interference to the amateur service. No amateur station transmitting in the 2400 - 2450 MHz segment is protected from interference due to the operation of industrial, scientific and medical devices on 2450 MHz.

(k) No amateur station transmitting in the 3.332 - 3.339 GHz and 3.3458 - 3525 GHz segments, the 2.5 mm band, the 144.68 - 144.98 GHz, 145.45 - 145.75 GHz and 146.82 - 147.12 GHz segments and the 343 - 348 GHz segment shall cause harmful interference to stations in the radio

[1]Waived July 2, 1990, to permit amateurs in the restricted areas to transmit in the following segments: 902.0-902.4, 902.6-904.3, 904.7-925.3, 925.7-927.3 and 927.7-928.0 MHz

astronomy service. No amateur station transmitting in the 300 - 302 GHz, 324 - 326 GHz, 345 - 347 GHz, 363 - 365 GHz and 379 - 381 GHz segments shall cause harmful interference to stations in the space research service (passive) or Earth exploration-satellite service (passive).

(l) In the 9 cm band:

(1) In ITU Regions 2 and 3, the band is allocated to the amateur service on a secondary basis.

(2) In the United States, the band is allocated to the amateur service on a co-secondary basis with the non-Government radiolocation service.

(3) In the 3.3 - 3.4 GHz segment, no amateur station shall cause harmful interference to, nor is protected from interference due to the operation of, stations authorized by other nations in the fixed and fixed-satellite service.

(4) In the 3.4 - 3.5 GHz segment, no amateur station shall cause harmful interference to, nor is protected from interference due to the operation of, stations authorized by other nations in the fixed and fixed-satellite service.

(m) In the 5 cm band:

(1) In the 5.650 - 5.725 GHz segment, the amateur service is allocated in all ITU Regions on a co-secondary basis with the space research (deep space) service.

(2) In the 5.725 - 5.850 GHz segment, the amateur service is allocated in all ITU Regions on a secondary basis. No amateur station shall cause harmful interference to, nor is protected from interference due to the operation of, stations authorized by other nations in the fixed-satellite service in ITU Region 1.

(3) No amateur station transmitting in the 5.725 - 5.875 GHz segment is protected from interference due to the operation of industrial, scientific and medical devices operating on 5.8 GHz.

(4) In the 5.650 - 5.850 GHz segment, no amateur station shall cause harmful interference to, nor is protected from interference due to the operation of, stations authorized by other nations in the radiolocation service.

(5) In the 5.850 - 5.925 GHz segment, the amateur service is allocated in ITU Region 2 on a co-secondary basis with the radiolocation service. In the United States, the segment is allocated to the amateur service on a secondary basis to the non-Government fixed-satellite service. No amateur station shall cause harmful interference to, nor is protected from interference due to the operation of, stations authorized by other nations in the fixed, fixed-satellite and mobile services. No amateur station shall cause harmful interference to, nor is protected from interference due to the operation of, stations in the non-Government fixed-satellite service.

(n) In the 3 cm band:

(1) In the United States, the 3 cm band is allocated to the amateur service on a co-secondary basis with the non-government radiolocation service.

(2) In the 10.00 - 10.45 GHz segment in ITU Regions 1 and 3, no amateur station shall cause interference to, nor is protected from interference due to the operation of, stations authorized by other nations in the fixed and mobile services.

(o) No amateur station transmitting in the 1.2 cm band is protected from interference due to the operation of industrial, scientific and medical devices on 24.125 GHz. In the United States, the 24.05 - 24.25 GHz segment is allocated to the amateur service on a co-secondary basis with the non-government radiolocation and Government and non-government Earth exploration-satellite (active) services.

(p) The 2.5 mm band is allocated to the amateur service on a secondary basis. No amateur station transmitting in this band shall cause harmful interference to, nor is protected from interference due to the operation of, stations in the fixed, inter-satellite and mobile services.

(q) No amateur station transmitting in the 244 - 246 GHz segment of the 1 mm band is protected from interference due to the operation of industrial, scientific and medical devices on 245 GHz.

The FCC has allocated portions of the 421 to 430 MHz band to the Land Mobile Service within a 50-mile radius centered on Buffalo New York, Detroit Michigan and Cleveland Ohio, so amateurs in these areas must not cause harmful interference to the Land Mobile or government radiolocation users.

Certain other restrictions may apply if your station is within specific geographical regions. For example, there is an area in Maryland, West Virginia and Virginia surrounding the National Radio Astronomy Observatory. This area is known as the **National Radio Quiet Zone**. The NRQZ serves to protect the interests of the National Radio Astronomy Observatory at Green Bank, WV and also Naval Research Laboratory at Sugar Grove, WV.

We must also remember that we share air space with others besides radio services. Amateurs cannot build an antenna tower or other structure that extends more than 200 feet into the air without first obtaining FCC and Federal Aviation Agency (FAA) approval (in addition to any local zoning regulations). Also, if your station will be located within 1 mile of an airport, your antenna height is restricted further.

§97.15 Station antenna structures.

(a) Unless the amateur station licensee has received prior approval from the FCC, no antenna structure, including the radiating elements, tower, supports and all appurtenances, may be higher than 61 m (200 feet) above ground level at its site.

(b) Unless the amateur station licensee has received prior approval from the FCC, no antenna structure, at an airport or heliport that is available for public use and is listed in the *Airport Directory* of the current *Airman's Information Manual* or in either the *Alaska* or *Pacific Airman's Guide and Chart Supplement*; or at an airport or heliport under construction that is the subject of a notice or proposal on file with the FAA, and except for military airports, it is clearly indicated that the airport will be available for public use; or at an airport or heliport that is operated by the armed forces of the United States; or at a place near any of these airports or heliports, may be higher than:

(1) 1 m above the airport elevation for each 100 m from the nearest runway longer than 1 km within 6.1 km of the antenna structure.

(2) 2 m above the airport elevation for each 100 m from the nearest runway shorter than 1 km within 3.1 km of the antenna structure.

(3) 4 m above the airport elevation for each 100 m from the nearest landing pad within 1.5 km of the antenna structure.

(c) An amateur station antenna structure no higher than 6.1 m (20 feet) above ground level at its site or no higher than 6.1 m above any natural object or existing manmade structure, other than an antenna structure, is exempt from the requirements of paragraphs (a) and (b) of this Section.

(d) Further details as to whether an aeronautical study and/or obstruction marking and lighting may be required, and specifications for obstruction marking and lighting, are contained in Part 17 of the FCC Rules, *Construction, Marking, and Lighting of Antenna Structures.* To request approval to place an antenna structure higher than the limits specified in paragraphs (a), (b), and (c) of this Section, the licensee must notify the FAA on FAA Form 7460-1 and the FCC on FCC Form 854.

(e) Except as otherwise provided herein, a station antenna structure may be erected at heights and dimensions sufficient to accommodate amateur service communications. [State and local regulation of a station antenna structure must not preclude amateur service communications. Rather, it must reasonably accommodate such communications and must constitute the minimum practicable regulation to accomplish the state or local authority's legitimate purpose. See PRB-1, 101 FCC 2d 952 (1985) for details.]

[Turn to Chapter 10 now, and study questions A1E01 through A1E04 and questions A1E10 and A1E11. Review this section as needed.]

BUSINESS COMMUNICATIONS

The Amateur Service is a personal communications service, and is not intended for any type of business communications. Nevertheless, amateurs have always wondered just what constitutes a business communication. Part of the answer is obvious. If you are the owner of (or work for) Joe's Oil Delivery Service, you can't use Amateur Radio to communicate between the main office and the delivery trucks.

In general, an Amateur Radio operator may not accept payment in any form in return for sending communications of any type over an Amateur Radio station. (There is one exception, which involves an amateur who is paid to operate a club station when it transmits Morse code practice or information bulletins, as long as certain conditions are met.)

This means you can't accept payment for sending a message to your neighbor's Uncle Sam over Amateur Radio. It also means if you are the control operator of a repeater, you can't accept payment for providing communications services to another party. For example, if your club provides safety and logistics communications for the sponsor of a local marathon or other public event, you can't accept payment.

What about some other less obvious types of communications? Can you "advertise" over the air that you want to sell or trade a piece of your amateur station equipment? This is okay as long as you don't do it on a regular basis. So if you have one or more different pieces of equipment for sale every week, it begins to look like you are operating an amateur equipment distributorship over the air.

There are even some conditions when you can send a message to a business by Amateur Radio. If neither the amateur nor his or her employer has any pecuniary (money or other payment) interest in the communication, it is okay. Can you use the repeater autopatch to call ahead to the local pizza emporium to order dinner? Yes, as long as you aren't calling your own business, so you are not being compensated to make the call.

Section 97.113 of the FCC Rules gives the specific details about the business communications rules.

§97.113 Prohibited transmissions.

(a) No amateur station shall transmit:

(1) Communications specifically prohibited elsewhere in this Part;

(2) Communications for hire or for material compensation, direct or indirect, paid or promised, except as otherwise provided in these rules;

(3) Communications in which the station licensee or control operator has a pecuniary interest, including communications on behalf of an employer. Amateur operators may, however, notify other amateur operators of the availability for sale or trade of apparatus normally used in an amateur station, provided that such activity is not conducted on a regular basis;

(4) Music using a phone emission except as specifically provided elsewhere in this Section; communications intended to facilitate a criminal act; messages in codes or ciphers intended to obscure the meaning thereof, except as otherwise provided herein; obscene or indecent words or language; or false or deceptive messages, signals or identification;

(5) Communications, on a regular basis, which could reasonably be furnished alternatively through other radio services.

(b) An amateur station shall not engage in any form of broadcasting, nor may an amateur station transmit one-way communications except as specifically provided in these rules; nor shall an amateur station engage in any activity related to program production or news gathering for broadcasting purposes, except that communications directly related to the immediate safety of human life or the protection of property may be provided by amateur stations to broadcasters for dissemination to the public where no other means of communication is reasonably available before or at the time of the event.

(c) A control operator may accept compensation as an incident of a teaching position during periods of time when an amateur station is used by that teacher as a part of classroom instruction at an educational institution.

(d) The control operator of a club station may accept compensation for the periods of time when the station is transmitting telegraphy practice or information bulletins, provided that the station transmits such telegraphy practice and bulletins for at least 40 hours per week; schedules operations on at least six amateur service MF and HF bands using reasonable measures to maximize coverage; where the schedule of normal operating times and frequencies is published at least 30 days in advance of the actual transmissions; and where the control operator does not accept any direct or indirect compensation for any other service as a control operator.

(e) No station shall retransmit programs or signals emanating from any type of radio station other than an amateur station, except propagation and weather forecast information intended for use by the general public and originated from United States Government stations and communications, including incidental music, originating on United States Government frequencies between a space shuttle and its associated Earth stations. Prior approval for shuttle retransmissions must be obtained from the National Aeronautics and Space Administration. Such retransmissions must be for the exclusive use of amateur operators. Propagation, weather forecasts, and shuttle retransmissions may not be conducted on a regular basis, but only occasionally, as an incident of normal amateur radio communications.

(f) No amateur station, except an auxiliary, repeater or space station, may automatically retransmit the radio signals of other amateur stations.

[Review your understanding of these business communications rules by turning to Chapter 10 and studying questions A1E05 through A1E09. Review this section if you have difficulty with any of those questions.]

VOLUNTEER EXAMINATIONS

All amateur exams are conducted under the guidance of a Volunteer Examiner Coordinator. The FCC has made formal agreements with several VECs throughout the US to conduct amateur exams. These VECs use Volunteer Examiners to conduct the actual exams. Sections 97.501 through 97.527 cover all aspects of this amateur volunteer examination system. As an Advanced class licensee, you become eligible to participate in this system with much wider responsibilities than was possible with a Technician, Technician Plus or General class license.

Examination Specifications

The standards for each examination element are given in §97.503. Instructions regarding the topics covered on amateur exams, and the number of questions from each topic are also given in §97.503. There are even directions about how to prepare the various examination elements.

For example, several paragraphs are devoted to Morse code telegraphy examinations.

§97.503 Element standards.

(a) A telegraphy examination must be sufficient to prove that the examinee has the ability to send correctly by hand and to receive correctly by ear texts in the international Morse code at not less than the prescribed speed, using all the letters of the alphabet, numerals 0-9, period, comma, question mark, slant mark and prosigns AR, BT and SK.

(1) Element 1(A): 5 words per minute;

(2) Element 1(B): 13 words per minute;

(3) Element 1(C): 20 words per minute.

§97.507 Preparing an examination.

(a) Each telegraphy message and each written question set administered to an examinee must be prepared by a VE who has been granted an Amateur Extra Class operator license. A telegraphy message or written question set, however, may also be prepared for the following elements by a VE who has been granted an FCC operator license of the class indicated:

(1) Element 3(B): Advanced Class operator.

(2) Elements 1(A) and 3(A): Advanced or General Class operator.

(3) Element 2: Advanced, General, Technician, or Technician Plus Class operator.

(c) Each telegraphy message and each written question set administered to an examinee for an amateur operator license must be prepared, or obtained from a supplier, by the administering VEs according to instructions from the coordinating VEC.

(d) A telegraphy examination must consist of a message sent in the international Morse code at no less than the prescribed speed for a minimum of 5 minutes. The message must contain each required telegraphy character at least once. No message known to the examinee may be administered in a telegraphy examination. Each 5 letters of the alphabet must be counted as 1 word. Each numeral, punctuation mark and prosign must be counted as 2 letters of the alphabet.

The FCC specifies what examination credit must be given to an applicant who already holds an unexpired amateur operator license in §97.505. We can summarize these regulations by saying the applicant must be given credit for at least the exam elements required for the license.

§97.505 Element credit.

(a) The administering VEs must give credit as specified below to an examinee holding any of the following documents:

(1) An unexpired (or expired but within the grace period for renewal) FCC-granted Advanced Class operator license document:

Elements 1(B), 2, 3(A), 3(B), and 4(A).

(2) An unexpired (or expired but within the grace period for renewal) FCC-granted General Class operator license document: Elements 1(B), 2, 3(A), and 3(B).

(3) An unexpired (or expired but within the grace period for renewal) FCC-granted Technician Plus Class operator (including a Technician Class operator license granted before February 14, 1991) license document: Elements 1(A), 2, and 3(A).

(4) An unexpired (or expired but within the grace period for renewal) FCC-granted Technician Class operator license document: Elements 2 and 3(A).

(5) An unexpired (or expired but within the grace period for renewal) FCC-granted Novice Class operator license document: Elements 1(A) and 2.

(6) A CSCE: Each element the CSCE indicates the examinee passed within the previous 365 days.

(7) An unexpired (or expired for less than 5 years) FCC-issued commercial radiotelegraph operator license document or permit: Element 1(C).

(8) An expired or unexpired FCC-issued Technician Class operator license document granted before March 21, 1987: Element 3(B).

(9) An expired or unexpired FCC-issued Technician Class license document granted before February 14, 1991: Element 1(A).

(10) An unexpired (or expired but within the grace period for renewal), FCC-granted Novice, Technician Plus (including a Technician Class operator license granted before February 14, 1991), General, or Advanced Class operator license document, and a FCC Form 610 containing:

(i) A physician's certification stating that because the person is an individual with a severe handicap, the duration of which will extend for more than 365 days beyond the date of the certification, the person is unable to pass a 13 or 20 words per minute telegraphy examination; and

(ii) A release signed by the person permitting the disclosure to the FCC of medical information pertaining to the person's handicap: Element 1(C).

(b) No examination credit, except as herein provided, shall be allowed on the basis of holding or having held any other license grant or document.

Volunteer Examiner Qualifications

If you want to participate in the Amateur Radio examination process as a Volunteer Examiner, you must meet a set of specific qualifications. The VEC that you plan to work with determines if you meet these qualifications through a certifica-

ARRL / VEC
VOLUNTEER EXAMINER APPLICATION FORM

Please type or print clearly in ink

Control Number
(ARRL / VEC will assign)

☐ General
☐ Advanced
Call: _____ ☐ Extra

License expiration date: _____

Name: _____
              (First)   MI   (Last)

Mailing Address: _____

City: _____ State: _____ ZIP: _____

Day Phone: (_____) _____ Night Phone: (_____) _____

Was your license ever suspended or revoked?   ☐ Yes   ☐ No
Have you ever been disaccredited by another VEC?   ☐ Yes   ☐ No
If yes, which VEC(s) and when?
Do you have a call sign change pending with FCC?   ☐ Yes   ☐ No
Do you have any kind of Form 610 pending action with the FCC?   ☐ Yes   ☐ No

Person to contact if you cannot be reached? _____ _____
                                             (name)   (phone)

Mailing address where UPS or daytime delivery is *reliably* possible:

_____   _____
              (name)   (street address)

_____   _____   _____
             (city)   (state)   (zip)

Please list any foreign countries that you will be serving: _____
For instant accreditation, have you participated as a VE in another VEC program,
and is your accreditation in that program current?   ☐ Yes   ☐ No
If yes, which VEC? _____

CERTIFICATION

By signing this Application Form, I certify that to the best of my knowledge the above information AND the following statements are true:

1) I am at least 18 years of age.
2) I agree to comply with the FCC Rules-(see especially Subpart F—Section 97.515 [b]).
3) I agree to comply with examination procedures established by the ARRL as Volunteer Examiner Coordinator.
4) I understand that violation of the FCC Rules or willful noncompliance with the VEC will result in the loss of my VE accreditation, and could result in loss of my Amateur Radio operator and/or station licenses, or both.
5) I understand that even though I may be accredited as a VE, if I am not able or competent to perform certain VE functions required for any particular examination, I should not administer that examination (Section 97.525[a][3]).

_____   _____   _____
           (signature)   (call sign)   (date)

(Please attach a photocopy of your Amateur Radio license, and if applicable a photocopy of any other VEC accreditation held, to this application.)

Figure 1-3 — You can copy this Volunteer Examiner Application Form, fill it out and send it to the ARRL/VEC, 225 Main Street, Newington, CT 06111. Include a note that this form came from *The ARRL Advanced Class License Manual*, and request a copy of the VE Manual and additional information about becoming a Certified ARRL Volunteer Examiner.

tion process. You must receive your certification credentials before you will be permitted to assist at any amateur exams. (To learn more about this certification process, and to become a Volunteer Examiner, contact the ARRL/VEC Office, 225 Main Street, Newington, CT 06111 (203) 666-1541. Figure 1-3 is a sample VE Application Form, which you can copy and send to the ARRL/VEC to request a copy of the VE Manual and additional information.)

You must be at least 18 years old to serve as a VE. Your amateur license must never have been revoked or suspended. You may not own a significant interest in (or be an employee of) any company that is involved with the manufacture or distribution of Amateur Radio equipment or supplies.

All Amateur Radio exams are conducted by a team of at least three accredited Volunteer Examiners. The examiners must have passed all exam elements for the next higher class of license in order to administer exams for a particular license class. For example, to conduct a Novice examination, the examiners must hold an Amateur Extra, Advanced or General class license to conduct examinations for the Novice, Technician or Technician Plus licenses.

---

§97.509 Administering VE requirements.

(a) Each examination for an amateur operator license must be administered by 3 administering VEs at an examination session coordinated by a VEC. Before the session, the administering VEs must make a public announcement stating the location and time of the session. The number of examinees at the session may be limited.

(b) Each administering VE must:

(1) Be accredited by the coordinating VEC;

(2) Be at least 18 years of age;

(3) Be a person who has been granted an FCC amateur operator license document of the class specified below:

(i) Amateur Extra, Advanced, or General Class in order to administer a Novice, Technician, or Technician Plus Class operator license examination;

(ii) Amateur Extra Class in order to administer a General, Advanced, or Amateur Extra Class operator license examination.

(4) Not be a person whose grant of an amateur station license or amateur operator license has ever been revoked or suspended.

(5) Not own a significant interest in, or be an employee of, any company or other entity that is engaged in the manufacture or distribution of equipment used in connection with amateur station transmissions, or in the preparation or distribution of any publication used in preparation for obtaining amateur operator licenses. (An employee who does not normally communicate with that part of an entity engaged in the manufacture or distribution of such equipment, or in the preparation or distribution of any publication used in preparation for obtaining amateur operator licenses, may be an administering VE.)

(c) Each administering VE must be present and observing the examinee throughout the entire examination. The administering VEs are responsible for the proper conduct and necessary supervision of each examination. The administering VEs must immediately terminate the examination upon failure of the examinee to comply with their instructions.

(d) No VE may administer an examination to his or her spouse, children, grandchildren, stepchildren, parents, grandparents, stepparents, brothers, sisters, stepbrothers, stepsisters, aunts, uncles, nieces, nephews, and in-laws.

(e) No VE may administer or certify any examination by fraudulent means or for monetary or other consideration including reimbursement in any amount in excess of that permitted. Violation of this provision may result in the revocation of the grant of the VE's amateur station license and the suspension of the grant of the VE's amateur operator license.

(f) No examination that has been compromised shall be administered to any examinee. Neither the same telegraphy message nor the same question set may be re-administered to the same examinee.

(g) Passing a telegraphy receiving examination is adequate proof of an examinee's ability to both send and receive telegraphy. The administering VEs, however, may also include a sending segment in a telegraphy examination.

Notice that Volunteer Examiners may not be compensated for their services under any circumstances. Being paid to conduct an examination is different from recovering "out-of-pocket" costs, however. The FCC has made provisions for examiners (and VECs) to recover at least part of their expenses involved with conducting an examination. This reimbursement is paid by a fee that may be charged applicants at an exam session. The FCC specifies a maximum fee that may be charged during the calendar year, and the VEC and VEs may decide whether to charge the fee. (Since the fee tends to increase each year based on inflation, we won't list the current fee here. The FCC bases the maximum fee on a $4.00 fee for 1984, and adjusts it each year based on the Department of Labor's Consumer Price Index.)

There are severe penalties for fraudulently administering amateur examinations, or for administering exams for money or other considerations. Your amateur station license may be revoked and your operator's license may be suspended.

Completing the Paperwork

As with every other step in the process, the FCC describes the procedures the Volunteer Examiners must follow at the conclusion of the exam. Since the applicants are sure to be anxious to learn if they passed the exam, the VEs must grade the papers right away. They can't take them home to grade later! Of course they must also issue any Certificates of Successful Completion of Examination (CSCEs) earned by the applicants.

Within 10 days of the exam session, the VEs must forward the applications and exam paperwork to the Volunteer Examiner Coordinator that was responsible for coordinating the session. Note that the paperwork does not go directly to the FCC. It is first sent to the VEC. The VEC then screens the application and reviews the paperwork from the exam session and forwards the information to the FCC. The VEC must forward the paperwork to the FCC within 10 days of receiving it.

§97.509 Administering VE Requirements

(h) Upon completion of each examination element, the administering VEs must immediately grade the examinee's answers. The administering VEs are responsible for determining the correctness of the examinee's answers.

(i) When the examinee is credited for all examination elements required for the operator license sought, the administering VEs must certify on the examinee's application document that the applicant is qualified for the license.

(j) When the examinee does not score a passing grade on an examination element, the administering VEs must return the application document to the examinee and inform the examinee of the grade.

(k) The administering VEs must accommodate an examinee whose physical disabilities require a special examination procedure. The administering VEs may require a physician's certification indicating the nature of the disability before determining which, if any, special procedures must be used.

(l) The administering VEs must issue a CSCE to an examinee who scores a passing grade on an examination element.

(m) Within 10 days of the administration of a successful examination for an amateur operator license, the administering VEs must submit the application document to the coordinating VEC.

[Congratulations! You have completed your study of the Commission's Rules section of the question pool. Before you go on to the next chapter, though, you should turn to Chapter 10 and study questions A1F01 through A1F14. Review this last section of Chapter 1 if you have any difficulty with these questions.

CHAPTER 2 KEYWORDS KEYWORDS KEYWORDS

Amplitude modulation (AM) — A method of combining an information signal and a radio-frequency (RF) carrier signal. The amplitude of the RF signal is varied in a way that is controlled by the information signal. For fax (A3C) or SSTV (J3F or A3F) transmissions it refers to a method of superimposing picture information on the RF signal.

Baud — A unit of signaling speed equal to the number of discrete conditions or events per second. (For example, if the duration of a pulse is 3.33 milli-seconds, the signaling rate is 300 bauds or the reciprocal of 0.00333 seconds.)

Bandwidth — The frequency range (measured in hertz — Hz) over which a signal is stronger than some specified amount below the peak signal level. For example, a certain signal is at least half as strong as the peak power level over a range of ±3 kHz, so it has a 3-dB bandwidth of 6 kHz.

Cathode-ray tube (CRT) — An electron-beam tube in which the beam can be focused on a luminescent screen. The spot position can be varied to produce a pattern on the screen. CRTs are used in oscilloscopes and as the "picture tube" in television receivers.

Direct sequence (DS) spread spectrum — A **spread-spectrum** technique in which a very fast binary bit stream is used to shift the phase of an RF carrier.

Facsimile (fax) — The process of scanning pictures or images and converting the information into signals that can be used to form a likeness of the copy in another location. The pictures are often printed on paper for permanent display.

Fast-scan television (ATV) — A television system used by amateurs that employs the same video-signal standards as commercial TV.

Frequency hopping (FH) spread spectrum — A **spread-spectrum** technique in which the transmitter frequency is changed rapidly according to a pseudorandom list of channels.

Frequency modulation (FM) — A method of combining an information signal and a radio-frequency (RF) carrier signal. The instantaneous frequency of the RF signal is varied by an amount that depends on the frequency of the information signal at that instant. For fax (F3C) or SSTV (J3F) transmissions it refers to a method of superimposing picture information on the radio-frequency carrier. G3C and G3F refer to a method of varying the phase of the carrier wave. There is no practical difference in the way you receive phase-modulated signals from frequency-modulated ones.

Gray scale — A photographic term that defines a series of neutral densities (based on the percentage of incident light that is reflected from a surface), ranging from white to black.

Horizontal synchronization pulse — Part of a TV signal used by the receiver to keep the **cathode-ray tube (CRT)** electron-beam scan in step with the camera scanning beam. This pulse is transmitted at the beginning of each horizontal scan line.

Photocell — A solid-state device in which the voltage and current-conducting characteristics change as the amount of light striking the cell changes.

Photodetector — A device that produces an amplified signal that changes with the amount of light striking a light-sensitive surface.

Phototransistor — A bipolar transistor constructed so the base-emitter junction is exposed to incident light. When light strikes this surface, current is generated at the junction, and this current is then amplified by transistor action.

Pseudonoise (PN) — A signal that *appears* to be noise because of its random properties. A pseudonoise signal is used to produce (or receive) spread-spectrum communication.

Slow-scan television (SSTV) — A television system used by amateurs to transmit pictures within a signal bandwidth allowed on the HF bands by the FCC. It takes approximately 8 seconds to send a single black and white SSTV frame, and between 12 seconds and 4½ minutes for the various color systems currently in use.

Spread-spectrum (SS) communication — A communications method in which the RF bandwidth of the transmitted signal is much larger than that needed for traditional modulation schemes, and in which the RF bandwidth is independent of the modulation content. The frequency or phase of the RF carrier changes very rapidly according to a particular pseudorandom sequence. SS systems are resistant to interference because signals not using the same spreading sequence code are suppressed in the receiver.

SSTV scan converter — A device that uses digital signal-processing techniques to change the output from a normal TV camera into an SSTV signal or to change a received SSTV signal to one that can be displayed on a normal TV.

Sync — Having two or more signals in step with each other, or occurring at the same time. A pulse on a TV or fax signal that ensures the transmitted and received images start at the same point.

Vertical synchronization pulse — Part of a TV signal used by the receiver to keep the CRT electron-beam scan in step with the camera scanning beam. This pulse returns the beam to the top edge of the screen at the proper time.

OPERATING PROCEDURES

As an Advanced class Amateur Radio licensee, you will be expected to know standard operating practices used on the amateur bands. For the Novice class license (FCC Element 2), you had to learn about a few procedures, such as the RST reporting system, how to tune a transmitter and how to zero beat a received signal. For the Technician and General exams, you learned more standard operating procedures, such as how to operate radiotelephone, how to use RTTY, and repeater techniques. These represented some of the new operating privileges that go with the higher-class license. As you increase your Amateur Radio knowledge and experience while working toward the Advanced class license, you will want to try some of the more exotic modes available to you.

There is one question on the Advanced class license exam from the Operating Procedures subelement. The exam questions in this section are grouped into a single syllabus point:

A2A Facsimile communications; slow-scan TV transmissions; spread-spectrum transmissions; HF digital communications (ie, PacTOR, CLOVER, HF packet); automatic HF forwarding.

FACSIMILE SYSTEMS

Facsimile (fax) is the earliest operating system used for image transmission over radio. Originally, fax systems used a printed picture or image, which was scanned to produce an electrical signal for transmission. The received signals were converted to a printed picture again, for permanent display. Today, many amateurs use computer systems to operate fax, and the pictures can be displayed on a computer monitor and stored digitally. The pictures have high resolution, which is achieved by using from 500 to several thousand lines per frame.

To keep the transmitted-signal **bandwidth** within the narrow limits allowed by the FCC on the HF bands, relatively long transmission times are used for fax pictures. Depending on the system in use, it may take from a bit more than three minutes up to 15 minutes to receive a single frame. Of course if it takes longer than 10 minutes for an amateur station to transmit the frame, there has to be a pause for station identification. That presents additional problems, and hams seldom use such systems.

Many amateurs use fax to receive weather-satellite photos that are transmitted direct from space. News services also use this mode to distribute "wirephotos" throughout the world, and fax systems are becoming increasingly common in business and law-enforcement offices to transfer documents, signatures, fingerprints and photographs.

Transmission

A typical modern fax system uses electronics to scan the image and produce an electrical signal that can be transmitted over a telephone line or used to modulate an RF signal for radio transmission. A small spot of light is focused on the printed material, and reflected light is picked up by a **photocell**, **photodetector** or **phototransistor**. The photodetector converts variations in picture brightness and darkness into voltage variations. The light source and sensor scan the entire picture area.

Voltage variations from the light pickup are amplified and used to modulate an audio subcarrier signal. Either **amplitude modulation (AM)** (A3C) or **frequency modulation (FM)** (F3C or G3C) methods may be used, with FM being the standard for amateur HF operation and some of the press services using AM.

The important characteristics for any fax system include the number of scan lines per frame, the scanning speed or transmission speed (which determines the scan density of the picture) and the modulation characteristics. These characteristics must be the same for the receiving and transmitting stations. Table 2-1 lists standards used by some of the common fax services. The 240 lines-per-minute speed

Table 2-1
Standards for Various Fax Services

Service	Transmission Speed (lines per minute)	Size (inches)	Scan Density (lines per inch)
WEFAX Satellite	240	11	75
APT Satellite	240	11	166
Weather Charts	120	19	96
Wire photos	90	11	96
Wire photos	180	11	166

is a good rate for amateur HF use because a detailed picture (800 lines) can be transmitted in about 3.3 minutes. If a speed of 120 lines-per-minute is used, then it takes approximately 6 minutes to transmit a complete picture.

A computerized fax system using 480 scan lines transmitted at a speed of 138.3 seconds per frame is becoming quite popular with amateurs. The 480-line system provides good detail on a computer VGA monitor set to 640 pixels per line, 480 scan lines and 16 levels of gray. A simple interface and free "FAX480" software (available on ARRL's phone-line BBS — 203-666-0578 — and from other sources) is all you need to experience this mode. The FAX480 system was developed by Ralph Taggart, WB8DQT, and first described in the February 1993 issue of *QST*.

Scan density is usually expressed as a number of lines per inch. This parameter determines the aspect ratio of the picture (image width/height) and the transmission time for one frame.

The FCC permits fax operation in the voice segments of all amateur bands. Note that there is no voice — or fax — operation allowed on 30 meters, though. The FCC rules require you to identify your station every 10 minutes during a transmission or series of transmissions, and at the beginning and end of the communications. You may give this identification by sending a fax picture that includes your call sign, although it is a good idea to also give your ID by voice or CW for those stations not copying your pictures.

Reception

To receive fax pictures, the characteristics of your system must match those of the transmitting station. Amateurs commonly use the 240 line-per-minute standard or the "FAX480" system.

There are several methods available for displaying a received fax picture. Most amateurs who operate fax use some type of computerized system. Some have dedicated microprocessor-based units while others have an interface and software for a personal computer of some type. The principle advantages over old-fashioned mechanical recorders are the ability to process the signals digitally during reception, and the ability to store the images digitally for later display or printing.

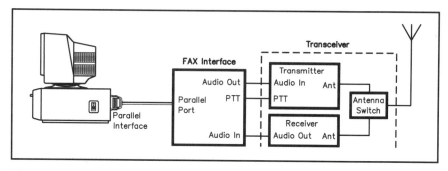

Figure 2-1 — This block diagram shows a computerized system for transmitting and receiving pictures by facsimile (fax).

Figure 2-1 shows a simplified block diagram of a computerized fax system.

Computer systems use high-resolution TV-monitor displays to provide a 16-level (or more) **gray scale** and good resolution. Permanent hard copy is possible with a graphics printer. The computer makes it possible to display inverted images, perform various types of image enhancement, and even produce "color" images. Figure 2-2 shows a fax picture received by Ralph Taggart, WB8DQT, using his FAX480 format.

[Before proceeding, study questions A2A01 through A2A05 in Chapter 10. Review this section as needed.]

Figure 2-2 — This is an example of a fax picture received by Ralph Taggart, WB8DQT, using his computerized FAX480 system.

SLOW-SCAN TELEVISION TRANSMISSION

A commercial TV signal has a bandwidth greater than 5 MHz. This is more frequency space than is available in all of the amateur bands below 6 meters combined. Obviously, if we want to send TV pictures over the air using the HF bands, we cannot use commercial-style techniques. Another stumbling block is the FCC's requirement that we limit our image signals to no more than the bandwidth of a single-sideband voice signal in certain portions of the HF bands.

Slow-scan television (SSTV) is a form of television that has a very slow scan rate. This makes it possible to keep the bandwidth within the required limitations. A regular fast-scan TV signal produces 30 frames every second, but a black and white SSTV system takes 8 seconds to send one frame! It takes considerably longer than that to send a frame of color SSTV. Amateurs can only use **fast-scan television (ATV)** on frequencies above 420 MHz. Because of the slow scan rate, SSTV can be used in portions of all HF amateur bands except 30 meters. Many DX stations are now equipped for this mode, and some amateurs have contacted more than 100 countries on SSTV.

If you think of ATV as watching home movies by radio, then SSTV is like a slide show over the air. Only still pictures are suitable for transmission by SSTV, at least for live transmission or immediate display as the pictures are being received. (It is possible to devise a system that would digitally store a series of frames, transmit them as slow-scan signals, and then store them at the receiving end for later display in a rapid sequence that would convey motion.)

Black and White SSTV Technical Details

The bandwidth of any radio signal is directly proportional to the data rate. The more information that is being transmitted in a given amount of time, the higher the bandwidth. This is true regardless of the signal type: CW, voice or other. Instead of the normal rate of 30 frames per second and 525 lines per frame, an SSTV picture takes 8 seconds for one black and white frame and has only 120 scan lines.

Table 2-2
Black and White SSTV Standards

Frame time	8 seconds
Lines per frame	120
Time to send one line	67 ms
Duration of horizontal sync pulse	5 ms
Duration of vertical sync pulse	30 ms
Horizontal and vertical sync frequency	1200 Hz
Black frequency	1500 Hz
White frequency	2300 Hz

This works out to 15 picture lines per second instead of 15,750 lines per second. It is easy to see that this slower data rate results in a greatly reduced bandwidth requirement.

The video information is normally transmitted as a frequency-modulated audio subcarrier. This would be designated as an F3F, J3F or G3F emission, although A3F is also permitted. For a black and white system, a 1500-Hz signal is used to produce black, and a 2300-Hz signal produces white on the TV screen. Frequencies between these two represent shades of gray. The **vertical synchronization pulse** and **horizontal synchronization pulses** must be added to keep transmitting and receiving systems in **sync**. These pulses are sent as bursts of 1200-Hz tones. Since this represents a shade "blacker than black," the sync pulses do not show up on the screen. Table 2-2 summarizes these SSTV standards. The horizontal sync pulse is included in the time to send one line, but the vertical sync pulse adds 30 ms of "overhead." So it takes a bit more than 8 seconds to actually transmit a black and white picture frame.

Color SSTV Technical Details

Early experiments with color transmission involved the use of red, green and blue filters between the camera and the image. Red, green and blue filters were also used between the **cathode-ray tube (CRT)** and a color Polaroid camera. A set of three frames of the same picture would be sent in succession, with the filter colors switched each time. The receiving station would record each new frame on the same piece of film, and if the camera alignment was not disturbed, a color picture would result. While this system worked, it was not very reliable or elegant. It is also interesting to note that CRTs with special, long-acting phosphors were needed to display the images because of the long scan times. The CRT had to "hold" the display at the top of the picture until it finished scanning the bottom, 8 seconds later.

With the advent of **SSTV scan-converters** to change a slow-scan image to a fast-scan format, new methods became feasible. Either a black-and-white camera with filters or a color camera can be used to produce the picture. The scan converter includes a memory area for each color, so three frames are transmitted and recombined for display on a color monitor at the receiving end.

Figure 2-3 is a block diagram of such a system. Various formats are being used. You can send a complete

Table 2-3
Color SSTV Standards

Format	Name	Time (sec)	Lines
Wraase SC-1	24	24	128
	48	48	256
	96	96	256
Martin	M1	114	256
	M2	58	256
	M3	57	128
	M4	29	128
Scottie	S1	110	256
	S2	71	256
	S3	55	128
	S4	36	128

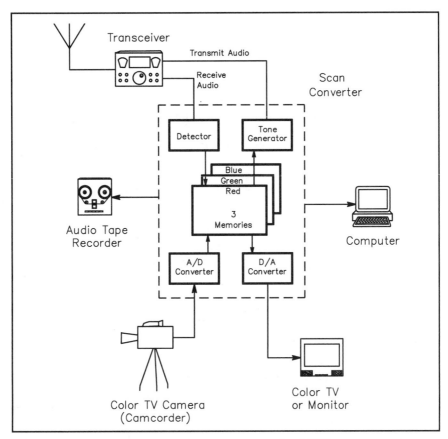

Figure 2-3 — This block diagram represents a color SSTV scan converter system. There are several formats for transmitting the red, green and blue picture information. A similar system with only one memory could be used to send and receive black-and-white pictures.

frame of each color, or a single line can be repeated for the red, green and blue information, then the next line sent and so on. If only one memory is used with the black-and-white camera, then you would have a basic SSTV scan-converter system.

There are several techniques being used to transmit color pictures by slow-scan TV. Most of these systems transmit either 120, 128, 240 or 256 scan lines, and can take from 12 seconds to 269 seconds to transmit a single frame. The 128-line and 256-line systems seem to be the most common. Table 2-3 summarizes the characteristics of some of the popular color SSTV formats.

Equipment

An SSTV camera produces a variable-frequency audio tone. Higher tones represent light areas, lower tones represent dark areas. To transmit the SSTV picture, simply feed this signal into the microphone input of your SSB transmitter. To receive a picture, just tune in a signal on an SSB receiver and feed the audio signal to your SSTV monitor or scan converter for display on a normal TV or computer monitor.

Even if you don't have a camera to start with, you can become involved in SSTV operation. Since the transmitted pictures are just audio tones, they can be recorded off the air using a normal audio tape recorder. You can play the signals back through your monitor for viewing again later and you can feed the audio into your microphone jack to retransmit the pictures. Another SSTVer may be willing to prerecord some pictures from his camera for your use.

With the popularity of home computers, there are some programs available to generate SSTV audio tones, so this is another way to get in on the SSTV fun. Computer graphics open up many new possibilities for creating your own SSTV pictures. You can store received pictures on disk for later retransmission, and you can also create images from photos and other video sources.

One very important equipment consideration is that SSTV is a 100%-duty-cycle transmission mode. This means your transmitter will be producing full power for the entire picture-transmission time. Most SSB transmitters and amplifiers will have to be run at reduced power output to avoid overheating the components.

Camera Procedures

Most amateurs use a fast-scan camera, and convert the signal to the slow-scan format with a scan converter. With this type of system, you can watch the picture on a normal TV screen or the camera viewfinder.

Lighting is critical with SSTV because the reproducible brightness range is rather limited. You should adjust the contrast and brightness controls on your monitor by using a test pattern. You might do this by asking another station to transmit a test pattern over the air for you, or by having such a pattern recorded on tape. Test patterns typically have several vertical bars of various shades of gray. Proper monitor calibration is achieved when the darkest bar is just dark enough to blacken the screen while the brightest bar just reaches maximum brightness.

Now you are ready to adjust the brightness and contrast controls on your camera. Set them so the darkest part of the image is black on your calibrated monitor screen and the brightest areas just reach maximum brightness.

Your subjects don't all have to be either live or on tape. Some SSTVers find it convenient to mount the camera in front of an easel. Various photographs, drawings or lettered signs can be placed on the easel for convenient display. The variety of subjects that you may choose to transmit is almost endless, but you will find that high-contrast pictures work best. They should not be cluttered with too much fine detail. Some possibilities are a close-in shot of you at your operating position, cartoons (commercial or homemade) and a couple of frames with your call and perhaps your name and QTH printed on them.

General Operating

SSTV has good potential for providing service to the general public by means of third-party traffic. For example, the scientists working on the Antarctic ice pack don't see their families for months — except by amateur slow-scan TV. People can write or call almost anywhere in the world today using commercial services. But how often can grandma and grandpa see their grandchildren from 3000 miles away? SSTV provides this opportunity, and without a charge!

SSTV is legal in the voice segments of all bands except on 30 meters, although it is primarily used on HF. Standard calling frequencies are 3.845, 7.171, 14.230, 21.340 and 28.680 MHz. The most popular bands are 20 and 75 meters.

Many operators send a couple of frames followed by voice comments concerning the picture, alternating between picture and voice signals on one frequency. It is important to announce the color SSTV format of the picture you are about to send, because some scan converters must be set to the proper mode before receiving the picture. You will probably hear an operator make an announcement such as, "Here comes a Scotty 1 picture," or "This will be in Martin 2." See Table 2-3 for the names of some of these popular color SSTV modes.

Some of the more elaborate stations are equipped to transmit video and audio simultaneously, with the voice on one sideband and the video on the other. These independent sideband (ISB) signals can be copied with two receivers or a single receiver with two IF filters, two detectors and two audio amplifiers.

An SSTV signal must be tuned in properly so that the picture will come out with the proper brightness and so the 1200-Hz sync pulses are detected properly. If the signal is not "in sync," the picture will appear wildly skewed. The easiest way to tune a signal is to wait for the operator to say something and then fine tune him or her. With some experience, you may find that you are able to zero in on an SSTV signal by listening to the sync pulses and watching for proper synchronization on the screen. Many SSTV monitors are equipped with some type of tuning aid.

If you want to record slow-scan pictures off the air, there are two ways of doing it. One is to tape record the audio signal for later playback. The other method is to store the image in digital form on a computer system after the received signal has been converted for display on a computer monitor.

In general, SSTV operating procedures are quite similar to those used on SSB. The FCC requires you to identify your station at the beginning and end of a transmission or series of transmissions and at not more than a 10-minute interval during a single long transmission. You may use SSTV pictures to identify your station in meeting this requirement. We recommend that you also use a CW or voice ID for the benefit of those who may be listening but are unable to copy your SSTV images. You may spark their interest in the mode when they hear you talking about the pictures you are sending.

[Before moving on to the next section, study those questions in Chapter 10 with numbers from A2A06 through A2A09. Review this section as needed.]

SPREAD-SPECTRUM COMMUNICATIONS

When Amateur Radio communication is disturbed by interference from other stations on nearby frequencies, the most common "solutions" are to increase transmitter power or use narrower bandwidth receive filters. The increased transmitter power is intended to "punch through" the interfering signals, while the narrower bandwidth receive filters are intended to reject more of the unwanted signals.

Amateurs can take a different approach to solving this problem, however, by using **spread-spectrum communication**. SS, as it is often called, actually reduces the average power transmitted on any one frequency, and distributes the signals over a much wider bandwidth. Figure 2-4 illustrates the effects of spreading the same transmitter power over a wider and wider bandwidth. The average power transmitted on any one frequency decreases, as long as the total transmitter power remains the same.

There are many ways to cause a carrier to spread over a range of frequencies, but all SS systems can be considered as a combination of two modulation processes. First, the information to be transmitted is applied to the carrier. A conventional form of analog or digital modulation is used for this step. Second, the carrier is modulated by a "spreading code." The spreading code distributes the carrier over

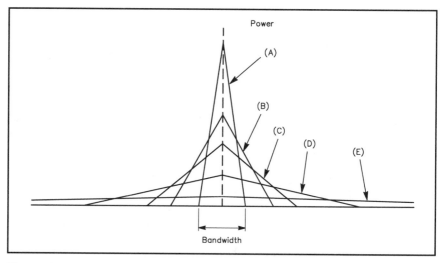

Figure 2-4 — This graph represents the average power distribution over a frequency range as the signal bandwidth increases. Signal A contains most of its energy in a narrow range around the center frequency. As the bandwidth increases and the transmitter power remains the same, the power at the center frequency decreases, as at B. Signals C and D show how the energy is distributed across the spread signal bandwidth. At E, the signal energy is spread over a very wide bandwidth, and there is little power at any one frequency.

a large bandwidth. Four spreading techniques are commonly used in military and space communications, but amateurs are only authorized to use two of them; frequency hopping and direct sequence.

Frequency Hopping

Frequency hopping (FH) spread spectrum is a form of spreading in which the center frequency of a conventional carrier is altered many times per second in accordance with a list of frequency channels. This list is called a *pseudorandom* list because the channel-selection order appears to be random. The order is not truly random, however, because the receiver has to be able to follow it. In other words, there doesn't appear to be a set pattern, such as going through the frequencies in direct increasing order. The frequency channels are selected to avoid interference to fixed-frequency users such as repeaters. FH is generally used to transmit single-sideband (SSB) signals.

As an example, suppose a transmitter can transmit on any of 100 discrete frequencies, which we can call F1 through F100. Now suppose the transmitter operates for 1 second on F1, then jumps to F62 and transmits for another second before going to F33 to transmit for a second and then on to F47, and so on. The frequency pattern seems random, but as long as the receiver knows the sequence, it can also jump from F1 to F62 to F33 to F47 in sync with the transmitter. There may be a signal on F33 that will interfere with the SS station, but that will only last for 1 second. Likewise, the SS signal may cause some interference to the station on F33, but only for a second. In either case the interference is barely perceptible.

Frequency hopping spread spectrum works because the frequency of the RF carrier is changed very rapidly according to a particular pseudorandom sequence of frequencies. There are many possible sequences for the same set of frequencies, so many SS contacts can take place simultaneously without interference. Signals from stations not using SS and from stations using a different spreading sequence are suppressed in the receiver, because the receiver is changing frequency in step with the transmitter. Spread spectrum signals present little or no interference to other stations because they only remain on any one frequency for a brief instant.

Direct Sequence

In **direct sequence (DS) spread spectrum**, a binary bit stream is used to shift the phase of an RF carrier very rapidly in a particular pattern. The binary sequence is designed to appear to be random (that is, a mix of approximately equal numbers of zeros and ones), but the sequence is generated by a digital circuit. The binary sequence can be duplicated and synchronized at the transmitter and receiver. This sequence is called **pseudonoise**, or **PN**, because the signal only *appears* to be random. The transmitting and receiving systems must both use the same spreading codes, so the receiver will know where to look next for the transmitted signal. DS spread spectrum is typically used to transmit digital information.

[Now turn to Chapter 10 and study questions A2A10 and A2A11. Review this section if either of these questions give you trouble.]

HF DIGITAL COMMUNICATIONS

Communications modes using personal computers have become very popular on all the amateur bands. You can hear many types of computerized communications on the HF bands. These modes all use similar operating procedures. In general, you will type your message on a computer keyboard or retrieve a message stored in a computer file. Your station will send the message and the receiving station will display the message on a computer screen or monitor, or store the message as a computer file.

Modes such as RTTY, ASCII, AMTOR, PacTOR, G-TOR, CLOVER and packet all transmit information that has been encoded as digital data. This means the text or other information is converted to a sequence of zeros and ones, the language of digital computers.

All these modes except RTTY provide some means of error detection or correction. The actual process of encoding, decoding, detecting and correcting errors is usually transparent to the operator. Many stations use a multimode communications processor between a computer and a radio to transmit and receive information. Figure 2-5 shows a block diagram of a typical system.

PacTOR was developed in Germany, and combines features from AMTOR and packet radio to produce a system that is superior to either. The CLOVER system uses a unique modulation waveform and data transfer protocol to provide high data rates on HF. CLOVER is a proprietary system of HAL Communications. G-TOR is another system developed by Kantronics to provide improved data flow over noisy HF communications circuits.

The FCC specifies that the maximum symbol rate for digital communications on the HF bands must not exceed 300 bauds. A **baud** is a unit of signaling speed equal to the number of discrete conditions or events per second. (For example, if the duration of a pulse is 3.33 milliseconds, the signaling rate is 300 bauds or the reciprocal of 0.00333 seconds.) The most common data rate used for HF packet communications is 300 bauds.

[Before you go on to Chapter 3, turn to Chapter 10 and study question A2A12. Review this section if that question gives you any difficulty.]

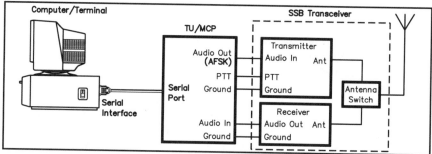

Figure 2-5 — This diagram shows a typical Amateur Radio HF digital communications station. Most multimode communications processors (MCP) can operate nearly all available data modes as well as SSTV and fax.

CHAPTER 3
KEY WORDS
KEY WORDS
KEY WORDS

Absorption — The loss of energy from an electromagnetic wave as it travels through any material. The energy may be converted to heat or other forms. Absorption usually refers to energy lost as the wave travels through the ionosphere.

Aurora — A disturbance of the atmosphere at high latitudes, which results from an interaction between electrically charged particles from the sun and the magnetic field of the Earth. Often a display of colored lights is produced, which is visible to those who are close enough to the magnetic-polar regions. Auroras can disrupt HF radio communication and enhance VHF communication. They are classified as visible auroras and radio auroras.

Equinoxes — One of two spots on the orbital path of the Earth around the sun, at which the Earth crosses a horizontal plane extending through the center of the sun. The *vernal equinox* marks the beginning of spring and the *autumnal equinox* marks the beginning of autumn .

Ground-wave propagation — An effect usually observed on the amateur 160 and 80-meter bands, in which the radio waves are bent slightly (diffracted) by the rounded edge of the surface of the Earth. Daytime propagation out to 120 miles or more is possible with ground-wave propagation.

K index — A geomagnetic-field measurement that is updated every three hours at Boulder, Colorado. Changes in the K index can be used to indicate HF propagation conditions. Rising values generally indicate disturbed conditions while falling values indicate improving conditions.

Multipath — A fading effect caused by the transmitted signal traveling to the receiving station over more than one path.

Pedersen ray — A high-angle radio wave that penetrates deeper into the F region of the ionosphere, so the wave is bent less than a lower-angle wave, and thus travels for some distance through the F region, returning to Earth at a distance farther than normally expected for single-hop propagation.

Polarization — A property of an electromagnetic wave that tells whether the electric field of the wave is oriented vertically or horizontally. The polarization sense can change from vertical to horizontal under some conditions, and can even be gradually rotating either in a clockwise (right-hand-circular polarization) or a counterclockwise (left-hand-circular polarization) direction.

Radio horizon — The position at which a direct wave radiated from an antenna becomes tangent to the surface of the Earth. Note that as the wave continues past the horizon, the wave gets higher and higher above the surface.

Selective fading — A variation of radio-wave field intensity that is different over small frequency changes. It may be caused by changes in the material that the wave is traveling through or changes in transmission path, among other things.

Solar wind — Electrically charged particles emitted by the sun, and traveling through space. The wind strength depends on how severe the disturbance on the sun was. These charged particles may have a sudden impact on radio communications when they arrive at the atmosphere of the Earth.

Sporadic-E propagation — A type of radio-wave propagation that occurs when *dense* patches of ionization form in the E layer of the ionosphere. These "clouds" reflect radio waves, extending the possible VHF communications range.

Summer solstice — One of two spots on the orbital path of the Earth around the sun at which the Earth reaches a point farthest from a horizontal plane extending through the center of the sun. With the North Pole inclined toward the sun, it marks the beginning of summer in the Northern Hemisphere.

Tropospheric ducting — A type of radio-wave propagation whereby the VHF communications range is greatly extended. Certain weather conditions cause portions of the troposphere to act like a duct or waveguide for the radio signals.

Winter solstice — One of two spots on the orbital path of the Earth around the sun at which the Earth reaches a point farthest from a horizontal plane extending through the center of the sun. With the North Pole inclined away from the sun, it marks the beginning of winter in the Northern Hemisphere.

RADIO-WAVE PROPAGATION

As you advance in your knowledge of Amateur Radio, and study for a higher license class, you earn the privilege of using additional frequencies and operating modes. You will also learn more about how radio waves travel from one place to another to carry the information you are transmitting. To pass examination Element 3B for a General class license, you learned many basic properties of radio-wave propagation. To pass the Element 4A exam for an Advanced class license, you must study some types of propagation that are a little more difficult to understand. Once you do, however, you will be ready to put that knowledge to use and enhance your enjoyment of our hobby.

The material in this chapter covers these areas of propagation: sporadic-E, auroral propagation, ground-wave propagation, selective fading, radio-path horizon, take-off angles over flat and sloping terrain and various Earth effects on propagation. Questions in the A3 section of the question pool test the information presented here. There will be two questions on your exam from the material covered in this chapter. Those questions will be taken from the two syllabus groups about Radio-Wave Propagation:

A3A   Sporadic-E; auroral propagation; ground-wave propagation (distances and coverage, and frequency versus distance in each of these topics).

Selective fading; radio-path horizon; take-off angle over flat or sloping terrain; earth effects on propagation.

You will be directed to the questions in Chapter 10 at appropriate places throughout this chapter. See *The ARRL Handbook* and *The ARRL Antenna Book* for further reading on propagation.

SPORADIC E

Sporadic-E propagation, also known as E-skip or E$_S$, occurs when radio waves are reflected by dense patches of ionization that form in the E region of the ionosphere, approximately 50 miles above the Earth. These ionized patches, or *clouds* as they are often called, form randomly and usually last up to a few hours at a time. Their transient nature, and the E-layer altitude, account for the name sporadic-E. Sporadic-E is, however, of a different origin and has different communications potential than E-layer propagation that affects mainly the 1.8 through 14-MHz bands.

E$_S$ Openings

Sporadic-E is an almost daily occurrence in the equatorial regions, but it is common in the temperate latitudes (roughly 15 to 45°) as well. Although it can occur at any time, sporadic-E propagation is most common in the Northern Hemisphere during May, June and July. This occurs around the **summer solstice**. There is a less-intense season at the end of December and early January, around the **winter solstice**. The long and short seasons are reversed in the Southern Hemisphere. Figure 3-1 shows the relative amount of sporadic-E activity during the year in northern latitudes. This propagation mode does not seem to depend on the sunspot cycle.

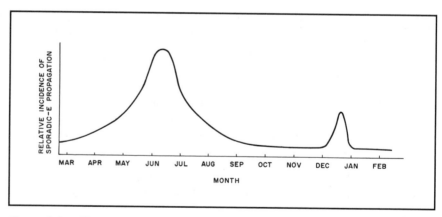

Figure 3-1 — There are two times during the year that sporadic-E propagation is most likely.

The duration and extent of E_S openings tends to be greater in the summer season. Late June and early July is usually the peak period. There are E_S openings almost every day then, and country-wide openings may last for many hours at a time. Midlatitude sporadic E is most likely to occur between about 9 AM and noon local time, and again in the early evening between about 5 and 8 PM. These events may last from a few minutes to several hours.

The MUF of intense E_S clouds is a function of their ionization density. The upper frequency limit for sporadic-E propagation is not known. It is observed fairly often up to about 100 MHz and amateur 222-MHz contacts have been confirmed using E_S. It is most common on the 28 and 50-MHz amateur bands. Although

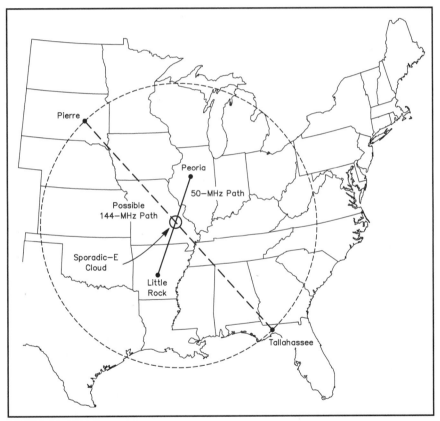

Figure 3-2 — Sporadic-E contacts of 450 miles or shorter (such as between Peoria, IL and Little Rock, AR) indicate that the MUF is above 144 MHz. Using the same sporadic-E reflecting point, 144 MHz contacts of 1300 miles (such as between Pierre, SD and Tallahassee, FL) should be possible.

sporadic-E is most often associated with 50-MHz work because it is the main DX mode there, it is often the only means of long-distance 28-MHz propagation during low-sunspot years, when F-layer propagation is not consistently available. E-skip is not nearly as common on 144 MHz; according to some estimates, only 5 to 10% of the 6-meter openings are accompanied by 2-meter openings. On the 2-meter band, openings are brief and extremely variable. Two-meter operators take full advantage of those occasions, however; thousands of 2-meter E_S contacts are made each summer.

Layer height and electron density determine the skip distance. On 28 and 50 MHz, sporadic-E propagation is most common over distances of about 300 to 1400 miles. If the skip distance on 10 meters is as short as 250 to 300 miles, it is an indication that the MUF has reached 50 MHz. Contacts out to the maximum of about 1400 miles should be possible on 6 meters. E-skip contacts as short as 450 miles on 6 meters may indicate that 1300 mile contacts on 2 meters are possible. See Figure 3-2.

Ionization develops rapidly, with effects first showing on lower frequencies. Amateurs interested in 6-meter sporadic-E work often monitor the lower VHF television channels and the 10-meter amateur band for signs of activity. They also monitor the FM broadcast band and TV channels 5 and 6 for indications of 2-meter sporadic-E propagation.

As ionization density increases and the MUF rises, the skip distance on a given band shortens. When the E-skip distance on 10 meters shortens to 500 miles or less, it is likely that E_S will be noted on 6 meters. Likewise, very short skip on 6 meters may signal a 2-meter opening.

On the air, signals arriving by sporadic E are often extremely strong. Signal strength varies as the E_S clouds shift, and at times signals may go from S-meter-pinning strength to barely audible in a matter of seconds. Also, the areas that are workable are very specific, changing as the clouds move. For example, a station in New England may work only stations in eastern Iowa, with excellent signal reports in both directions and then not hear them at all a few minutes later as the propagation shifts to stations in Minnesota. During intense E_S openings, minimal power is required for reliable communications. Often stations using a few watts and a dipole will work nearly the same range as stations running the legal power limit and a large Yagi array.

[Study the examination questions A3A01 through A3A04 in Chapter 10. Review this section as needed.]

AURORAL PROPAGATION

Auroral propagation occurs when VHF radio waves are reflected from ionization associated with an auroral curtain. It is a VHF and UHF propagation mode that allows contacts up to about 1400 miles. Auroral propagation occurs for stations near the northern and southern polar regions, but the discussion here is limited to auroral propagation in the Northern Hemisphere.

Aurora results from a large-scale interaction between the magnetic field of the Earth and electrically charged particles. During times of enhanced solar activity, electrically charged particles are ejected from the surface of the sun. These particles form a **solar wind**, which travels through space. If this solar wind travels toward Earth, then the charged particles interact with the magnetic field around the Earth. A visible aurora, often called the northern lights, or aurora borealis, is caused by the collision of these solar-wind particles with oxygen and nitrogen molecules in the upper atmosphere.

When the oxygen and nitrogen molecules are struck by the electrically charged particles in the solar wind, they are ionized. When the electrons that were knocked loose recombine with the molecules, light is produced. The extent of the ionization determines how bright the aurora will appear. At times, the ionization is so strong that it is able to reflect radio signals with frequencies above about 20 MHz. This ionization occurs at an altitude of about 70 miles, very near the E layer of the ionosphere. Not all auroral activity is intense enough to reflect radio signals, so a distinction is made between a visible aurora and a radio aurora.

The number of auroras (both visible and radio) varies with geomagnetic latitude. Generally, auroral propagation is available only to stations in the northern states, but, on occasion, extremely intense auroras reflect signals from stations as far south as the Carolinas. Auroral propagation is most common for stations in the northeastern states and adjacent areas of Canada, which are closest to the north magnetic pole. This mode is rare below about 32° north latitude in the southeast and about 38° to 40° N in the southwest. See Figure 3-3.

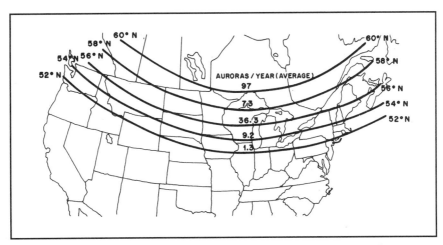

Figure 3-3 — The possibility of auroral propagation decreases as distance from the geomagnetic North Pole increases.

The number and distribution of auroras are related to the solar cycle. Auroras occur most often during sunspot peaks, but the peak of the auroral cycle appears to lag the solar-cycle peak by about two years. Intense auroras can, however, occur at any point in the solar cycle.

Auroras also follow seasonal patterns. Although they may occur at any time, they are most common around the **equinoxes** in March and September. Auroral propagation is most often observed in the late afternoon and early evening hours, and it usually lasts from a few minutes to many hours. Often, it will disappear for a few hours and reappear around midnight. Major auroras often start in the early afternoon and last until early morning the next day.

Using Aurora

Most common on 10, 6 and 2 meters, some auroral work has been done on 222 and 432 MHz. The number and duration of openings decreases rapidly as the operating frequency rises.

The reflecting properties of an aurora vary rapidly, so signals received via this mode are badly distorted by multipath effects. CW is the most effective mode for auroral work. CW signals have a fluttery tone. The tone is distorted, and is most

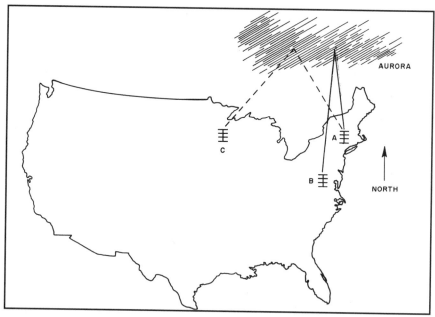

Figure 3-4 — To work the aurora, stations point their antennas north. Station A may have to beam west of north to work station C.

often a buzzing sound rather than a pure tone. For this reason, auroral propagation is often called the "buzz mode." SSB is usable for 6-meter auroral work if signals are strong; voices are often intelligible if the operator speaks slowly and distinctly. SSB is rarely usable at 2 meters or higher frequencies.

In addition to scattering radio signals, auroras have other effects on world-wide radio propagation. Communication below 20 MHz is disrupted in high latitudes, primarily by absorption, and is especially noticeable over polar and near-polar paths. Signals on the AM broadcast band through the 40-meter band late in the afternoon may become weak and "watery" sounding. The 20-meter band may close down altogether. At the same time, the MUF in equatorial regions may temporarily rise dramatically, providing transequatorial paths at frequencies as high as 50 MHz.

All stations point their antennas north during the aurora, and, in effect, "bounce" their signals off the auroral zone. The optimum antenna heading varies with the position of the aurora and may change rapidly, just as the visible aurora does. Constant probing with the antenna is recommended to peak signals, especially if the beamwidth is narrow. Usually, an eastern station will work the greatest distance to the west by aiming as far west of north as possible. The opposite applies for western stations working east. This does not always follow, however, so you should keep your antenna moving to find the best heading. See Figure 3-4.

You can observe developing auroral conditions by monitoring signals in the region between the broadcast band and 5 MHz or so. If, for example, signals in the 75-meter band begin to waver suddenly (flutter or sound "watery") in the afternoon

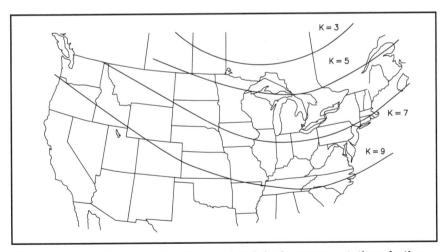

Figure 3-5 — As the intensity of auroral activity increases, stations farther south are able to take advantage of it. The numbers refer to the K index. See Figure 3-6.

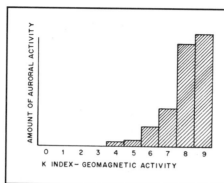

Figure 3-6 — It is possible to predict auroral activity by monitoring the K index during WWV broadcasts at 18 minutes past the hour. A K index of 3 or greater, and rising, may indicate auroral activity. The most auroral activity occurs with K-index values of 8 or 9.

or early evening hours, a radio aurora may be beginning. Since auroras are associated with solar disturbances, you can often predict one by listening to the WWV Geo alert broadcasts at 18 minutes after each hour. In particular, the **K index** may be used to indicate auroral activity. K-index values of 3, and rising, indicate that conditions associated with auroral propagation are present in the Boulder, Colorado, area. Timing and severity may be different elsewhere, however. Maximum occurrence of radio aurora is for K-index values of 7 to 9. See Figures 3-5 and 3-6.

On 6 and 2 meters, the buzzing sound that is characteristic of signals reflected by the aurora may be heard even on local signals when both beams are aimed north. The range of stations workable via aurora extends from local out to 1400 miles, but distances of a few hundred miles are most common. Range depends to some extent on transmitter power, antenna gain and receiver sensitivity, but as with most modes of operation, patience and operator skill are important.

[At this point, you should proceed to Chapter 10 and study examination questions A3A05 through A3A09. Review this section as needed.]

GROUND-WAVE PROPAGATION

We usually describe radio signals as waves that travel in straight lines. Signals that travel into the ionosphere can be *refracted* (bent) by the ionized gases, and return to Earth some distance away. This refraction occurs because the area of ionized gases cause the radio wave to slow down, and this bends the wave. Refraction is primarily an HF propagation mode.

We also know that radio signals can be reflected from a building, mountain or even areas of dense ionization such as sporadic-E clouds, aurora and meteor trails. The reflected signals allow communications between stations that could not otherwise hear each other. Such reflections can occur with radio signals of any frequency.

Radio signals can also *diffract* as they pass over the edge of an obstruction. *Knife-edge diffraction* occurs when VHF, UHF or microwave signals pass over the edge of a mountain or other obstruction and are bent into an area that would normally be in the shadow. Figure 3-7 shows how knife-edge diffraction can make communication possible between two stations that would not normally expect to hear each other.

Refraction and diffraction both involve bending of the waves. The primary difference is that refraction occurs because of changes in properties of the material through which the wave passes. Diffraction occurs because the wave passes over an edge of an obstruction in the path of the wave.

There is a special form of diffraction that primarily affects longer-wavelength radio waves that have vertical **polarization**. This diffraction produces **ground-wave propagation**. Ground-wave propagation is most noticeable on the AM broadcast band and the 160-meter and 80-meter amateur bands. Practical ground-wave communications distances on these bands often extend to 120 miles or more. Ground-wave loss increases significantly with shorter wavelengths (higher frequencies), so its effects are not noticeable even at 40 meters. As the frequency of radio signals increases, the ground-wave propagation distance decreases. Although the term *ground-wave propagation* is often applied to any short-distance communication, the actual mechanism is unique to the longer wavelength bands.

Radio waves are bent slightly (diffracted) as they pass over a sharp edge, but even rounded edges can diffract waves. At medium and long wavelengths, the curvature of the Earth looks like a rounded edge. Bending results when the lower part of the wave loses energy because of currents induced in the ground. This slows the lower portion of the wave, causing the entire wave to tilt forward slightly. This

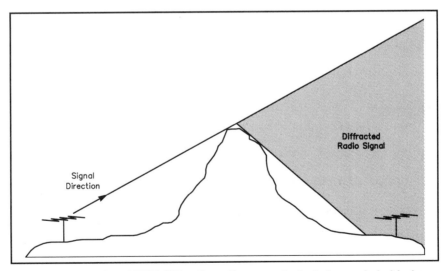

Figure 3-7 — VHF and UHF diffraction allows contacts to be made behind a mountain, where you may not normally expect signals to be heard.

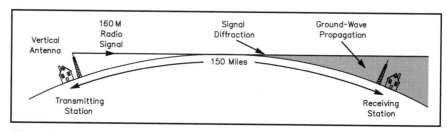

Figure 3-8 — Even the curved surface of the Earth horizon can diffract long-wavelength (low frequency) radio waves. The waves can follow the curvature of the Earth for as much as several hundred miles. This is called ground-wave propagation.

tilting follows the curvature of the Earth, allowing low-frequency signals to be heard over distances well beyond line of sight. See Figure 3-8.

Ground-wave propagation is most useful during the day at 1.8 and 3.5 MHz, when D-region **absorption** makes any skywave propagation impossible. Vertically polarized antennas with excellent ground systems provide the best results. Ground-wave losses are reduced considerably over saltwater and are highest over dry and rocky land.

One simple way to observe the effects of ground-wave propagation is to listen to stations on the AM broadcast band. During the day you will regularly hear stations 100 to 150 miles away. You won't hear stations much farther than 200 miles, however. At night, when the D-region absorption decreases, skywave propagation becomes possible, and you will begin to hear stations several hundred miles away. Of course AM broadcast stations usually have vertical antennas with excellent ground systems!

[Now turn to Chapter 10 and study questions A3A10 and A3A11. Review this section if these ground-wave propagation questions give you trouble.]

FADING

Fading is a general term used to describe variations in the strength of a received signal. It may be caused by natural phenomena such as constantly changing ionospheric-layer heights, variations in the amount of **absorption**, or random **polarization** shifts when the signal is refracted. Fading may also be caused by man-made phenomena such as reflections from passing aircraft and ionospheric disturbances caused by exhaust from large rocket engines.

Multipath

A common cause of fading is an effect known as **multipath**. Several components of the same transmitted signal may arrive at the receiving antenna from different directions, and the phase relationships between the signals may cancel or

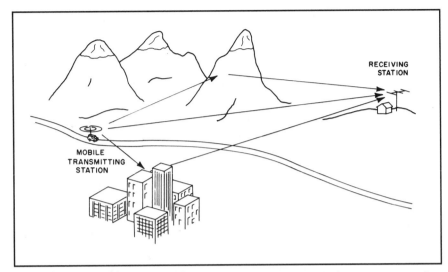

Figure 3-9 — If a signal travels from a transmitter to a receiver over several different paths, the signals may arrive at the receiver slightly out of phase. The out-of-phase signals alternately cancel and reinforce each other, and the result is a fading signal. This effect is known as multipath fading.

reinforce each other. Multipath fading is responsible for the effect known as "picket fencing" in VHF communications, when signals from a mobile station have a rapid fluttering quality. This fluttering is caused by the change in the paths taken by the transmitted signal to reach the receiving station as the mobile station moves. This effect is illustrated in Figure 3-9.

Multipath effects can occur whenever the transmitted signal follows more than one path to the receiving station. Some examples of this with HF propagation would be if part of the signal goes through the ionosphere and part follows a ground-wave path, or if the signal is split in the ionosphere and travels through different layers before reaching the receiving station. It is even possible to experience multipath fading if part of the signal follows the long path around the Earth to reach the receiver, while part of the signal follows the direct short-path route. When the transmitted signal reaches the receiver over several paths, the end result is a variable-strength signal.

Selective Fading

Selective fading is a type of fading that occurs when the wave path from a transmitting station to a receiving station varies with very small changes in frequency. It is possible for components of the same signal that are only a few kilohertz apart (such as the carrier and the sidebands in an AM signal) to be acted upon differently by the ionosphere, causing modulation sidebands to arrive at the receiver out of phase. Selective fading occurs because of phase differences between radio-wave components of the same transmission, as experienced at the receiving station. The result is distortion that may range from mild to severe.

Wideband signals, such as FM and double-sideband AM, suffer the most from selective fading. The sidebands may have different fading rates from each other or from the carrier. Distortion from selective fading is especially bad when the carrier of an FM or AM signal fades while the sidebands do not. In general, the distortion from selective fading is more pronounced for signals with wider bandwidths. It is worse with FM than it is with AM. SSB and CW signals, which have a narrower bandwidth, are affected less by selective fading.

[Now study examination questions A3B01 through A3B04 in Chapter 10. Review this section as needed.]

RADIO PATH HORIZON

In the early days of VHF amateur communications, it was generally believed that space-wave communications depended on direct line-of-sight paths between the communicating station antennas. After some experiments with good equipment and antennas, however, it became clear that radio waves are bent or scattered in several ways, and that reliable VHF and UHF communications are possible with stations beyond the visual horizon. The farthest point to which space waves will travel directly is called the **radio horizon**.

Under normal conditions, the structure of the atmosphere near the Earth causes radio waves to bend into a curved path that keeps them nearer to the Earth than true straight-line travel would. This bending of the radio waves is why the distance to

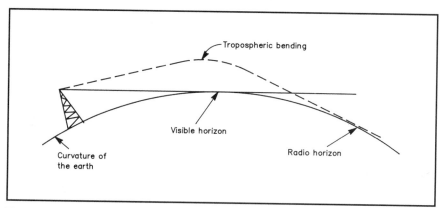

Figure 3-10 — Under normal conditions, tropospheric bending causes VHF and UHF radio waves to be returned to Earth beyond the visible horizon.

the radio horizon exceeds the distance to the visual, or geometric, horizon. See Figure 3-10. The distance to the radio horizon can be approximated by assuming that the waves travel in straight lines, but that the radius of the Earth is increased by one third. On this assumption, the distance from the transmitting antenna to the horizon is given by:

$$D(mi) = 1.415 \times \sqrt{H\,(ft)}$$ (Equation 3-1)

or

$$D(km) = 4.123 \times \sqrt{H\,(m)}$$ (Equation 3-2)

where D is the distance to the radio horizon and H is the height above average terrain of the transmitting antenna. The formula assumes that the Earth is perfectly smooth out to the horizon. Of course, any obstructions that rise along any given path must be taken into consideration. An antenna that is on a high hill or tall building well above any surrounding obstructions has a much farther radio horizon than an antenna located in a valley or shadowed by other obstructions.

With these equations, the point at the horizon is assumed to be on the ground. If the receiving antenna is also elevated, the maximum space-wave distance between the two antennas is equal to D + D1; that is, the sum of the distance to the horizon from the transmitting antenna plus the distance to the horizon from the receiving antenna. Figure 3-11 illustrates this principle.

Radio-horizon distances are shown graphically in Figure 3-12. The radio horizon is approximately 15% farther than the geometric horizon. To make best use of the space wave, the antenna must be as high as possible above the surroundings. This is why stations located high and in the clear on hills or mountaintops have a substantial advantage on the VHF and UHF bands, compared with stations in lower areas.

[It is time for another question review to check your understanding of the material covered in this section. Turn to Chapter 10 now and study questions A3B05 and A3B06. Review the material in this section if necessary.]

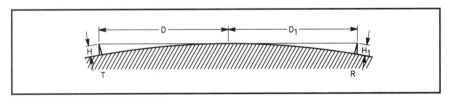

Figure 3-11 — The distance, D, to the radio horizon from an antenna of height H is given by the formula in the text. The maximum distance over which two stations may communicate by space wave is equal to the sum of their distances to the horizon.

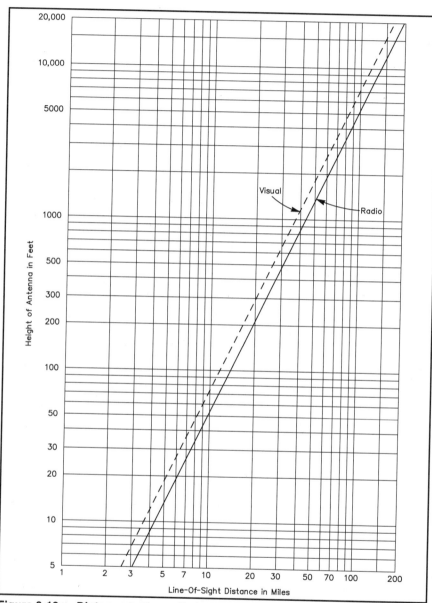

Figure 3-12 — Distance to the radio horizon from an antenna of given height above average terrain is indicated by the solid line. The broken line indicates the distance to the visual, or geometric, horizon. The radio horizon is approximately 15% farther than the visual horizon.

EARTH EFFECTS ON PROPAGATION

The space wave goes essentially in a straight line between the transmitter and the receiver. Antennas that are low-angle radiators (that is, antennas that concentrate the energy toward the horizon) are best. Energy radiated at angles above the horizon may pass over the receiving antenna.

In general, the radiation takeoff angle from a Yagi antenna with horizontally mounted elements decreases as the antenna height increases above flat ground. If you can raise the height of your antenna, the takeoff angle will decrease.

If the ground under your antenna is not flat, you may be able to use the sloping ground to your advantage. For example, a Yagi antenna mounted on the side of a hill, and pointed *away* from the hill, will have a lower radiation takeoff angle. The steeper the slope, the lower the takeoff angle will be. See Figure 3-13. Of course if the antenna is pointed *toward* the hill, the takeoff angle will be increased.

The polarization of both the receiving and transmitting antennas should be the same for VHF and UHF operation because the polarization of a space wave remains constant as it travels. There may be as much as 20 dB of signal loss between two stations that are using antennas with opposite polarizations.

As a radio wave travels in space, it collides with air molecules and other particles. When it collides with these particles, the radio wave gives up some energy. This is why there is a limit to distances that may be covered by space-wave communications.

VHF propagation is usually limited to distances of approximately 500 miles. This is the normal limit for stations using high-gain antennas, high power and sensitive receivers. At times, however, VHF communications are possible with stations up to 2000 or more miles away. Certain weather conditions cause ducts in the troposphere, simulating propagation within a waveguide. Such ducts cause VHF radio waves to follow the curvature of the Earth for hundreds, or thousands, of

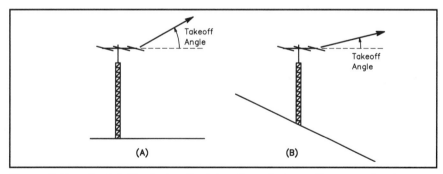

Figure 3-13 — Part A illustrates the takeoff angle for radio waves leaving a Yagi antenna with horizontal elements over flat ground. Higher antenna elevations result in smaller takeoff angles. Part B shows the takeoff angle for a similar antenna over sloping ground. For steeper slopes away from the front of the antenna, the takeoff angle gets smaller.

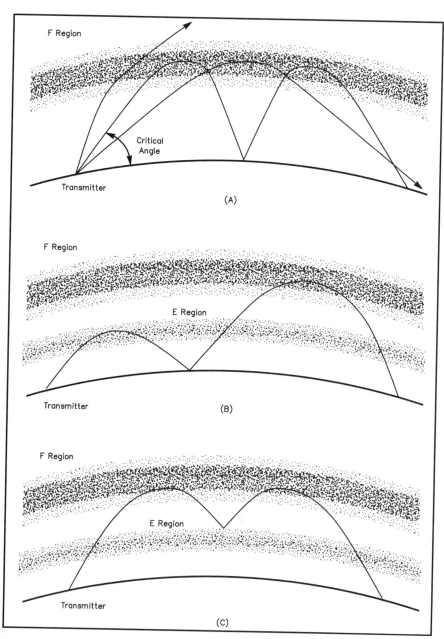

Figure 3-14 — Parts A, B and C show various multihop paths for radio signals that reach farther than 2500 miles.

miles. This form of propagation is called **tropospheric ducting**.

The possibility of propagating radio waves by tropospheric ducting increases with frequency. Ducting is rare on 50 MHz, fairly common on 144 MHz and more common on higher frequencies. Gulf Coast states see it often, and the Atlantic Seaboard, Great Lakes and Mississippi Valley areas see it occasionally, usually in September and October.

[Now turn to Chapter 10 and study examination questions A3B07, A3B08, A3B10, A3B11 and A3B12. Review this section if any of those questions give you difficulty.]

F-REGION PROPAGATION

The maximum one-hop skip distance for high-frequency radio signals is usually considered to be about 2500 miles. Most HF communication beyond that distance takes place by means of several ionospheric hops. Radio signals return to Earth from the ionosphere, and the surface of the Earth reflects them back into the ionosphere. It is also possible that signals may reflect between the E and F regions, or even be reflected several times within the F region. Figure 3-14 shows several possible paths for the signals to take in reaching some distant location.

There is a propagation theory that suggests radio waves may at times propagate for some distance through the F region of the ionosphere. This theory is supported by the results of propagation studies showing that a signal radiated at a medium elevation angle sometimes reaches the Earth at a greater distance than a lower-angle wave. This higher-angle wave, called the **Pedersen ray**, is believed to

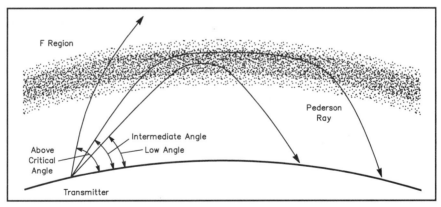

Figure 3-15 — This diagram shows a radio wave entering the F region at an intermediate angle, which penetrates higher than normal into the F region and then follows that region for some distance before being bent enough to return to Earth. A signal that travels for some distance through the F region is called a Pedersen ray.

penetrate the F region farther than lower-angle rays. In the less densely ionized upper edge of the region, the amount of refraction is less, nearly equaling the curvature of the region itself as it encircles the Earth. Figure 3-15 shows how the Pedersen ray could provide propagation beyond the normal single-hop distance.

This Pedersen-ray theory is further supported by studies of propagation times and signal strengths for signals that travel completely around the Earth. The time required is significantly less than would be necessary to hop between the Earth and the ionosphere 10 or more times while circling the Earth. Return signal strengths are also significantly higher than should be expected otherwise. There is less attenuation for a signal that stays in the F region than for one that makes several additional trips through the E and D regions, in addition to the loss produced by reflections off the surface of the Earth.

[Before going on to Chapter 4, turn to Chapter 10 and study examination question A3B09. Review this section if you do not understand the Pedersen ray.]

Absorption wavemeter — A device for measuring frequency or wavelength. It takes some power from the circuit under test when the meter is tuned to the same resonant frequency.

Capacitive coupling (of a dip meter) — A method of transferring energy from a dip-meter oscillator to a tuned circuit by means of an electric field.

Capture effect — An effect especially noticed with FM and PM systems whereby the strongest signal to reach the demodulator is the one to be received. You cannot tell whether weaker signals are present.

Cathode-ray tube — An electron-beam tube in which the beam can be focused on a luminescent screen. The spot position can be varied to produce a pattern on the screen.

Cavity — A high-Q tuned circuit that passes energy at one frequency with little or no attenuation but presents a high impedance to another nearby frequency.

Circulator — A passive device with three or more ports or input/output terminals. It can be used to combine the output from several transmitters to one antenna. A circulator acts as a one-way valve to allow radio waves to travel in one direction (to the antenna) but not in another (to the receiver).

Cross modulation — A type of intermodulation caused by the carrier of a desired signal being modulated by an unwanted signal in a receiver.

D'Arsonval meter movement — A type of meter movement in which a coil is suspended between the poles of a permanent magnet. DC flowing through the coil causes it to rotate an amount proportional to the current. A pointer attached to the coil indicates the amount of deflection on a scale.

Desensitization — A reduction in receiver sensitivity caused by the receiver front end being overloaded by noise or RF from a local transmitter.

Dip meter — A tunable RF oscillator that supplies energy to another circuit resonant at the frequency that the oscillator is tuned to. A meter indicates when the most energy is being coupled out of the circuit by showing a dip in indicated current.

Duplexer — A device, usually employing cavities, to allow a transmitter and receiver to be connected simultaneously to one antenna. Most often, as in the case of a repeater, the transmitter and receiver operate at the same time on different frequencies.

Dynamic range — The ability of a receiver to tolerate strong signals outside the band-pass range. Blocking dynamic range and intermodulation distortion (IMD) dynamic range are the two most common dynamic range measurements used to predict receiver performance.

Electromagnetic radiation — Another term for electromagnetic waves, consisting of an electric field and a magnetic field that are at right angles to each other.

Frequency counter — A digital-electronic device that counts the cycles of an electromagnetic wave for a certain amount of time and gives a digital readout of the frequency.

Frequency standard — A circuit or device used to produce a highly accurate reference frequency. The frequency standard may be a crystal oscillator in a marker generator or a radio broadcast, such as from WWV, with a carefully controlled transmit frequency.

Image signal — An unwanted signal that mixes with a receiver local oscillator to produce a signal at the desired intermediate frequency.

Inductive coupling (of a dip meter) — A method of transferring energy from a dip-meter oscillator to a tuned circuit by means of a magnetic field between two coils.

Intermodulation distortion (IMD) — A type of interference that results from the unwanted mixing of two strong signals, producing a signal on an unintended frequency. The resulting mixing products can interfere with desired signals on those frequencies. "Intermod" usually occurs in a nonlinear stage or device

Isolator — A passive attenuator in which the loss in one direction is much greater than the loss in the other.

Lissajous figure — An oscilloscope pattern obtained by connecting one sine wave to the vertical amplifier and another sine wave to the horizontal amplifier. The two signals must be harmonically related to produce a stable pattern.

Marker generator — An RF signal generator that produces signals at known frequency intervals. A marker generator is often a crystal oscillator that is rich in harmonics, usually for the purpose of calibrating a receiver dial.

Minimum discernible signal (MDS) — The smallest input signal level that can just be detected above the receiver internal noise. Also called **noise floor**.

Noise figure — A ratio of the noise output power to the noise input power when the input termination is at a standard temperature of 290 K. It is a measure of the noise generated in the receiver circuitry.

Noise floor — The smallest input signal level that can just be detected above the receiver internal noise. Also called **minimum discernible signal (MDS)**.

Oscilloscope — A device using a cathode-ray tube to display the waveform of an electric signal with respect to time or as compared with another signal.

Phase noise — Undesired variations in the phase of an oscillator signal. Phase noise is usually associated with phase-locked loop (PLL) oscillators.

Sensitivity — A measure of the minimum input signal level that will produce a certain audio output from a receiver.

Signal-to-noise ratio — Signal input power divided by noise input power or signal output power divided by noise output power.

Surface-mount package — An electronic component without wire leads, designed to be soldered directly to copper-foil pads on a circuit board.

White noise — A random noise that covers a wide frequency range across the RF spectrum. It is characterized by a hissing sound in your receiver speaker.

Zero beat — The condition that occurs when two signals are at exactly the same frequency. The beat frequency between the two signals is zero. When two operators in a QSO are transmitting on the same frequency the stations are zero beat.

AMATEUR RADIO PRACTICE

No Amateur Radio operator will be able to do much operating without using some test equipment occasionally. Elaborate equipment is required only for the most advanced techniques; most operators will not need such equipment for many years after they begin operating. This chapter describes some commonly used frequency-measurement devices and other items of test equipment such as the frequency counter, dip meter, oscilloscope and D'Arsonval meter movement. Performance limitations that effect the accuracy, stability and frequency response of this test equipment are also included. You will learn about the various component-mounting techniques commonly used by amateurs who build their own equipment. This chapter also includes information about receiver performance limitations such as phase noise, desensitization, capture effect, intercept point, noise floor, dynamic range, image rejection, minimum discernible signal (MDS) and signal-to-noise ratio. There is also a discussion about a few commonly encountered interference problems (such as intermodulation and cross modulation) and their solutions.

Your Advanced class license exam will include four questions from the Amateur Radio Practice subelement. These questions will come from the four syllabus points:

A4A   Frequency measurement devices (ie, frequency counter, oscilloscope Lissajous figures, dip meter); component mounting techniques (ie, surface, dead bug {raised}, circuit board).

A4B   Meter performance limitations; oscilloscope performance limitations; frequency counter performance limitations.

A4C   Receiver performance characteristics (ie, phase noise, desensitization, capture effect, intercept point, noise floor, dynamic range {blocking and IMD}, image rejection, MDS, signal-to-noise-ratio).

A4D   Intermodulation and cross-modulation interference.

TEST EQUIPMENT

This section describes the use of marker generators (including frequency standards), frequency counters, dip meters and oscilloscopes. Included are discussions on the performance limitations of these test-equipment items. Uses of other test equipment, such as multimeters, vacuum-tube voltmeters and FET multimeters are covered in *Now You're Talking!* and *The ARRL Technician Class License Manual*.

Marker Generator

A **marker generator** is sometimes referred to as a crystal calibrator. In Amateur Radio, it is used almost exclusively to calibrate receiver frequency readouts and locate band edges. A marker generator usually includes a **frequency standard**, or highly accurate reference frequency.

A marker generator emits a signal, usually a harmonic of the main oscillator frequency, at every 25, 50, or 100-kHz point through the LF, MF and HF bands, and often well into the VHF and UHF bands. These harmonic signals are used to verify that the frequency readout on the receiver or transceiver is correct or, if incorrect, what adjustments must be made. Most marker generators use a 100-kHz frequency standard (quartz crystal) and a trimmer capacitor to effect minor changes in the actual marker frequency. The quartz-crystal frequency standard is an accurate and stable reference. If 50- or 25-kHz marker signals are required, a frequency-divider circuit is employed. See Figure 4-1.

Several nations, including the United States, operate radio broadcast stations that transmit on accurately controlled transmit frequencies 24 hours per day. These radio signals serve as a **frequency standard**, which can be used to compare the accuracy of a receiver marker generator or frequency display. In the US, these standards are broadcast from WWV in Fort Collins, Colorado, and from WWVH in Kekahu (Kauai), Hawaii. Thus, the first step in the use of a marker generator is to **zero beat** its signal with that of WWV or WWVH, usually on 5, 10 or 15 MHz. The trimmer capacitor is used to zero-beat the signals. Then the marker generator can be used to check where the band edges appear on the radio dial.

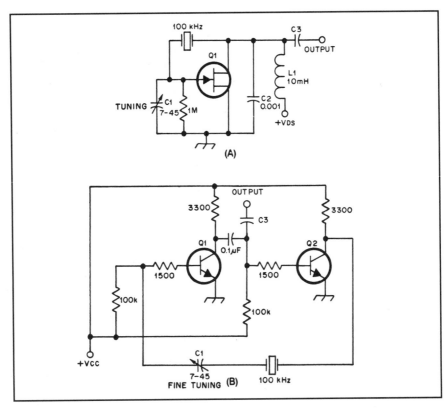

Figure 4-1 — Two simple 100-kHz crystal-oscillator circuits. B is the most suitable for use as a marker generator. In both circuits, C_1 is for fine frequency adjustment. The output coupling capacitor, C_3, is generally small — 20 to 50 pF — a compromise to avoid loading the oscillator by the receiver antenna input while maintaining adequate coupling for good harmonic strength.

A marker generator might be built into a transceiver or receiver, or it might be used as a separate external accessory. If it is used with a transceiver, you must be sure the receiver incremental tuning (RIT) control is set at zero before marking band edges. If a separate receiver and transmitter are in use, it is a good idea to monitor the transmitted signal on a calibrated receiver (preferably while using a dummy antenna), to be sure the transmitter frequency readout is correct. In almost all equipment, it is the carrier frequency that is displayed, whether the carrier is transmitted or suppressed. So, be certain that all transmitted sidebands, regardless of emission mode, are within the band.

Many manufacturers of frequency standards specify the maximum percentage error that can be expected. You should assume that at least that much error will be present. It is good practice to operate far enough from the band edge that, even in the worst possible case, the sidebands still will be inside the band.

[Now turn to Chapter 10 and study question A4A01. Review this section if you don't understand that question.]

Frequency Counter

A **frequency counter**, once considered almost a luxury item, is now an integral part of most commercially made amateur transceivers. The name is completely descriptive: A frequency counter is used to make frequency measurements. It counts the number of cycles per second (hertz) of a signal, and displays that number on a digital readout. Some frequency counters even incorporate voice synthesizers that announce the frequency on command.

Modern transceivers often have a built-in frequency counter to measure and display the operating frequency, although not all digital readouts employ a frequency counter. If a counter is used, then the frequency readout will be as accurate as the counter itself. Some rigs sample some information from the control circuitry and then calculate the operating frequency for display. In such a case, if some part of the radio circuitry is not working properly, you may get an erroneous display.

You can also use a frequency counter to measure signal frequencies throughout a piece of equipment, and to make fine adjustments to tuned circuits. In this case, a frequency counter becomes a valuable piece of test equipment.

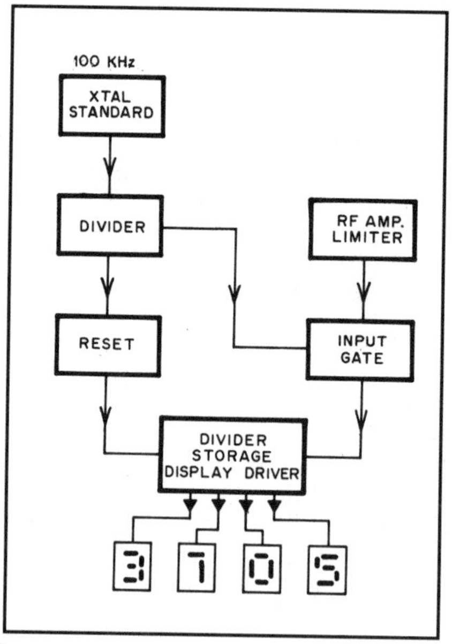

Although usually quite accurate, a frequency counter should be checked regularly against WWV, WWVH or some other frequency-standard broadcast. Frequency counters that operate well into the gigahertz range are available. Counters that operate at VHF or UHF, and sometimes those that operate near the top of the HF range, usually employ a prescaler. The prescaler uses logic circuitry to divide the frequency prior to counting, greatly extending the useful range of the frequency counter. A typical counter is illustrated in block-diagram form in Figure 4-2.

A frequency counter circuit can only measure time as accurately as the crystal reference oscillator, or time base, built into the circuit. The more accurate this crystal is, the more accurate the readings

Figure 4-2 — Frequency-counter block diagram.

will be. Close-tolerance crystals are used, and there is usually a trimmer capacitor across the crystal so the frequency can be set exactly once it is in the circuit. One way to increase the accuracy of a frequency counter is to increase the accuracy of the time base oscillator. (Accuracy refers to the closeness of the measured value to the actual value.)

The accuracy of frequency counters is often expressed in parts per million (ppm). Even after checking the counter against WWV, you must take this possible error into account. The readout error can be as much as:

$$\text{Error} = f(\text{Hz}) \times \frac{\text{counter error}}{1,000,000} \qquad \text{(Equation 4-1)}$$

Suppose you are using a frequency counter with a time base accuracy of 10 parts per million to measure the operating frequency of a 146.52-MHz transceiver. You can use Equation 4-1 to calculate the maximum frequency-readout error:

$$\text{Error} = 146,520,000 \text{ Hz} \times \frac{10}{1,000,000} = 1465.20 \text{ Hz}$$

This means actual operating frequency of this radio could be as high as 146.5214652 MHz or as low as 146.5185348 MHz. You should always take the maximum frequency-readout error into consideration when deciding how close to a band edge you should operate. In addition, you must be sure that the modulation sidebands also stay inside the band edge.

The stability of the time-base oscillator is also very critical to the accuracy and precision of the counter. (Precision refers to the repeatability of a measurement.) Any variation of the time base oscillator frequency affects the counter accuracy and precision. The counter keeps track of the number of RF cycles that occur in a given time interval. Even a slight increase in the time interval could result in a significant increase in the measured frequency. As a simple example, suppose you are using a frequency counter that counts RF cycles for 1 second, and displays the frequency as a result of that count. If the time-base oscillator slows down, it will count RF cycles for a longer period, so the displayed frequency will increase. If the oscillator speeds up, it will count RF cycles for a shorter period, so the displayed frequency decreases. The displayed frequency on this counter will change even if the signal it is counting does not vary!

The speed of the digital logic in the frequency counter limits the upper frequency response of the counter. In an extreme example, if the signal you are trying to measure has two RF cycles in the time it takes the logic in your counter to respond to a single pulse, then the displayed count will only be half the actual frequency. A basic frequency counter may have an upper frequency limit of 10 MHz. By using faster digital logic in the circuit it may be possible to build a similar counter with an upper frequency limit of 100 MHz. The most common way to increase the useful measurement range of a counter is to include a *prescaler*. A prescaler divides the input signal by some value such as 10 or 100, so the counter logic doesn't have to be as fast.

Several factors limit the accuracy, frequency response and stability of a frequency counter. These factors are the time-base oscillator accuracy, the speed of the digital logic and the stability of the time-base oscillator.

[Before you go on to the next section, study the examination questions in Chapter 10 with numbers A4A02 and A4B04 through A4B11. Review this section as needed.]

Dip Meter

Once called a grid-dip meter because it employed a vacuum tube with a meter to indicate grid current, this handy device is actually an RF oscillator. Modern solid-state **dip meters** use FETs.

The principle of operation of a dip meter is that when the meter is brought near a circuit resonant at the meter oscillation frequency, that circuit will take some power from the dip meter. Power taken from the oscillator results in a slight drop in the meter reading of the feedback circuit current. A dip-meter, then, gives an indication of the resonant frequency of a circuit. This will not be a highly accurate frequency measurement, but often all you need is a general indication.

Figure 4-3 is the schematic diagram of a simple dip meter. The plug-in coil assembly includes an inductor (L_1) and capacitors (C_1 and C_2) to form an LC oscillator. The main tuning capacitor (C_3) varies the oscillator frequency over a range set by the plug-in coil assembly. Most dip meters have several plug-in coil assemblies to cover a wide frequency range. A meter in the feedback circuit

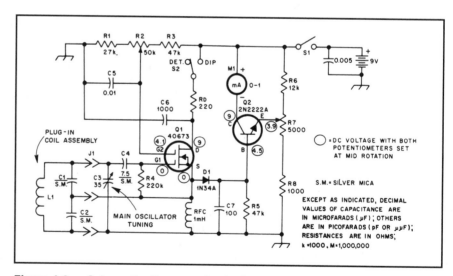

Figure 4-3 — Schematic diagram of a dual-gate MOSFET dip meter. L_1, C_1 and C_2 make up plug-in tuned circuits to change the operating frequency.

measures the current to give an indication that power is being coupled out of the oscillator.

The most frequent amateur use of a dip meter is to determine the resonant frequency of an antenna or antenna traps. Dip meters can also be used to determine the resonant frequency of other circuits. The circuit under test should have no signal or power applied to it while you couple the dip meter to the circuit. Sometimes you simply need a small signal at some specific frequency to inject into a circuit. The tunable RF oscillator of a dip meter is ideal for this task.

Some dip meters are equipped with a switch to disable the power supply and insert a diode into the circuit. Then the meter can be used as an **absorption wavemeter**. In such a case, of course, resonant frequency would be indicated by a slight increase (rather than a decrease) in the meter reading. The circuit under test must have normal supply voltage and signals applied to it to use an absorption wavemeter. Be careful when you work around live circuits.

Although dip meters are relatively easy to use, the touchiest part of their use is coupling the oscillator coil to the circuit being tested. This coupling should be as loose as possible and still provide a definite, but small, dip in the current when coupled to a circuit resonant at the dip-meter-oscillator frequency. Coupling that is too loose will not give a dip sufficient to be a positive indication of resonance.

Whenever two circuits are coupled, however loosely, each circuit affects the other to some extent. If the coupling is loose, the effect will be small and will not create a significant change in the resonant frequency of either circuit. Too tight a coupling, however, almost certainly will yield a false reading on the dip meter.

Dip meters are usually coupled to a circuit by allowing the oscillator-coil field to cut through a coil in the circuit under test. This is called **inductive coupling**. The energy is transferred through the magnetic fields of the coils. Sometimes it is not possible to couple the meter to an inductor, so **capacitive coupling** is used. In this case, the dip-meter coil is simply brought close to an element in the circuit, and the capacitance between the components couples a signal to the circuit. This means that it is the electric fields between the components that transfers the energy. Figure 4-4 indicates the methods of coupling a dip meter to a circuit.

The procedure for using a dip meter is to bring the dip-meter coil within a few inches of the circuit to be tested and then sweep the oscillator through the frequency band until the meter needle indicates a dip. This dip should be symmetrical — that is, the needle should move downward and upward at about the same speed when the oscillator is tuned through resonance. A jumpy needle may indicate that the coupling is too tight, or that there is a problem with the dip meter itself or in the circuit being tested. A jumpy meter needle may also indicate there is a strong influence from yet another circuit active in the vicinity of the test area. The test should be repeated several times, to be sure the dip occurs at the same point each time, and thus accurately indicates the resonance point.

If no dip appears during this check, try another coil on the dip meter. The actual resonant frequency of the circuit being tested might be far from that

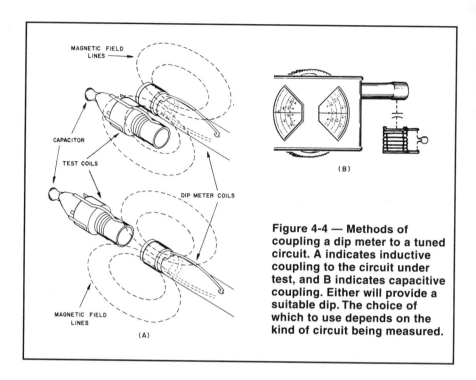

MAGNETIC FIELD LINES

CAPACITOR

TEST COILS

DIP METER COILS

MAGNETIC FIELD LINES

(A)

(B)

Figure 4-4 — Methods of coupling a dip meter to a tuned circuit. A indicates inductive coupling to the circuit under test, and B indicates capacitive coupling. Either will provide a suitable dip. The choice of which to use depends on the kind of circuit being measured.

expected — too far even to be within the resonant-frequency range of the coil originally used on the dip meter. Some experience with dip meters usually is required before you can be sure you are not getting a false reading, caused by coupling that is too tight or by another resonant circuit near or connected to the one you are testing.

If you want to check the resonant frequency of an antenna, a noise bridge and receiver can also be used to indicate resonance. This setup will also give an indication of antenna impedance. If a noise bridge is available, it may be better to learn how to use it for checking your antenna resonant frequency.

Another problem with dip meters is the possibility of reading a harmonic rather than the fundamental frequency. The dip for a harmonic, of course, will not be as deep as that for the fundamental. Nevertheless, it is a good idea to take a reading at double, triple, one-half and one-third the apparent resonant frequency. By doing so, you can be sure the original dip actually occurs at the true resonant frequency.

[Go to the examination questions in Chapter 10 and study questions A4A04 through A4A10 before proceeding. Review this section as needed.]

Oscilloscope

An **oscilloscope** is built around a **cathode-ray tube**. The cathode-ray tube in an oscilloscope differs greatly from that in a television receiver. In fact, about the only thing the tubes have in common is that both use an electron beam focused on a fluorescent screen. The oscilloscope electron beam is controlled by electrostatic charges on the vertical and horizontal deflection plates.

When a voltage is applied to the plates, the beam is pulled toward the plate with the positive charge and repelled by the one with the negative charge. (Remember that electrons are negatively charged ions.) With two pairs of plates at right angles, the beam can be moved from side to side and up and down. Frequencies far into the RF region can be applied to the plates, with the use of very little power.

In the absence of deflection voltages, the oscilloscope controls are adjusted so a small bright spot appears in the center of the screen. Then, an ac voltage applied to the horizontal plates causes the spot to move from side to side (Figure 4-5). Usually, the time base signal or horizontal sweep voltage is applied to these plates, causing the spot to move toward the right at a steady speed. The sweep-voltage frequency (a voltage with a sawtooth waveform) is selected by the user. The signal to be analyzed is applied to the vertical deflection plates. The speed at which the spot moves in any direction is exactly proportional to the rate at which the voltage is changing.

In the course of one ac cycle, the spot will move upward while the voltage applied to the vertical plates is increasing from zero. When the positive peak of the cycle is reached, and the voltage begins to decrease, the spot will reverse its direction and move downward. It will

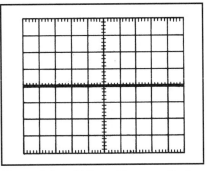

Figure 4-5 — If the voltage applied to the vertical plates of the oscilloscope is zero, only the horizontal line created by the sweep oscillator will be visible.

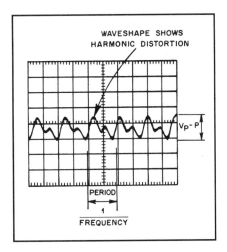

Figure 4-6 — A typical pattern resulting when a complex waveform is applied to the vertical plates, and the sweep-oscillator output is applied to the horizontal plates.

continue in this direction until the negative peak is reached. Then, as the voltage increases toward zero, the spot moves upward again. If the horizontal sweep voltage is moving the spot horizontally at a uniform speed at the same time, it is easy to analyze the signal waveform applied to the vertical deflection plates. See Figure 4-6.

There are many uses for an oscilloscope in an amateur station. This instrument is often used to display the output waveform of a transmitter during a two-tone test. Such a test can help you determine if the amplifier stages in your rig are operating in a linear manner. An oscilloscope can also be used to display signal waveforms during troubleshooting procedures.

Another frequent use of the oscilloscope in amateur practice is the comparison of two signals, one of which is applied to the vertical deflection plates, and the other to the horizontal plates. This procedure produces a **Lissajous figure** on the scope. When sinusoidal ac voltages are applied to both sets of deflection plates, the resulting pattern depends on the relative amplitudes, frequencies and phases of the two voltages. If the ratio between the two frequencies is constant and can be expressed as two integers, the Lissajous pattern will be stationary. If the frequency ratio is not an exact integer value, the pattern may seem to shift or rotate around a vertical or horizontal axis.

Examples of some simple Lissajous patterns are shown in Figure 4-7. The frequency ratio is found by counting the number of loops along two adjacent edges. Thus, in the pattern shown at C, there are three loops along a horizontal edge and only one along the vertical, so the ratio of the vertical frequency to the horizontal frequency is 3:1. In part E, there are four loops along the horizontal edge and three along the vertical edge, producing a ratio of 4:3. Assuming that a known frequency is applied to the horizontal plates and an unknown frequency is applied to the vertical plates, the relationship is given by:

$$\frac{f_H}{f_V} = \frac{n_V}{n_H}$$

(Equation 4-2)

where:

f_H is the known frequency, on the horizontal axis
f_V is the unknown frequency, on the vertical axis
n_V is the number of loops along a vertical edge
n_H is the number of loops along a horizontal edge.

If you know the two signal frequencies being supplied to the oscilloscope, you can predict the Lissajou pattern that will be produced. Reduce the fraction f_H / f_V to the lowest denominator. To do this, divide the top (numerator) and bottom (denominator) of the fraction by the same number, until further division does not give integer (whole number) values. The resulting fraction is the ratio of vertical to horizontal loops in the Lissajous pattern (n_V / n_H). An example will make this procedure easier to understand: Suppose you feed a 100-Hz signal to the horizontal input of an oscilloscope and a 150-Hz signal to the vertical input. What type of Lissajous pattern will this produce? Use Equation 4-2.

$$\frac{f_H}{f_V} = \frac{100 \text{ Hz}}{150 \text{ Hz}} = \frac{n_V}{n_H}$$

We can reduce this fraction by dividing both the numerator and denominator by 50. Notice that we could divide both by 2, 5 or 10 and still get integer values for the numerator and denominator, but by choosing 50 we get the smallest possible integer values.

$$\frac{f_H}{f_V} = \frac{\dfrac{100}{50}}{\dfrac{150}{50}} = \frac{2}{3} = \frac{n_V}{n_H}$$

From this calculation, you can see that the Lissajous pattern will have two loops on the vertical axis and three loops on the horizontal axis. The pattern will look like the one shown in Figure 4-7D.

Lissajous figures are used more often to measure a signal with an unknown frequency by comparing it with a signal of known frequency. In that case, you would look at the Lissajous pattern produced on the oscilloscope and then solve Equation 4-2 for the unknown frequency. Equation 4-3 is the result of cross multiplying the terms of Equation 4-2, solving for the vertical frequency, which is usually the unknown.

$$f_V = \frac{n_H}{n_V} f_H \qquad \text{(Equation 4-3)}$$

For example, suppose the signal applied to the horizontal input has a frequency of 200 Hz. When an unknown signal is applied to the vertical input, a Lissajous pattern like the one shown in Figure 4-7C results. What is the frequency of the signal applied to the vertical input? Equation 4-3 gives the result.

$$f_V = \frac{n_H}{n_V} f_H = \frac{3}{1} \times 200 \text{ Hz} = 600 \text{ Hz}$$

An important application of Lissajous figures is the calibration of audio-frequency signal generators. For very low frequencies, the 60-Hz power-line frequency is used as a reasonably good standard for comparison. This frequency is accurately controlled by the power companies, so although the signal frequency may vary somewhat, the long-term average is very accurate. The medium AF range can be covered by comparison with the 440- and 600-Hz audio modulation

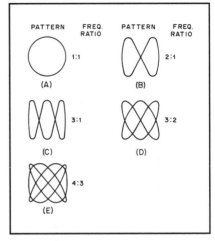

Figure 4-7 — Lissajous figures and corresponding frequency ratios for a 90° phase relationship between the voltages applied to the two sets of deflection plates.

on WWV transmissions. It is possible to calibrate over a 10:1 range both upward and downward from these frequencies and thus cover the audio ranges useful for voice communication.

An oscilloscope has both a horizontal and a vertical amplifier (in addition to the horizontal sweep-oscillator circuitry). This is desirable because it is convenient to have a means for adjusting the voltages applied to the deflection plates to secure a suitable pattern size. Several hundred volts is usually required for full-scale deflection in either the vertical or horizontal direction, but the current required is usually somewhere in the microampere range. Thus, the actual power needed for full-scale deflection is extremely low.

One important limitation to the accuracy, frequency response and stability of an oscilloscope is the bandwidth (frequency response) of the scope deflection amplifiers. Another important limitation is the accuracy and linearity of the time base, or horizontal sweep oscillator. If the horizontal sweep oscillator only works in the audio-frequency range, the scope won't be very useful for making RF measurements. Unless the sweep oscillator is stable and the frequency can be set accurately, frequency measurements made with the scope will not be very accurate.

Scopes with a horizontal sweep rate that is limited to audio frequencies are relatively inexpensive and serve many amateur needs. Increasing the bandwidth of the horizontal and vertical amplifier, and increasing the horizontal sweep-oscillator frequency, will increase the useful frequency response of the scope. Of course, it is a distinct advantage for amateurs to have a scope that will handle RF through their most-used range; that is, to 30 MHz, to 150 MHz or even higher. Scopes with such RF capabilities are usually expensive, however.

One way to extend the useful frequency range of a narrow-bandwidth oscilloscope is with an adapter circuit. The idea is to use a mixer and an oscillator set to a frequency that will provide a sum or difference frequency within the useful range of your scope. With a 25-MHz oscillator, for example, you can display signals in the 20- to 30-MHz range on a 5-MHz scope. Other oscillator frequencies will enable you to display a signal with almost any desired frequency. This will not improve the transient-signal response of the narrowband scope, however.

Other limitations amateurs should be aware of are the stability of the scope sweep-frequency oscillator, absence of a horizontal deflection-plate amplifier (which would make it impossible to produce a Lissajous pattern), the ease of varying the horizontal sweep rate, the upper and lower limits of the available sweep rates, and the amount of voltage required for full-scale deflection of the electron beam. All of these factors will affect the frequency response and stability of a scope.

[Now turn to the examination questions in Chapter 10 and study question A4A03. Also study questions A4B02 and A4B03 before proceeding. Review this section if you have difficulty with any of these questions.]

D'Arsonval Meters

Almost all high-quality ammeters, voltmeters and multimeters that use an analog dial and a moving needle pointer use a **D'Arsonval meter movement** for the

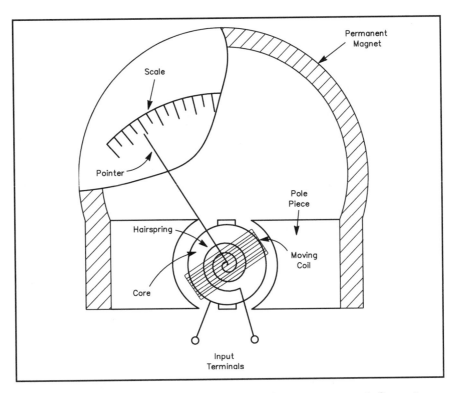

Figure 4-8 — This diagram shows a D'Arsonval meter movement. Current through the coil creates a magnetic field in opposition to the permanent magnet, causing the coil to rotate and the pointer to move across the scale.

measuring device. This type of meter movement has a coil suspended between the poles of a permanent magnet. An indicator needle is attached to this moveable coil. When current flows through the coil, the magnetic field will cause the coil to move. It is important to note that it is the current through the coil that causes the meter needle to move. The whole movement is carefully balanced, and a spring is used to oppose the needle movement when a current flows through the coil. This provides equal needle deflections for equal current changes, which means that the scale is linear. The D'Arsonval meter movement is named for its inventor, Jacques Arsène D'Arsonval, a 19th century physicist. Figure 4-8 illustrates the construction and operating principles of the D'Arsonval meter movement.

The accuracy of most meters is specified as a percentage of full scale. If the specification states that the meter accuracy is within two percent of full scale, the possible error anywhere on a scale of 0 to 10 V is two percent of 10 V, or 0.2 V. The actual value read anywhere on that scale could be as much as 0.2 V above or below the indicated meter reading. If you are using a 0- to 100-mA scale, the possible error would be ± 2 mA anywhere on that scale. Besides the limitations of the basic meter movement, any other components, such as current- or voltage-multiplying resistors, will affect the accuracy of the instrument.

The actual meter-scale calibration and the mechanical tolerance of the movement are two factors that limit the accuracy and stability of a particular meter. In addition, the coil impedance controls how much current will flow through the movement when a certain voltage is applied across it, and that will affect how sensitive the meter is.

D'Arsonval meters are dc-operated devices. In order to measure ac, you must use a diode to rectify the ac. A bridge rectifier is often used for this purpose. Then a scale can be calibrated to read effective (RMS) voltage. Such a scale is only useful for a sine-wave signal, however.

Instruments that operate directly from ac to measure current and voltage are available. They are relatively expensive, however, and are seldom used in an Amateur Radio station. Most such meters are useful only at power-line frequencies.

[Turn to Chapter 10 and study question A4B01 now. Review this section if you are uncertain of the answer to this question.]

COMPONENT MOUNTING TECHNIQUES

Sooner or later, most amateurs try building some piece of station equipment or accessory. They may use any of a variety of construction techniques to build this equipment.

Most electronics experimenters are familiar with printed circuit (PC) boards. This is often the technique used for a first project. An insulating board with copper circuit traces forms the basis for the project. Component leads are inserted through holes in the circuit-board material, and the leads are soldered to the copper traces. Some amateurs trace the circuit pattern onto the copper-clad circuit-board material and etch away the excess copper, while others prefer to purchase circuit boards prepared by commercial suppliers.

You can also build a project on "perf board," which is an insulating material with a grid of holes for component leads. Connecting wires may be soldered to the component leads or wire-wrap construction may be used. Others may prefer to build their projects on "universal" or "experimenters' circuit boards." These boards usually have holes and circuit pads for one or more dual in-line package (DIP) integrated circuits, with copper traces to additional pads for mounting other components.

Still another circuit-construction technique that is popular with many hams is variously called "dead bug construction," "ugly construction" or "ground-plane construction." With this construction technique, a piece of copper-clad circuit-board material is used for a base. Components that connect to ground are soldered directly to the copper foil. Ungrounded connections are made directly to the component leads. Transistors and integrated circuits are often glued to the circuit board upside down, so their leads stick up for connections to other components. (This is where the name "dead bug" comes from.) Figure 4-9 is a photo of a circuit built using this construction technique.

To reduce the effects of stray inductance and capacitance, and to aid robotic circuit construction, many components are available in **surface-mount packages**. These devices are soldered directly to circuit-board pads, and mount flush against the board. Amateurs can take advantage of these surface-mount packages, especially for UHF and microwave circuits, where stray capacitance and inductance must be considered. Surface-mount resistors and capacitors are small rectangles with metal end caps so they can be soldered to the circuit-board pads. These tiny components without wire leads are sometimes called "chips." Figure 4-10 is a photo of a circuit that uses some surface-mounted components.

[Study question A4A11 in Chapter 10 before you go on to the next section. Review this material if you have difficulty with that question.]

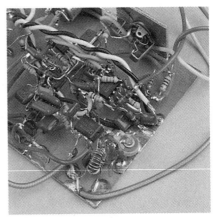

Figure 4-9 — This photo shows how the "dead bug" construction method can be used to build a circuit that requires little or no advance preparation.

Figure 4-10 — Surface-mount components are used for circuits that require minimal stray inductance and capacitance between the component leads. Amateurs use such components especially for UHF and microwave circuits.

RECEIVER PERFORMANCE CHARACTERISTICS

When you begin to evaluate a receiver or transceiver for your Amateur Radio station, you can read about a wide array of measurements and tests in *QST* Product Reviews. The results of those measurements can present a dizzying array of numbers unless you understand how they were made, and what performance characteristics they represent. You will learn about some of these measurements in this section, as you prepare for your Advanced class license exam.

Two of the basic receiver specifications are **noise figure** and **sensitivity**, or

minimum discernible signal (MDS). The MDS is also called the receiver **noise floor**, because it represents the weakest, or smallest signal that can just be detected above the receiver noise. You can think of noise figure as a "figure of merit" for the receiver. The higher the noise figure, the more noise that is generated in the receiver itself. This also means a higher noise floor. Lower noise figures are more desirable.

The noise figure of a receiver is related to the signal-to-noise ratio of the input and output signals. Signal-to-noise (S/N) ratio is defined as signal power divided by noise power. Input S/N ratio uses input signal and noise powers while output S/N ratio uses output signal and noise powers. Signal-to-noise ratios are often expressed in decibels.

It is useful to know that the theoretical noise power at the input of an ideal receiver, with an input-filter bandwidth of 1 hertz, is − 174 dBm. (*The ARRL Handbook* contains more detailed information about how to calculate this number.) This is considered to be the theoretical best (lowest) noise floor a receiver can have. In other words, for this ideal receiver, the strength of any received signal would have to be slightly more than − 174 dBm. A receiver bandwidth of 1 Hz is impractical, and undesirable for receiving information on the signal. A more typical receiver might have a 500-Hz bandwidth for CW operation, or even wider for SSB or FM voice. The 500-Hz bandwidth increases the receiver noise by a factor of 500, which corresponds to 27 dB (10 log 500). The theoretical noise floor (minimum discernible signal, or MDS) of this receiver will be − 174 dBm + 27 dB = − 147 dBm. If the receiver has a wider bandwidth, you can calculate the theoretical MDS for that receiver by finding the log of the bandwidth and multiplying that value by ten. Add the result to the 1-Hz bandwidth value of − 174 dBm.

The receiver noise figure degrades the noise floor, or raises the power that actual signals must have to be heard. You can calculate the actual noise floor of a receiver by adding the noise figure (expressed in dB) to the theoretical best MDS value.

Noise Floor = Theoretical MDS + noise figure   (Equation 4-4)

For example, suppose our 500-Hz-bandwidth receiver has a noise figure of 8 dB. We can use Equation 4-4 to calculate the actual noise floor of this receiver.

Noise Floor = − 147 dBm + 8 dB = − 139 dBm

Dynamic range is another important receiver parameter. This refers to the ability of the receiver to tolerate strong signals outside the normal passband. Intermodulation distortion (IMD) dynamic range measures the impact on the receiver of the production of spurious signals that result when two or more signals mix in the receiver. When the IMD dynamic range is exceeded, false signals begin to appear along with the desired signal. (Undesired signals are strong enough to mix with other signals and produce spurious signals that show up in the receiver passband along with the desired signal.)

You can calculate the third-order intercept point by multiplying the IMD dynamic range by 1.5, and then adding the receiver noise floor value to this result.

Suppose the example receiver just described has an IMD dynamic range of 94 dB. The third-order intercept point is:

Third-order intercept = 1.5(94 dB) + (– 139 dBm) = 141 dB – 139 dBm = 2 dBm

A larger value for the third-order intercept point would represent a better receiver.

Blocking dynamic range refers to the difference in signal powers between the noise floor and a signal that causes 1 dB of gain compression in the receiver. When the blocking dynamic range is exceeded, the receiver begins to lose the ability to amplify weak signals.

Figure 4-11 shows the basic test setup to measure blocking dynamic range. One signal generator is used to feed a weak signal into the receiver input at about 20 dB above the noise floor. The receiver is tuned to this first signal. A second signal generator is set to a frequency 20 kHz away, and the strength of this signal is increased until the receiver audio output of the desired signal drops by 1 dB.

As Figure 4-11 shows, you can calculate the receiver blocking level by adding the undesired signal power (expressed in dBm) with the hybrid combiner loss and the step attenuator loss. The blocking dynamic range is then calculated by subtracting the blocking level from the receiver noise floor, and taking the absolute value of the result. (The absolute value means you take the final result as a positive value, even if the calculation results in a negative value. Keep in mind that when you subtract a negative value, the sign changes, so the operation is the same as adding a positive value. Equation 4-5 gives this expression in mathematical terms.

Blocking Dynamic Range = | Noise Floor – Blocking Level | (Equation 4-5)

The vertical bars around the calculation indicate that the result is taken as an absolute value.

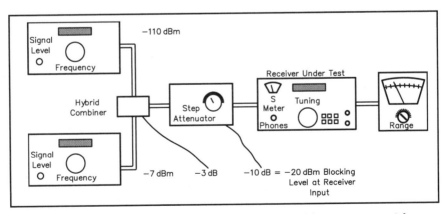

Figure 4-11 — This diagram shows the equipment and its arrangement for measuring receiver blocking dynamic range. Measurements shown are for the example discussed in the text.

Suppose we have a receiver that has a 500-Hz bandwidth, and we measure the noise figure as 8-dB. We find the blocking level is – 20 dBm. What will the blocking dynamic range of this receiver be?

The first step is to calculate the receiver noise floor, using Equation 4-4. We know that a theoretically ideal receiver with a 500-Hz bandwidth has an MDS of –147 dBm.

Noise Floor = Theoretical MDS + noise figure   (Equation 4-4)

Noise Floor = – 147 dBm + 8 dB = – 139 dBm

Next we will use Equation 4-5 to calculate the blocking dynamic range.

Blocking Dynamic Range = | Noise Floor – Blocking Level | (Equation 4-5)

Blocking Dynamic Range = | – 139 dBm – (– 20 dBm) | = | – 139 dBm + 20 dBm |
= 119 dB

This blocking dynamic range measurement tells us that signals more than 119 dB above the noise floor will block reception of weak signals, even though the undesired signal is well outside the normal pass band of the receiver.

Phase noise is a problem that has become more apparent as other improvements have reduced the noise floor and improved dynamic range measurements. Most modern commercial transceivers use phase-locked loop (PLL) frequency synthesizers. These oscillators use a feedback loop to detect the output frequency, compare it to the desired frequency, and make automatic corrections to the oscillator frequency. Many factors can cause the output signal frequency to vary a bit, so the circuit is constantly adjusting itself. On average, the output frequency is very close to the desired value, but at any instant the actual frequency is likely to be a little higher or lower than desired. These slight fluctuations result in a signal in which the phase of any cycle is likely to be slightly different from the phase of adjacent cycles. Figure 4-12 exaggerates these phase variations of a PLL oscillator.

One result of receiver phase noise is that as you tune towards a strong signal, the receiver noise floor appears to increase. In other words, you hear an increasing amount of noise in an otherwise quiet receiver as you tune towards the strong signal.

Excessive phase noise in a receiver local oscillator allows strong signals on nearby frequencies to interfere with the reception of a weak desired signal. You can understand this if you think about tuning in a weak signal that is barely above the receiver noise floor. While listening to this signal, another station begins transmitting a strong signal on a nearby frequency. The new signal is outside the receiver passband, so you don't hear the interfering signal directly, but there is a sudden increase in receiver noise. This increased receiver noise can cover a weak desired signal, or at least make copying it more difficult. Figure 4-13 illustrates how this phase noise can cover a weak signal.

Excessive phase noise can cause interference in two ways. The first is the result of receiver phase noise just discussed. The second is caused by a transmitter

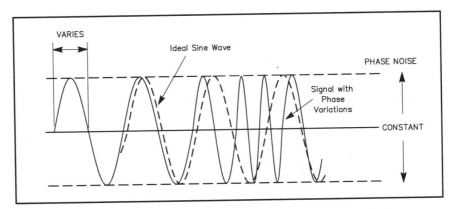

Figure 4-12 — This diagram exaggerates the effects of phase noise from a phase-locked loop oscillator. The dashed line represents the output of an ideal oscillator as it might be viewed on an oscilloscope. The solid line represents the output of an oscillator with a large amount of phase noise.

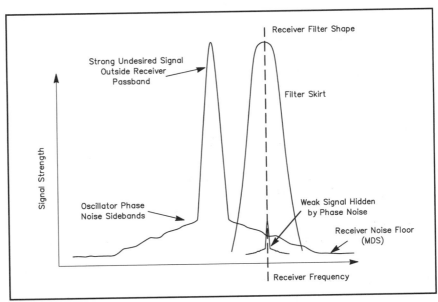

Figure 4-13 — In a receiver with excessive phase noise, a strong signal near the receiver passband can raise the apparent receiver noise floor in the passband. This increased noise can cover a weak signal that you are trying to receive.

with excessive phase noise on its transmitted signal. The effect is virtually the same. The phase-noisy signal "splatters" up and down the band for some range around the desired transmit frequency, and this additional noise can fall within the passband of a receiver tuned to some weak signal. So even if you have a receiver with very low phase noise, you can be bothered by this type of interference!

One final receiver-performance limitation that you need to understand for your Advanced class license exam is the concept of images and image-rejection ratio. You learned earlier that the process of mixing two signals produces a new signal at the sum of their frequencies and another new signal at the difference between their frequencies. One of these mixing products is the desired one, and the other is undesired. Suppose we want to use the signal at the difference between the two frequencies for further processing in our receiver. For example, a receiver uses a 15-MHz local oscillator frequency to tune a signal at 25 MHz. The difference signal is 10 MHz, which represents an intermediate frequency for this receiver. Now suppose a signal at 5 MHz also gets into the mixer. The sum of the LO frequency and the unwanted signal is also 10 MHz! Both signals are amplified in the IF stage. The unwanted signal is known as an **image signal**.

The input filter to the RF amplifier must reject the unwanted signals and pass the desired signals. Most receivers have separate input filters for each band, which are selected by the bandswitch. Proper filter design is the key to solving an image problem with any receiver. It is not difficult to design a filter for the receiver just described, that will pass the 25-MHz signal and reject the 5-MHz signal. If the 5-MHz signal doesn't reach the mixer input, then it can't produce an image response in the receiver.

Other signals will also produce a response in this receiver. For example, a signal at 27 MHz will mix with the LO to produce a signal at 12 MHz. The RF amplifier input filter may not be able to adequately reject the 27-MHz signal, so the 12-MHz image may be passed to the first IF stage. If that signal were to make it through the first IF filter, and be passed on to the product detector or a second mixer, an image response could be produced. For example, the second IF may be at 1 MHz, and use an 11-MHz local oscillator. Both the desired 10-MHz signal and the undesired 12-MHz signal would produce signals at the 1-MHz IF, in this case. From this example, you can see that the RF amplifier stage must include adequate filtering to pass the desired signal while blocking the undesired one.

[Turn to Chapter 10 now and study question A4C01 and questions A4C08 through A4C11. If you have difficulty with any of these questions, review this section.]

Receiver Desensitization

A problem often encountered with repeater systems is **desensitization** of the receiver, almost invariably by a strong signal from a nearby transmitter (regardless of that transmitter's frequency). This signal so overloads the receiver that it becomes relatively insensitive to the signals it is supposed to receive. Desensitization can also result when a strong signal on a frequency near the received frequency causes a reduction in receiver sensitivity. In that case strong adjacent-channel signals cause the receiver desensitization.

In the case of a desensitized repeater receiver, the offending transmitter is often that of the repeater itself, so the transmitter is physically close to the receiver *and* operating on a frequency near the receive frequency. The key to eliminating desensitization is isolation — that is, isolating the receiver from any transmitters that might be causing or contributing to the problem.

Obviously, the repeater receiver and transmitter must be carefully shielded from each other. Such shielding usually involves physically separating the receiver from the transmitter and enclosing each in a metal box. All connections between the two boxes must be carefully shielded. Double-shielded coaxial cable is recommended. Separation of the transmitting and receiving antennas also is helpful, but often is impractical.

When both receiver and transmitter must use the same antenna, a series of **cavities** called a **duplexer** is employed to provide the isolation. The duplexer acts as a notch filter at the receive frequency to effectively attenuate the transmitter signal at the receiver. Total attenuation can easily be as much as 35 dB.

Often, additional capacitance is placed across the transmitter tank circuits to raise the circuit Q and make the bandwidth as small as possible. This reduces the amount of **white noise** generated by the transmitter that reaches the receiver. White noise is a constant hissing, buzzing sound in a receiver that is characteristic of unsquelched FM.

If the desensitization is the result of powerful nearby transmitters not operating on the same band, however, the use of low-pass, band-pass or high-pass filters might solve the problem. Which kind of filter is used depends on the frequency of the offending transmitter. If it is a commercial AM transmitter, a high-pass filter on the repeater receiver input might solve the problem. If not, rearrangement of the receiver antenna might also be necessary. If the problem should be a nearby commercial UHF television transmitter, a low-pass filter on the receiver input probably would help, assuming the receiver input frequency is lower than the TV frequency. If the problem is caused by a nearby amateur transmitter on the same band as the receiver, the best solution is to use a high-dynamic-range front end in your receiver.

[Go to question numbers A4C02 through A4C04 in Chapter 10, and study them before proceeding. Review this section as needed.]

Capture Effect

One of the most notable differences between an amplitude-modulated (AM) receiver and a frequency-modulated (FM) receiver is how noise and interference affect an incoming signal.

From the time of the first spark transmitter, "rotten QRM" has been a major problem for amateurs. The limiter and discriminator stages in an FM receiver can eliminate most of the impulse-type noise, except any noise that has frequency-modulation characteristics. For good noise suppression, the receiver IF system and detector phase tuning must be accurately aligned.

FM receivers perform quite differently from AM, SSB and CW receivers when QRM is present, exhibiting a characteristic known as the **capture effect**. The loudest signal received, even if it is only two or three times stronger than other signals on the same frequency, will be the only signal demodulated. On the other hand, an S9 AM, SSB or CW signal suffers noticeable interference from an S2 signal. Capture effect can be an advantage if you are trying to receive a strong station and there are weaker stations on the same frequency. At the same time, this phenomenon will prevent you from receiving one of the weaker signals if that is your desire.

Capture effect is most pronounced with FM emissions. The stronger signal blocks weaker signals, preventing them from being detected.

[Now turn to Chapter 10 and study examination questions A4C05 through A4C07. Review this section as needed.]

ELECTROMAGNETIC COMPATIBILITY

We live in a world surrounded by **electromagnetic radiation**, including frequencies that range from just above dc through visible light! There is an even wider variety of electronic devices that can respond to this radiation in some way. Some devices are designed to receive specific frequencies, and others are not supposed to respond to electromagnetic waves at all. Some electronic devices generate radio signals either as their main function or as an incidental part of their operation. Very often, some of this equipment will not function as it was intended to, and radio interference will result.

As an Amateur Radio operator, you may cause interference to others, you may be interfered with or you may be blamed for interference that you do not think you could be causing. This section deals with a few of those cases, and should help you understand some of the interference, as well as prepare you to pass your Advanced class license exam.

Intermodulation Distortion

Intermodulation distortion (IMD) or intermodulation interference occurs when signals from several transmitters, each operating on a different frequency, are mixed in a nonlinear manner. This produces signals at mixing products and may cause severe interference in a nearby receiver. Harmonics can also be generated in the nonlinear stage, and those frequencies will add to the possible mixing combinations. The intermod, as it is called, is transmitted along with the desired signal. Intermodulation distortion can also be produced in a receiver. Nonlinear circuits or devices cause intermodulation distortion in electronic circuits.

For example, suppose an amateur repeater receives on 144.85 MHz. Nearby, are relatively powerful nonamateur transmitters operating on 181.25 MHz and on

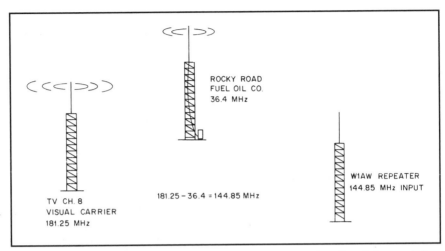

Figure 4-14 — This diagram shows a potential intermod situation. It is possible for some of the 36.4-MHZ Rocky Road transmitter carrier to mix with the channel 8 carrier signal at 181.25 MHz in the output stage of the TV transmitter. This mixing produces intermodulation distortion products at the sum and difference frequencies (181.25 + 36.4 = 217.65 MHz, 181.25 – 36.4 = 144.85 MHz). One of these intermod products is the input frequency of the W1AW repeater, and may key up the repeater. An isolator or circulator (see text) should be installed at the Channel 8 transmitter to solve this problem.

36.4 MHz (Figure 4-14). Neither of these frequencies is harmonically related to 144.85. The difference between the frequencies of the two nonamateur transmitters, however, is 144.85 MHz. If the signals from these transmitters are somehow mixed, the difference frequency (intermod) could be picked up by the amateur repeater and retransmitted over its own antenna. In this example, the signals may be mixing in one of the transmitters, and the intermod signal may actually be transmitted with the desired signal from that transmitter. Intermodulation interference can be produced when two transmitted signals mix in the final amplifiers of one or both transmitters, and unwanted signals at the sum and difference frequencies of the original signals are generated.

All transmitter operators are morally and legally required to ensure that their stations emit only clean signals. The problem often is that the operators of the other transmitters do not know their equipment is causing intermod with your club's repeater. When they become aware of it, they can take several steps to eliminate the problem.

Push-pull amplifiers are quite effective in eliminating even-numbered harmonics, and it is often the second harmonic of the frequency that is causing intermod. Another solution would be to use a linear (class-AB1) power amplifier instead of a class-C amplifier. (Linear amplifiers present fewer intermod problems than do non-linear amplifiers.) Should this remedy be impractical, at the very least, the class-C amplifier should be operated with the minimum grid current possible for efficient operation. (Chapter 7 contains detailed explanations of all these types of amplifiers and how they work.) All stages in the offending transmitter should be neutralized, to eliminate parasitic oscillations. (Low-pass and band-pass filters usually are ineffective in reducing intermod problems, because at VHF and UHF they are seldom sharp enough to suppress the offending signal without also weakening the wanted one.)

Two other devices that usually are highly effective in eliminating intermod are **isolators** and **circulators**. A terminated circulator is a precisely engineered ferrite component that functions like a one-way valve. It is a three-port device that can combine two or more transmitters for operation on one antenna. Very little transmitter energy is lost as RF travels to the antenna, but a considerable loss is imposed on any energy coming down the feed line to the transmitter. Thus, the circulator effectively reduces intermod problems. Another advantage to its use is that it provides a matched load at the transmitter output regardless of what the antenna-system SWR might be.

A two-port device called an isolator also helps reduce intermod; this device incorporates a built-in termination or load. Figure 4-15 illustrates how circulators or isolators may be included in your repeater system.

Circulators and isolators are available for 144- and 450-MHz use. They come in power levels up to a few hundred watts. Typical bandwidth for a 150-MHz unit is 3 MHz. Insertion loss is roughly 0.5 dB, and rejection of unwanted energy coming down the feeder is usually 20 to 28 dB. The 450-MHz types have greater band-

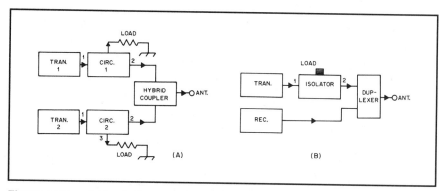

Figure 4-15 — A block diagram showing the use of circulators and an isolator. Two circulators may be used to connect one antenna to two transmitters, as in A. Often, duplexers also are used, along with the circulators. In B, an isolator is placed between the transmitter and the duplexer to reduce inter-modulation products.

width — about 20 MHz — but otherwise perform about the same as the 150-MHz versions. Isolators and circulators are rather expensive, but, when needed, they are worth considerably more than their price.

Intermod, of course, is not limited to repeaters. Anywhere two relatively powerful and close-by transmitter fundamental-frequency outputs or their harmonics combine to create a sum or a difference signal at the frequency on which any other transmitter or receiver is operating, an intermod problem can develop.

A two-tone IMD test on a receiver is a measure of the range of signals that can be tolerated at the receiver input before it begins to generate spurious mixing products. This specification gives an idea of how the receiver will perform in the presence of strong signals.

A receiver can experience intermodulation distortion interference when it receives two strong signals. In that case, the interference is actually generated in the receiver, so there is nothing the transmitter operators can do to solve the problem. (Well, nothing short of changing frequency or turning it off, at least.) As an example, suppose you have a receiver tuned to 146.70 MHz, to monitor a local repeater. You begin to hear a distorted signal, which you eventually identify as another local amateur operating on 146.52 MHz. The interference is intermittent, and by listening on another receiver tuned to 146.52 MHz, you determine that the interference does not occur every time the other ham transmits. What are the two most likely frequencies for the second signal that may be causing this intermod in your receiver?

This interference is an example of *third-order IMD*. The interfering signals occur at frequencies corresponding to the sum and difference of two times one signal frequency and a second signal frequency. This statement may be easier to understand if we express it as four equations.

$$f_{IMD\ 1} = 2f_1 + f_2 \qquad \text{(Equation 4-6)}$$

$$f_{IMD\ 2} = 2f_1 - f_2 \qquad \text{(Equation 4-7)}$$

$$f_{IMD\ 3} = 2f_2 + f_1 \qquad \text{(Equation 4-8)}$$

$$f_{IMD\ 4} = 2f_2 - f_1 \qquad \text{(Equation 4-9)}$$

It turns out that only the subtractive products (given by Equations 4-7 and 4-9) are close enough to the desired signal to cause significant interference. In our example, we know the IMD frequency of interest is 146.70 MHz, and we have found one of the signals producing the interference at 146.52 MHz, and want to find the two most likely frequencies for the second signal. Use Equation 4-7 to calculate one possible unknown frequency (f_2). The first step is to rearrange the terms to solve the equation for f_2.

$$f_{IMD\ 2} = 2f_1 - f_2$$

$$f_2 = 2f_1 - f_{IMD\ 2} = 2 \times 146.52\ \text{MHz} - 146.70\ \text{MHz}$$

$$f_2 = 293.04\ \text{MHz} - 146.70\ \text{MHz} = 146.34\ \text{MHz}$$

Next we will use Equation 4-9 to calculate another possible frequency for the second interfering signal. Again, the first step is to rearrange terms to solve for f_2.

$$f_{\text{IMD 4}} = 2f_2 - f_1$$

$$2f_2 = f_{\text{IMD 4}} + f_1$$

$$f_2 = \frac{f_{\text{IMD 4}} + f_1}{2} = \frac{146.70 \text{ MHz} + 146.52 \text{ MHz}}{2} = \frac{293.22 \text{ MHz}}{2}$$

$$f_2 = 146.61 \text{ MHz}$$

The two most likely frequencies to be mixing with the 146.52 MHz signal and producing this receiver intermodulation distortion interference are 146.34 MHz and 146.61 MHz. Higher-order IMD products can also occur, and they could be causing the interference, but the higher-order products are usually weaker, so they are less likely to be the cause.

Another IMD topic has to do with transmitter spectral-output purity. When several audio signals are mixed with the carrier signal to generate the modulated signal, spurious signals will also be produced. These are normally reduced by filtering after the mixer, but their strength will depend on the level of the signals being mixed, among other things, and they will be present in the transmitter output to some extent. You can perform a transmitter two-tone test by putting two equal-amplitude audio tones into the microphone circuit. Then you can view the transmitter output on an oscilloscope (or spectrum analyzer) to get an indication of how linear the amplifier stages are. This is important if you are to be sure that your transmitted signal is clean.

Excessive intermodulation distortion of an SSB transmitter output signal results in splatter being transmitted over a wide bandwidth. The transmitted signal will be distorted, with spurious (unwanted) signals on adjacent frequencies.

[Before proceeding, turn to Chapter 10 and study questions A4D01, A4D02, A4D03, A4D08, A4D09 and A4D11. Review this section as needed.]

Cross Modulation

Cross modulation interference occurs when an unwanted signal is heard in addition to the desired signal. Sometimes signals from a very strong station are superimposed on other signals being received.

Cross modulation often results in one or more of several kinds of RFI. For instance, if a TV receiver is overloaded by a strong nearby signal, the result can be cross modulation. If a 14-MHz amateur signal mixes with the signal from a 92-MHz commercial FM station to produce a beat at 78 MHz, interference to VHF television channel 5 can result. The 14-MHz signal also might mix with that of a TV station on channel 5 to cause TVI on channel 3. Neither of these channels is harmonically related to 14 MHz, and in each case, both signals must be on the air for the interference to occur. Eliminating the unwanted signal at the TV receiver will cure the interference problem. Attaching a high-pass filter to the TV receiver input is the most likely cure for this problem.

Many combinations of this type can occur, depending on the band in use and on the local frequency assignments to FM and TV stations. The interfering frequency is equal to the amateur fundamental frequency either added to or subtracted from the frequency of some local station. Should the amateur station unintentionally (but nevertheless illegally) be emitting a strong harmonic of its fundamental frequency, this harmonic also can combine with the signal from another transmitter to produce cross modulation. When interference occurs on a TV channel or a commercial frequency that is not harmonically related to the amateur transmitting frequency, the possibility of cross modulation should be investigated.

Yet another type of cross modulation can occur with amateur phone transmitters. Occasionally a voice from an amateur transmission is heard whenever a broadcast receiver is tuned to a station, but there is no interference when tuning between stations. This type of cross modulation results from rectification of the amateur signal in one of the early receiver stages. Receivers that are susceptible to this kind of trouble usually also get a similar type of interference if there is a strong local BC station and the receiver is tuned to some other station.

The remedy for cross modulation in the receiver is the same as for images and oscillator-harmonic response — reduce the strength of the amateur signal at the receiver by using a filter. You may also be able to reduce the interference by tuning your receiver slightly off frequency. The interference level should drop, while the desired signal is still strong.

The trouble is not always in the receiver. Cross modulation can occur in any nearby rectifying circuit such as a poor contact in water or steam piping, rain gutters, and other conductors in the strong field of the transmitting antenna, external to both receiver and transmitter. Locating the cause may be difficult. It is best attempted with a battery-operated portable BC receiver. The receiver should be used as a probe to find the spot where the interference is most intense. When such a spot is located, inspection of the metal structures in the vicinity may indicate the cause. The remedy is to make a good electrical bond between the two conductors, eliminating the rectification.

[Turn to Chapter 10 and study examination questions A4D04, A4D05, A4D06, A4D07 and A4D10. Review this section as needed.]

**CHAPTER 5
KEYWORDS
KEYWORDS
KEYWORDS**

Apparent power — The product of the RMS current and voltage values in a circuit without consideration of the phase angle between them.

Average power — The product of the RMS current and voltage values associated with a purely resistive circuit, equal to one half the peak power when the applied voltage is a sine wave.

Back EMF — An opposing electromotive force (voltage) produced by a changing current in a coil. It can be equal to the applied EMF under some conditions.

Bandwidth — The frequency range (measured in hertz) over which a signal is stronger than some specified amount below the peak signal level. For example, if a certain signal is at least half as strong as the peak power level over a range of ± 3 kHz, the signal has a 3-dB bandwidth of 6 kHz.

Decibel (dB) — One tenth of a bel, denoting a logarithm of the ratio of two power levels — dB = 10 log (P2/P1). Power gains and losses are expressed in decibels.

Effective radiated power (ERP) — The relative amount of power radiated in a specific direction from an antenna, taking system gains and losses into account.

Electric field — A region through which an electric force will act on an electrically charged object.

Field — The region of space through which any of the invisible forces in nature, such as gravity, electric force or magnetic forces, act.

Half-power points — Those points on the response curve of a resonant circuit where the power is one half its value at resonance.

Joule — The unit of energy in the metric system of measure.

Magnetic field — A region through which a magnetic force will act on a magnetic object.

Parallel-resonant circuit — A circuit including a capacitor, an inductor and sometimes a resistor, connected in parallel, and in which the inductive and capacitive reactances are equal at the applied-signal frequency. The circuit impedance is a maximum, and the current through the circuit is a minimum at the resonant frequency.

Peak envelope power (PEP) — The average power of the RF envelope during a modulation peak. (Used for modulated RF signals.)

Peak power — The product of peak voltage and peak current in a resistive circuit. (Used with sine-wave signals.)

Phase — A representation of the relative time or space between two points on a waveform, or between related points on different waveforms. Also the time interval between two events in a regularly recurring cycle.

Phase angle — If one complete cycle of a waveform is divided into 360 equal parts, then the phase relationship between two points or two waves can be expressed as an angle.

Potential energy — Stored energy. This stored energy can do some work when it is "released." For example, electrical energy can be stored as an electric field in a capacitor or as a magnetic field in an inductor. This stored energy can produce a current in a circuit when it is released.

Power — The time rate of transferring or transforming energy, or the rate at which work is done. In an electric circuit, power is calculated by multiplying the voltage applied to the circuit by the current through the circuit.

Power factor — The ratio of real power to apparent power in a circuit. Also calculated as the cosine of the phase angle between current and voltage in a circuit.

Q — A quality factor describing how closely a practical coil or capacitor approaches the characteristics of an ideal component

Reactive power — The apparent power in an inductor or capacitor. The product of RMS current through a reactive component and the RMS voltage across it. Also called wattless power

Real power — The actual power dissipated in a circuit, calculated to be the product of the apparent power times the phase angle between the voltage and current.

Rectangular coordinates — A graphical system used to represent length and direction of physical quantities for the purpose of finding other unknown quantities.

Resonant frequency — That frequency at which a circuit including capacitors and inductors presents a purely resistive impedance. The inductive reactance in the circuit is equal to the capacitive reactance.

Series-resonant circuit — A circuit including a capacitor, an inductor and sometimes a resistor, connected in series, and in which the inductive and capacitive reactances are equal at the applied-signal frequency. The circuit impedance is at a minimum, and the current is a maximum at the resonant frequency.

Skin effect — A condition in which ac flows in the outer portions of a conductor. The higher the signal frequency, the less the electric and magnetic fields penetrate the conductor and the smaller the effective area of a given wire for carrying the electrons.

Thevenin's Theorem — Any combination of voltage sources and impedances, no matter how complex, can be replaced by a single voltage source and a single impedance that will present the same voltage and current to a load circuit.

ELECTRICAL
PRINCIPLES

To pass the Technician and General class exams, you studied basic electronic principles dealing with dc circuits, along with the ac theory for some elementary components such as resistors, capacitors and inductors. The Advanced class examination will emphasize ac circuit theory. You will be expected to have a good working knowledge of reactance and impedance, be able to perform power calculations, and to know what reactive power and power factor are. The Element 4A exam also covers energy storage in capacitors and inductors, resonant circuits, bandwidth, phase angle and circuit Q. Thevenin's Theorem is a powerful tool that you can use to simplify the solution to many circuit problems. You will have ten questions on your exam about these important electronics topics.

This chapter explains all of these topics so you will understand how to perform the required calculations and will know the information covered by the exam questions. There are many sample problems worked out in the text. Follow those explanations carefully, being sure to work out the arithmetic and manipulate the equations as indicated. As you complete each section of text, you will be instructed to turn to Chapter 10 and study the appropriate group of examination questions. If you have difficulty with any of the questions, be sure to review the material in this chapter. If you would like to delve deeper into the theory behind the topics covered here, you may want to read some more-advanced texts. *The ARRL Handbook* is always on the recommended reading list, of course. There are many good books about electronics theory and principles, and by reading some of the material in another book you may gain additional insight into the electronics theory covered in this chapter.

There will be ten exam questions taken from the Electrical Principles sub-element. These questions will come from the ten syllabus groups covering this material:

A5A   Characteristics of resonant circuits.
A5B   Series resonance (capacitor and inductor to resonate at a specific frequency).
A5C   Parallel resonance (capacitor and inductor to resonate at a specific frequency).
A5D   Skin effect; electrostatic and electromagnetic fields.
A5E   Half-power bandwidth.
A5F   Circuit Q.
A5G   Phase angle between voltage and current.
A5H   Reactive power; power factor.
A5I   Effective radiated power; system gains and losses.
A5J   Replacement of voltage source and resistive voltage divider with equivalent voltage source and one resistor (Thevenin's Theorem).

ELECTRICAL ENERGY

Before you can understand what electrical energy is, you must know some important definitions. Let's use some simple examples from your everyday experience to build those definitions. Pick up a stone, and carry it to an upstairs window of your house. You are doing work against the force of gravity exerted by the Earth as you move the stone farther away from the surface of the Earth. Gravity is an invisible force that the Earth exerts on the stone. There are many invisible forces in nature. These invisible forces work through space, so there is no physical contact required for the force to act. The space through which these invisible forces act is often referred to as a **field**. (No, this is not the same as a farmer's field!) What we mean here is a region of space through which a force acts without actual contact. When you pick up the stone, there is physical contact, so you are not exerting an invisible force — you are not a field, then. The gravitational field of the Earth pulls every object toward the center of the Earth.

Okay, so how much work did you do on the stone? Well, you have to multiply the distance you moved it through the gravitational field (let's say to a height of 10 feet above the Earth) times the force you had to exert, which was equal to the force of gravity on the stone (say 1 pound). So you have done 10 foot-pounds of work against the gravitational field. Now place the stone on a windowsill. By doing work on the stone, you have stored some energy in it. That energy is equal to the amount of work that you did, and is called **potential energy**. In effect, you are storing the energy by the position of the stone in the gravitational field of the Earth. If you push the stone out the window, it will fall back to the Earth, and while it is falling, the stored potential energy is being converted to kinetic energy, or energy of motion.

In electronics we are interested mainly in two types of invisible forces. Those are the *electric force* and the *magnetic force*. The space through which these forces act are the **electric field** and the **magnetic field**. These fields make up an electromagnetic wave. (A wave of this type is a field in motion.) We can store electrical **potential energy** as a voltage in an electric field and as a current in a magnetic field. In either case, that potential energy can be released in the form of an electric current in the circuit.

Storing Energy in an Electric Field

You can store energy in a capacitor by applying a dc voltage across the terminals. There will be an instantaneous inrush of current to charge the capacitor plates. The only thing limiting the current at the instant the voltage is connected is any resistance there may be in the circuit. (Of course there will always be some resistance in the wires connecting the components, and in the components themselves.) The capacitor builds up an electric charge as one set of plates accumulates an excess of electrons and the other set loses an equal number. The voltage across the capacitor rises as this charge builds up. Eventually, the voltage at the capacitor terminals is equal to the source voltage, and the current stops. If the voltage source is disconnected, the capacitor will remain charged to that voltage. The charge will stay on the capacitor plates as long as there is no path for the electrons to travel from one plate to the other.

At this point we should be careful to point out that our discussion in this section deals with ideal components. We are thinking of resistors that have no stray capacitance or inductance associated with the leads or composition of the resistor itself. Ideal capacitors exhibit no losses, and there is no resistance in the leads or capacitor plates. Ideal inductors are made of wire that has no resistance, and there is no stray capacitance between turns. Of course, in practice we do not have ideal components, so the conditions described here may be modified a bit in real-life circuits. Even so, components can come pretty close to the ideal conditions. For example, a capacitor with very low leakage will hold a charge for days or even weeks.

Stored electric potential energy produces an **electric field** in the capacitor. Since the charge is not moving, this field is sometimes called an *electrostatic field*. If a resistor or some other circuit is connected across the capacitor terminals, that field will return the stored energy by creating a current in the circuit.

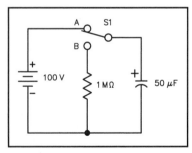

Figure 5-1 — A simple circuit for charging a capacitor and then discharging it through a resistor.

Figure 5-1 illustrates a simple circuit for charging and discharging a capacitor, depending on the switch setting. The electric energy will be converted to heat energy in a resistor, or into other forms of energy, such as sound energy in a speaker. If the capacitance is high, and you connect a large-value resistance across the terminals, it may take a long time for all the energy to be dissipated and the capacitor voltage to drop to zero. On the other hand, by touching a wire across the terminals to short circuit the capacitor, you can discharge it very quickly. If you short a capacitor of several microfarads that has been charged to several hundred volts, you can produce quite a spark. If your skin touches the terminals instead of a piece of wire, you can get a dangerous shock. Large-value filter capacitors in a power supply have bleeder resistors connected across them to drain this charge when the supply is turned off.

The basic unit for expressing energy in the metric system, which is the system used to express all common electrical units, is the **joule** (pronounced with a long u sound, similar to jewel). Electrical energy (or work) is the product of the force needed to move a single electron through an electric potential, times the distance it moves. So if it takes a force of 1 newton to move an electron 1 meter through an electric field, 1 joule of work is done, and 1 joule of energy is stored in the electron.

We are not usually concerned with single electrons, however. We want large numbers of electrons to flow in a circuit. You probably remember from your Novice days that if one coulomb of electrons (6.24×10^{18} or 6 quintillion, 240 quadrillion electrons) flow past a point in one second, then we have a current of one ampere. So even a current of one microampere represents over 6 trillion electrons flowing past a point per second! If 1 joule of work is done on an entire coulomb of electrons, then we say the electric potential is 1 volt. Stating this in mathematical terms, we can write:

$$1 \text{ volt} = \frac{1 \text{ joule}}{1 \text{ coulomb}} \qquad \qquad \text{(Equation 5-1)}$$

The actual work done in charging a capacitor, or the energy stored in the charged capacitor, can be found by:

$$W = \frac{V^2 C}{2} \qquad \qquad \text{(Equation 5-2)}$$

where:
   W = work or energy in joules (or watt-seconds).
   V = potential in volts across the capacitor.
   C = capacitance in farads.

Joules measure the electrical energy stored in an electrostatic field across a capacitor. This electrical potential energy is returned to the circuit as a current when the capacitor discharges.

Storing Energy in a Magnetic Field

When electrons flow through a conductor, a **magnetic field** is produced. This magnetic field exists in the space around the conductor, and a magnetic force acts through this space. This can be demonstrated by bringing a compass near a current-carrying wire and watching the needle deflect. Figure 5-2A illustrates the magnetic field around a wire connected to a battery. If the wire is wound into a coil, so the fields from adjacent turns add together, then a much stronger magnetic field can be produced. The direction of the field, which points to the magnetic north pole, can be found using a "left hand rule." For a straight wire, point the thumb of your left hand in the direction of the *electron flow* and your fingers curl in the direction of the north pole of the magnetic field around the wire. In a similar manner, if you curl the fingers of your left hand around the coil in the direction of electron flow, your thumb points in the direction of the north pole of the field. Parts B and C of Figure 5-2 illustrate the fields around two coils wound in opposite directions.

Notice that this *left-hand rule* describes the direction of the magnetic field in terms of the direction of *electron flow*. Older texts often describe the magnetic-field direction with a *right-hand rule*, using *conventional current*, which flows from positive to negative. Don't be confused by this difference if you read an older text.

The strength of the magnetic field depends on the amount of current, and is stronger when the current is larger. Electrical energy from the voltage source is transferred to the magnetic field in the process of creating the field. So we are storing energy by building up a magnetic field, and that means work must be done against some opposing force. That opposing force is the result of a voltage induced in the circuit whenever the magnetic field (or current) is changing. If the current remains constant, then the magnetic field remains a constant, and there is no more energy being stored.

When you first connect a dc source to a coil of wire, a current begins to flow, and a magnetic field begins to build up. The field is changing very rapidly at that

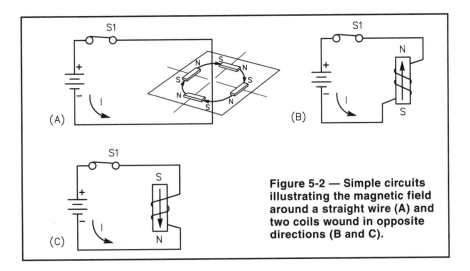

Figure 5-2 — Simple circuits illustrating the magnetic field around a straight wire (A) and two coils wound in opposite directions (B and C).

time, so a large opposing voltage is created, preventing a large current from flowing. As a maximum amount of energy is stored and the magnetic field reaches its strongest value, the opposing voltage will decrease to zero, so the current increases gradually to a maximum value. That maximum value is limited only by the resistance of the wire in the coil, and by any internal resistance of the voltage source.

If the current decreases, then a voltage is induced in the wire that will try to prevent the decrease. The stored energy is being returned to the circuit in this case. As the magnetic field collapses, and the stored energy is returned to the circuit, current continues to flow.

The work done in producing a magnetic field in an inductor (or the energy stored in the field) can be found by:

$$W = \frac{I^2 L}{2}$$
(Equation 5-3)

where:
  W = work in joules (or watt-seconds).
  I = current through the coil in amperes.
  L = inductance in henrys.

This induced voltage or EMF is sometimes called a **back EMF**, since it is always in a direction to oppose any change in the amount of current. When the switch in the circuit of Figure 5-2B is first closed, this back EMF will prevent a sudden surge of current through the coil. Notice that this is just the opposite of the condition when a capacitor is charging. Likewise, if you open the switch to break the circuit, a back EMF will be produced in the opposite direction. This time the EMF tries to keep the current going, again preventing any sudden change in the magnetic-field strength.

The magnitude of the induced back EMF depends on how rapidly the current is changing. If you have a strong magnetic field built up in a coil, and then suddenly break the circuit at some point, a large voltage is induced in the coil, which tries to maintain the current. It is quite common to have a spark jump across the switch contacts as they open. In fact, this is exactly the principle used in the induction coil of an automobile. A large current flows through the coil, then at the proper instant a set of contacts open, inducing a much larger voltage in a second coil. The voltage in this second coil causes a spark to jump across a spark-plug gap, and the gasoline in the cylinder explodes.

[Turn to Chapter 10 at this time, and study questions A5D06 through A5D11. Review this section if any of these questions give you difficulty.]

PHASE ANGLE BETWEEN CURRENT AND VOLTAGE

Now that you understand how capacitors and inductors store energy in electric and magnetic fields, we can learn how those devices react when an alternating voltage is applied to their terminals. While the resistance of a pure resistor does not vary with the frequency, you already know that the reactance of both a coil and a capacitor do change with frequency. That should tell us that coils and capacitors will behave differently with an ac voltage than a dc one. Remember that a pure inductor does not impede direct current, but does impede alternating current, and the higher the frequency the more it opposes the current. Also a pure capacitor will not allow dc to pass through it, but will hamper ac less and less as the frequency increases.

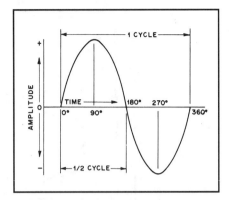

Figure 5-3 — An ac cycle is divided into 360° that are used as a measure of time or phase.

To understand the variations in voltage and current through inductors and capacitors, we must look at the amplitudes of these ac signals at certain instants of time. The relationship between the current and voltage waveforms at a specific instant is called the **phase** of the waveforms. Phase essentially means time, or a time interval between when one event occurs and the instant when a second, related, event takes place. The event that occurs first is said to lead the second, while the second event lags the first.

Since each ac cycle takes exactly the same time as any other cycle of the same frequency, we can use the cycle as a basic time unit. This makes the phase measurement independent of the waveform frequency. If two or more different frequencies are being considered, phase measurements are usually made with respect to the lowest frequency.

It is convenient to relate one complete cycle of the wave to a circle, and to divide the cycle into 360 equal parts or degrees. So a phase measurement is usually specified as an angle. In fact, we often refer to the **phase angle** between two waveforms. Figure 5-3 shows one complete cycle of a sine-wave voltage or current, with the wave broken into four quarters of 90° each.

AC Through a Capacitor

As soon as a voltage is applied across the plates of an ideal capacitor, there is a sudden inrush of current as the capacitor begins to charge. That current tapers off as the capacitor is charged to the full value of applied voltage. By the time the applied voltage is reaching a maximum, the capacitor is also reaching full charge, and so the current through the capacitor goes to zero. A maximum amount of energy has been stored in the electric field of the capacitor at this point. Figure 5-4 shows the voltage across a capacitor as it charges and the charging current that flows into a capacitor with a dc voltage applied.

The situation is a bit different when an ac voltage is applied. Figure 5-5 graphs

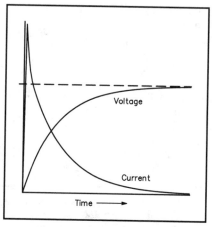

Figure 5-4 — This graph illustrates how the voltage across a capacitor changes as it charges with a dc voltage applied. The charging current is also shown.

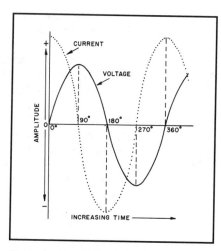

Figure 5-5 — Voltage and current phase relationships when an altering voltage is applied to a capacitor.

the relative current and voltage amplitudes over time, when an ac sine wave signal is applied. The two lines are shown differently to help you distinguish between them; the scale is not intended to show specific current or voltage values. When the applied voltage passes the peak and begins to decrease, the capacitor starts returning some of its stored energy to the circuit. Electrons are now flowing in a direction opposite to the direction they were flowing when the capacitor was charging. By the time the applied voltage reaches zero, the capacitor has returned all of its stored energy to the circuit, and the current is a maximum value in the reverse direction. Now the applied voltage direction has also reversed, and a large charging current is applied to the capacitor, which decreases to zero by the time the applied voltage reaches a maximum value. The second half of the applied voltage cycle is exactly the same as the first half, except the relative directions of current and voltage are reversed.

Study Figure 5-5 to understand the current and voltage relationships for a capacitor over an entire cycle. Notice that the current reaches each point on a cycle 90° ahead of the applied-voltage waveform. We say that the current through a capacitor *leads* the applied voltage by 90°. You could also say that the voltage applied to a capacitor *lags* the current through it by 90°. To help you remember this relationship, think of the word ICE. This will remind you that the current (I) comes before (leads) the voltage (E) in a capacitor (C). Notice also that the applied-voltage waveform and the stored electric-field waveform are in phase — that is, similar points on those waveforms occur at the same instant (Figure 5-6).

[To check your understanding of the current and voltage relationships in a capacitor, you should turn to Chapter 10 now and study question A5G10. Review this section if you are uncertain of the answer to that question.]

AC Through an Inductor

The situation with ac through an inductor is a little more difficult to understand than the capacitor case. As we go through the conditions for a single current cycle, study the graph in Figure 5-7. That should make it easier to follow the changing conditions as they are described.

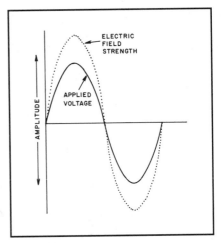

Figure 5-6 — The applied voltage and electric field in a capacitor are in phase.

Let's apply an alternating current to an ideal inductor and observe what happens. At the instant when the current is zero and starts increasing in a positive direction, a magnetic field will start to build up around the coil, storing the applied energy. The current is increasing at a maximum rate, so the magnetic field strength is increasing at a maximum rate. As the current through the coil reaches a positive peak and begins to decrease, a maximum amount of energy has been stored in the magnetic field, and then the coil will begin to return energy to the circuit as the magnetic field collapses. When the current is crossing zero on the way down, it is again changing at a maximum rate, so the magnetic field is also changing at a maximum rate. As the current direction changes, it will begin to build a magnetic field in the direction opposite to the first field, and energy is being stored once again.

When the current reaches a maximum negative value, a maximum magnetic field has been built up, and then the stored energy begins to return to the circuit again as the current increases toward zero. The second half of the cycle is the same as the first, but with all polarities reversed. You should realize by now that the current and the magnetic field in an inductor are in phase — similar points on both waveforms occur at the same instant.

In the section on storing energy in an inductor, you learned about the back EMF that is induced in the coil. That EMF is greatest when the magnetic field is changing the fastest. Furthermore, it is in a direction that opposes the change in current or magnetic-field strength. So when the current is crossing zero on the way to a positive peak, the induced EMF is at its greatest negative value. When the current is at the positive peak, the back EMF is zero, and so on. Figure 5-8 shows the phase relationship between the current through the inductor and the back EMF across it.

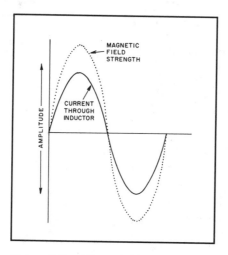

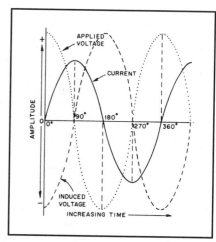

Figure 5-7 — The current through an inductor is in phase with the magnetic field strength.

Figure 5-8 — Phase relationships between voltage and current when an alternating voltage is applied to an inductance.

Since we know that the back EMF opposes the effects of an applied voltage, we can draw in the waveform for applied voltage as shown in Figure 5-8. The applied voltage is 180° out of phase with the induced voltage. From this fact, Figure 5-8 shows that the voltage across an inductor *leads* the current through it, or the current *lags* the voltage. The phase relationship between applied voltage and current through an inductor is just the opposite from the relationship for a capacitor. A useful memory technique to remember these relationships is the little saying, "ELI the ICE man." The L and C represent the inductor and capacitor, and the E and I stand for voltage and current. Right away you can see that E (voltage) comes before (leads) I (current) in an inductor and that I comes before E in a capacitor.

[To check your understanding of the relationship between the voltage across an inductor and the current through it, turn to Chapter 10 and study question A5G11. Review this section if you are uncertain of the answer to this question.]

Phase Angle With Real Components

Up to this point we have been talking about ideal components in discussing the phase relationships between voltage and current in inductors and capacitors. Of course, any real components will have some resistance associated with the inductance or capacitance. Since the voltage across a resistor is in phase with the current through it, the overall effect is that the phase difference between voltage and current will be less than 90° in both cases.

Solving Problems Involving Inductors

Probably the easiest way to illustrate the effect of adding resistance to the circuit is by means of a calculation. Figure 5-9A shows a simple circuit with a resistor and an inductor in series with an ac signal source. Let's pick a frequency of 10 kHz for the signal generator, and connect it to a 20-mH inductor in series with a 1-kΩ resistor. The question is, what is the phase angle between the voltage and current in this circuit? To aid our solution, draw a set of **rectangular coordinates**, and label the X (horizontal) axis R (for resistance) and the Y (vertical) axis X (for reactance). The degree indications shown on Figure 5-9B show the standard way of relating the coordinate system to degrees around a circle. This coordinate system can represent either voltage or current, as needed. In this case, it represents voltage, since the current is the same in all parts of a series circuit. Next, we must calculate the inductive reactance of the coil:

$$X_L = 2\pi \, f \, L = 6.28 \times (10 \times 10^3 \text{ Hz}) \times (20 \times 10^{-3} \text{ H}) = 1257 \ \Omega$$

Now calculate the voltage across the inductor using Ohm's Law. Since the actual current is not important, as long as we know it is the same through each part of the circuit, choose a simple value, such as 1 A:

$$E_L = I \, X_L = 1 \text{ A} \times 1257 \ \Omega = 1257 \text{ V}$$

The voltage across the resistor is also found easily using Ohm's Law:

$$E_R = I \, R = 1 \text{ A} \times 1000 \ \Omega = 1000 \text{ V}$$

Notice that the voltage is numerically equal to the reactance or resistance in these calculations. Wow! That leads to a nifty trick. By assuming a current of 1 A when the actual current value is not important to the problem, we can eliminate a step in the solution. There is no need to actually calculate voltages for a series-circuit problem. Just use the resistance and reactance values.

Okay, so now what? Well, we can't just add the two voltages to get the total voltage across the circuit, because they are not in phase. Remember that the voltage across an inductor *leads* the current, and the voltage across a resistor is *in phase* with the current. This means that the peak voltage across the inductor occurs 90° before the peak voltage across the resistor. On your graph, draw a line

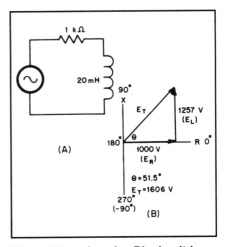

Figure 5-9 — A series RL circuit is shown at A. B shows the right triangle used to calculate the phase angle between the circuit current and voltage.

along the R (horizontal) axis to represent the 1000 V across the resistor. At the end of that line draw another line to represent the 1257 V across the inductor. Since this voltage leads the voltage in the resistor by 90°, it must be drawn vertically upward, along the 90° line. Now complete the figure by drawing a line from the origin (the point where the two axes cross) to the end of the E_L line. This right triangle represents the solution to our problem. Figure 5-9B shows a complete drawing of the triangle. The last line is the total voltage across the circuit, and the angle measured up from the E_R line to the E_T line is the phase angle between the voltage and current. If you use graph paper and make up a suitable scale of divisions per volt on the graph, you can actually solve the problem with no further calculation. Graphical solutions are usually not very accurate, however, and depend a great deal on how carefully you draw the lines.

Let's complete the mathematical solution to our problem. If you are not familiar with right triangles and their solutions (a branch of mathematics called trigonometry), just follow along. You should be able to pick up the techniques from a few examples, but if you continue to have trouble, there is an excellent discussion about the basics of trigonometry in *Understanding Basic Electronics*, published by ARRL. Alternatively, go to your library and check out a book on elementary high school trigonometry. You will find the tools of trigonometry to be very helpful for solving many electronics problems involving ac signals. The various sides and angles are identified by their positions in relation to the right (90°) angle. The side opposite the right angle is always the longest side of a right triangle, and is called the hypotenuse. The other sides are either opposite or adjacent to the remaining angles. It is important to realize that these methods of trigonometry apply only to triangles that contain a 90° angle.

We will find the phase angle first, using the tangent function. Since the angle of interest is the one between the E_R line and the hypotenuse, E_L is the side opposite and E_R is the side adjacent to the angle. Tangent (often abbreviated as *tan*) is defined as:

$$\tan \theta = \frac{\text{side opposite}}{\text{side adjacent}} \qquad \text{(Equation 5-4)}$$

Then for our problem,

$$\tan \theta = \frac{E_L}{E_R} = \frac{1257 \text{ V}}{1000 \text{ V}} = 1.257$$

where θ is the angle between the E_R and E_T lines.

Now that we know the tangent of the angle, it is a simple matter to refer to a table of trigonometric functions to find what angle has a tangent of that value. If you have an electronic calculator that is capable of doing trig functions, then you simply ask it to find the angle using the inverse tangent or arc tangent function, often written $\tan^{-1}$:

$$\tan^{-1}(1.257) = 51.5°$$

If the resistance and reactance of our components had been equal, then we would have found tan θ to be 1, and the phase angle would be 45°. If the reactance were many times larger than the resistance, then the phase angle would be close to 90°, and if the resistance were many times larger than the reactance, then the phase angle would be close to 0°.

There is one more question we would like an answer to here, and that is the total assumed voltage across the circuit, given our assumed current of 1 ampere. We could find the actual voltage in the same manner, given an actual current through the circuit. There are two common methods available to find this, so let's look at both of them. The first method uses the sine function. (Sine is often abbreviated as *sin*.)

$$\sin \theta = \frac{\text{side opposite}}{\text{hypotenuse}} \qquad \text{(Equation 5-5)}$$

We know that the phase angle is 51.5°, so we find the value of the sine function for that angle (using a trig table, calculator or slide rule). Sin (51.5°) = 0.7826. By solving Equation 5-5 for the hypotenuse, we can find the answer:

$$\text{hypotenuse} = E_r = \frac{\text{side opposite}}{\sin \theta} = \frac{E_L}{\sin (51.5°)} = \frac{1257 \text{ V}}{0.7826}$$

$$E_T = 1606 \text{ V}$$

The second method for finding E_T involves the use of an equation known as the Pythagorean Theorem (named after Pythagoras, a Greek mathematician who discovered this important relationship). The sides of a right triangle are related to each other by the equation:

$$C^2 = A^2 + B^2 \qquad \text{(Equation 5-6)}$$

where:
   C = the length of the hypotenuse.
   A and B = the lengths of the other two sides.

We can solve this equation for the length of the hypotenuse, and rewrite it:

$$C = \sqrt{A^2 + B^2} \qquad \text{(Equation 5-7)}$$

Since the side we want to find is the hypotenuse of our triangle (side E_T), we can use Equation 5-7.

$$E_T = \sqrt{(1257 \text{ V})^2 + (1000 \text{ V})^2} = \sqrt{1580049 \text{ V}^2 + 1000000 \text{ V}^2}$$

$$E_T = \sqrt{2580049 \text{ V}^2} = 1606 \text{ V}$$

Electrical Principles 5-13

Notice that in this example when we squared the value of the voltage, we also show the units of volts being squared. We carried the units along with the calculations, and when we took the square root of the number, we also took the square root of the units. This technique, called *dimensional analysis*, can be useful for showing that the result has the units you expected. If the units don't work out, it may be because you selected an improper equation or followed an incorrect mathematical procedure!

It doesn't matter which method you use to find the total voltage, so pick whichever one seems easier, then stick with it. After you have worked a few of these problems, they will begin to seem much easier. It is very important to follow an organized, systematic approach, however, or you will become easily confused, and any slight change in the wording of the problem will throw you off.

Solving Problems Involving Capacitors

Let's try another problem, to see how well you understood that solution. Figure 5-10A shows a circuit with a capacitor instead of an inductor. For simplicity, we will keep the same signal-generator frequency and the same resistor in the circuit.

The 12660-pF capacitor is a strange value, but let's use it anyway, just for the sake of example.

After drawing a set of coordinates to draw the triangle on, we have to calculate the capacitive reactance.

$$X_C = \frac{1}{2\pi fc} = \frac{1}{2 \times 3.14 \times (10 \times 10^3 \text{ Hz}) \times (12660 \times 10^{-12} \text{ F})}$$

$$X_C = \frac{1}{7.95 \times 10^{-4}} = 1258 \ \Omega$$

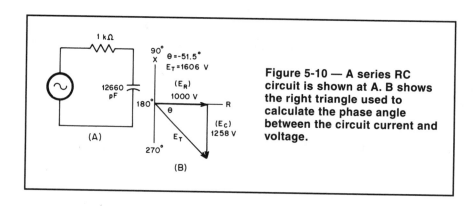

Figure 5-10 — A series RC circuit is shown at A. B shows the right triangle used to calculate the phase angle between the circuit current and voltage.

Assume a current of 1 A and use Ohm's Law to calculate the voltages across the resistor and capacitor. (Remember the trick we learned with the inductor problem. The voltages are numerically equivalent to the reactance and resistance when we assume a current of 1 A.) $E_R = 1000$ V and $E_C = 1258$ V. Draw the resistor-voltage line on your diagram, as before. Keep in mind that the voltage across a capacitor *lags* the current through it, so the capacitor voltage is 90° behind the resistor voltage. Show this on your graph by drawing the capacitor-voltage line vertically downward at the end of the E_R line. Our graph shows this as 270° but if you go clockwise from the resistance axis — the opposite direction from the way we normally measure angles — it is equivalent to –90°. Actually, we should refer to the capacitor voltage as a negative value, –1258 V, then. Complete the triangle by drawing the E_T line, and proceed with the solution as in the previous problem.

Tan θ = – 1.258, so $\tan^{-1} (- 1.258) = - 51.5°$ and $E_T = 1606$ V (1607 V by the Pythagorean Theorem solution). Notice that the phase angle came out negative this time, because the capacitor voltage was taken as a negative value. This indicates that the total voltage across the circuit is 51.5° behind the current — a result of the fact that the voltage across a capacitor *lags* the current through it. Of course we could also say that the current *leads* the voltage by 51.5°, and that would mean the same thing.

Solving Problems Involving Both Inductors and Capacitors

Series Circuits

The problem illustrated in Figure 5-11 adds a slight complication, but don't get too confused. Start out the same way we did for the first two problems, by drawing a set of coordinates, labeling the axes, and then calculating the reactance values. Since this is still a series circuit, the current will be the same through all components, so we are still interested in calculating the voltages. The inductive reactance of a 1.60-µH inductor with a 10-MHz signal applied to it is 100 Ω, and the capacitive reactance of a 637-pF capacitor with the same signal is 25 Ω. Now write a value for the assumed voltages across each component: $E_R = 100$ V, $E_L = 100$ V and $E_C = 25$ V.

(Don't forget that the voltage across the capacitor is negative, 180° out of phase with the voltage across the inductor.) When we go to add these lines to make up our "triangle" you will notice that the lines for the inductor and capacitor voltages go in opposite directions. Just subtract them before drawing the line. In this example, the inductor voltage is the larger, so the total reactive voltage is 75 V, and is represented by a line drawn upward on the diagram. If the capacitor voltage were greater, then the total reactive voltage would be negative, and would be represented by a line drawn downward on the diagram. That just tells us if the phase angle between circuit voltage and current will be positive (upward) or negative (downward).

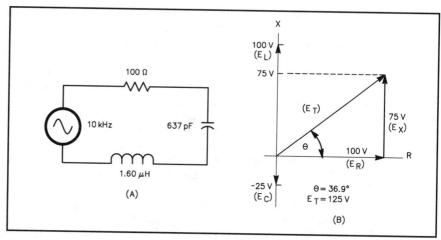

Figure 5-11 — A series RLC circuit is shown at A. B shows the right triangle used to calculate the phase angle between circuit current and voltage.

Okay, so now you have a triangle drawn to look like the one in Figure 5-11B, and the solution is straightforward. Tan θ = 0.75, $\tan^{-1}$ (0.75) = 36.9° and E_T = 125 V. The total voltage across our circuit leads the current by 36.9°.

Parallel Circuits

Now for the ultimate complication in this type of circuit problem! All of our work so far has been with series circuits. But what happens if the components are connected in parallel across the signal source? Let's handle this in one big jump, with a circuit including a resistor, an inductor and a capacitor all connected in parallel with a 10-kHz signal generator. Figure 5-12 gives a circuit diagram, with component values.

Again, start by drawing a set of coordinates, labeling the axes R for resistance and X for reactance, and then calculate the inductive and capacitive reactances.

X_L = 628 Ω and X_C = 159 Ω.

Now the problem becomes a little different than all the others we have done so far. Because this is a parallel circuit, we must recognize that it is the applied voltage that is the same for all components, and that the current through each will have a different phase. Of course the current through a resistor is in phase with the voltage across it, as always. The current through the capacitor leads the voltage across it, and the current through an inductor lags the voltage across it.

Since the actual value of applied voltage is not important, we can assume a value that will make our problem as easy as possible. Let's pick a value of 1000 V.

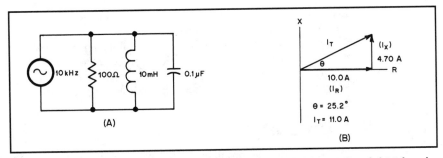

Figure 5-12 — A parallel RLC circuit is shown at A. B shows the right triangle used to calculate the phase angle between circuit voltage and current.

Why 1000? Well, as always, Ohm's Law applies to the circuit, so we can solve it for the individual currents through each component:

$$I_R = \frac{E}{R}, I_L = \frac{E}{X_L} \text{ and } I_C = \frac{E}{X_C} \hspace{2cm} \text{(Equation 5-8)}$$

Since the largest value of resistance or reactance is 628 Ω, all of our currents will be larger than 1 by selecting E = 1000 V. Will the solution be the same if you pick 10 V or 1 V? Yes! In fact, it would be a good exercise for you to also work out the problem with those (or other) values of applied voltage.

With the applied voltage of 1000, $I_R = 10$ A, $I_L = -1.59$ A and $I_C = 6.29$ A. Why is the current through the inductor negative? Because it lags the voltage by 90°. The minus sign tells us that when we combine the current through the inductor with the current through the capacitor, we have to subtract the numbers because they go in opposite directions. Whichever number is larger determines if the current is leading or lagging the voltage, and whether to draw the line up or down on your triangle. For this problem, $I_X = 4.70$ A. The current through the capacitor is larger, so the total current leads the voltage. After drawing the line for resistor current on your graph, draw a line upward to represent the capacitive current in the circuit. Complete the triangle and continue with the solution, as before. Tan θ = 0.470, $\tan^{-1}$ (0.470) = 25.2° and $I_T = 11.0$ A. That's all there is to it!

How would you find the phase angle in a series or parallel circuit that contains several inductors, capacitors and resistors? Well, it really is no more difficult than any of the problems we have done so far. After calculating all the reactances, you can find an assumed voltage across the components in a series circuit or an assumed current through the components in a parallel circuit. Then the combined voltage or current is calculated by adding the values for like components (inductive, capacitive and resistive values added separately). Finally, draw a triangle with the resistive value on the horizontal axis and the reactive value on the vertical axis.

Just remember that a leading voltage or current is drawn upward and a lagging voltage or current is drawn downward.

[If you worked all of these problems along with the text, you should have no trouble with the phase-angle questions on the Advanced class exam. Just to prove that to yourself, turn to Chapter 10 now. Study questions A5G01 through A5G09. Review the examples in this section if needed.]

RESONANT CIRCUITS

With all of the problems so far, we have used inductor and capacitor values that give different inductive and capacitive reactances. So there was always a bigger voltage across one of the series components or a larger current through one of the parallel ones. But you may have wondered about the condition if both reactances turned out to be equal, with equal (but opposite) voltages or currents. You have probably realized that in a series circuit with an inductor and a capacitor, if the inductive reactance is equal to the capacitive reactance, then the voltage drop across each component is the same, but they are 180° out of phase. The two values cancel, and the only remaining voltage drop is across any resistance in the circuit.

It is a common practice to say that the two reactances cancel in this case, and to talk about inductive reactance as a positive value and capacitive reactance as a negative value. Just keep in mind that this terminology is a simplification that applies only to series circuits, and comes about because the voltage across the inductor is positive or leads the current and that the voltage across the capacitor is negative or lags the current.

With a parallel circuit, if the inductive and capacitive reactances are equal, then the currents through the components will be equal, but 180° out of phase. The two currents cancel, and the only current in the circuit is a result of a parallel resistance. Here again, it is common practice to say that the reactances cancel. In the parallel circuit, however, it is the inductive component that is considered to be negative and the capacitive component is considered to be positive. Of course that terminology comes from the fact that the current through a parallel inductor lags the applied voltage and the current through a parallel capacitor leads the applied voltage.

Whether the components are connected in series or parallel, we say the

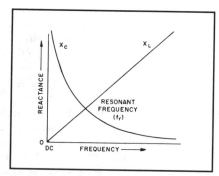

Figure 5-13 — A graph showing the relative change in inductive reactance and capacitive reactance as the frequency increases. For a specific inductor and a specific capacitor, the point where the two lines cross ($X_L = X_C$) is called the resonant frequency.

circuit **resonant frequency** occurs when the inductive-reactance value is the same as the capacitive-reactance value. Remember that inductive reactance increases as the frequency increases and that capacitive reactance decreases as frequency increases. Figure 5-13 is a graph of inductive and capacitive reactances for two general components.

Neither the exact frequency scale nor the exact reactance scale is important. The two lines cross at only one point, and that point represents the resonant frequency of a circuit using those two components. Every combination of a capacitor and an inductor will be resonant at some frequency.

Since resonance occurs when the reactances are equal, we can derive an equation to calculate the resonant frequency of any capacitor-inductor pair:

$$X_L = 2\pi f_r L = X_C = \frac{1}{2\pi f_r C} \qquad \text{(Equation 5-9)}$$

$$1 = \frac{1}{(2\pi f_r L)(2\pi f_r C)}$$

$$f_r^2 = \frac{1}{(2\pi)^2 LC}$$

$$f_r = \frac{1}{2\pi\sqrt{LC}}$$

Series Resonant Circuits

Figure 5-14 shows a signal generator connected to a series RLC circuit. The signal generator produces a current through the circuit, which will cause a voltage drop to appear across each component. The voltage drops across the inductor and capacitor are always 180° out of phase. When the signal generator produces an output signal at the resonant frequency of the circuit, the voltage drops across the inductor and capacitor can be many times larger than the voltage applied to the circuit. In fact, those voltages are sometimes at least 10 times as large, and may be as much as a few hundred times as large as the applied voltage in a practical circuit. With perfect components and no resistance in the circuit, there would be nothing to restrict the current in the circuit. An ideal **series-resonant circuit**, then, "looks like" a short circuit to the signal generator. There is always some resistance in a circuit, but if the total resistance is small, the current will be large, by Ohm's Law.

See Figure 5-14B. The large current that results from this condition produces large, but equal, voltage drops across each reactance. The phase relationship between the voltages across each reactance means that the voltage across the coil reaches a positive peak at the same time that the voltage across the capacitor reaches a negative peak. We can say that the impedance of an ideal series-resonant circuit is zero, and is approximately equal to the circuit resistance in an RLC circuit. The main purpose of the series resistor is to prevent the circuit from overloading the generator. The current at the input to a series RLC circuit is a maximum at resonance.

We can think of the resonance condition in another way, which may help you understand these conditions a little better. The large voltages across the reactances develop because of energy stored in the electric and magnetic fields associated with the components. The energy going into the magnetic field on one half cycle is coming out of the electric field of the capacitor. Then, when all of the energy has been transferred to the magnetic field, it is returned to the circuit and is stored in the electric field of the capacitor again. A large amount of energy can be handed back and forth between the inductor and the capacitor without the source supplying any additional amount. The source only has to supply the actual power dissipated in the resistance of imperfect inductors and capacitors plus what is used by the resistor.

Let's see how you do with a series-resonant-circuit problem. What frequency should the signal generator in Figure 5-14 be tuned to for resonance if the resistor is 47 Ω, the coil is a 50-μH inductor and the capacitor has a value of 40 pF? Probably the biggest stumbling block on these problems will be remembering to convert the inductor value to henrys and the capacitor value to farads. After you have done that, use Equation 5-9 to calculate the resonant frequency.

$$50 \ \mu H = 50 \times 10^{-6} \ H = 0.000050 \ H$$

$$40 \ pF = 40 \times 10^{-12} \ F = 0.000000000040$$

$$f_r = \frac{1}{2\pi \sqrt{LC}} = \frac{1}{2\pi \sqrt{(50 \times 10^{-6})(40 \times 10^{-12})}}$$

$$f_r = \frac{1}{2\pi \sqrt{(2000 \times 10^{-18})}} = \frac{1}{(6.28)(44.7 \times 10^{-9})} = \frac{1}{2.81 \times 10^{-8}}$$

$$f_r = 3.56 \times 10^6 \ Hz = 3.56 \ MHz$$

[Turn to Chapter 10 and study questions A5A01 through A5A05, A5A07 and questions A5B01 through A5B10. Review this section if any of these questions give you trouble.]

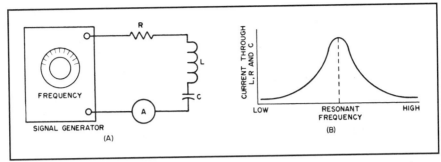

Figure 5-14 — A series-connected LC or RLC circuit behaves like a very low resistance at the resonant frequency. Therefore, at resonance, the current passing through the components reaches a peak.

Parallel-Resonant Circuits

With a **parallel-resonant circuit** there are several current paths, but the same voltage is applied to the components. Figure 5-15 shows a parallel LC circuit connected to a signal generator. The series resistor is just a precaution to prevent the circuit from overloading the generator. The applied voltage will force some current through the branches. The current through the coil will be 180° out of phase with the current in the capacitor, and again they add up to zero. So the total current into or out of the generator is very small. Part B is a graph of the relative generator current. It is a mistake to assume, however, that because the generator current is small there is a small current flowing through the capacitor and inductor. In a parallel resonant circuit, the current through the inductor and capacitor is very large. Part C is a graph of the voltage across the inductor and capacitor. Because the voltage can be very large near resonance, the current through the components can also be very large. We call this current the *circulating current*, or *tank current*.

The current at the input of a parallel RLC circuit at resonance is at a minimum. The circulating current within the components, however, is at a maximum.

The parallel resonant circuit has a high impedance, and can appear to be an open circuit to the signal generator, because the current from the signal generator is quite small. A resistor placed in parallel with the inductor and capacitor will draw current from the signal generator that depends on the generator output voltage and the resistance value. At resonance, the magnitude of the impedance of a circuit with a resistor, inductor and capacitor all connected in parallel will be approximately equal to the circuit resistance.

If we use points 1 and 2 in Figure 5-15A as reference points and examine the branch currents, we observe an interesting phenomenon. Because the current through the capacitor and inductor are 180° out of phase, the branch currents appear to flow in opposite directions. For example, in the figure the current through the inductor

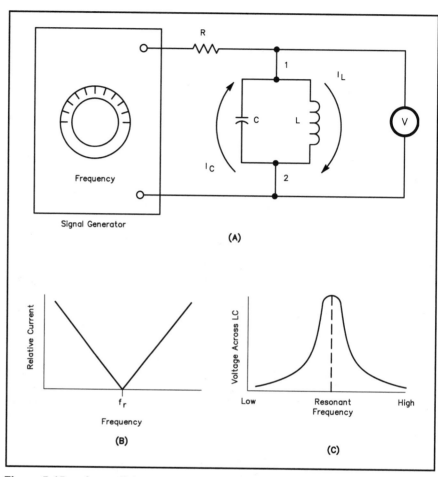

Figure 5-15 — A parallel-connected LC or RLC circuit behaves like a very high resistance at the resonant frequency. Therefore, the voltage measured across the circuit reaches a peak at the resonant frequency.

may appear to flow in the direction from 1 to 2, while the current through the capacitor is flowing from 2 to 1. When the applied current from the generator reverses polarity, the branch currents reverse direction. From the point of view of the circulating current, the two reactive components are actually in series. Current flows from point 1 through the inductor to point 2, and from point 2 through the capacitor to point 1. At resonance, this circulating current will only be limited by resistive losses in the components and in the wire connecting them.

We can again think about the energy in the circuit being handed back and forth between the magnetic field of the inductor and the electric field of the capacitor. Large amounts of energy are being transferred, but the generator only has to supply a small amount to make up for the losses in imperfect components. While the total current from the generator is small at resonance, the voltage measured across the tank reaches a maximum value at resonance.

It is also interesting to consider the phase relationship between the voltage across a resonant circuit and the current through that circuit. Because the inductive reactance and the capacitive reactance are equal but opposite, their effects cancel each other. The current and voltage associated with a resonant circuit are in phase, then. This is true whether we are considering a series resonant circuit or a parallel resonant circuit.

Calculating the resonant frequency of a parallel circuit is exactly the same as for a series circuit. So you should have no trouble finding the resonant frequency of the circuit shown in Figure 5-15 if R = 4.7 kΩ, L = 1 μH and C = 10 pF.

$$f_r = \frac{1}{2\pi\sqrt{LC}}$$ (Equation 5-9)

$$f_r = \frac{1}{2\pi\sqrt{\left(1 \times 10^{-6}\right)\left(10 \times 10^{-12}\right)}}$$

$$f_r = \frac{1}{2\pi\sqrt{\left(10 \times 10^{-18}\right)}}$$

$$f_r = \frac{1}{\left(6.28\right)\left(3.16 \times 10^{-9}\right)} = \frac{1}{1.99 \times 10^{-8}}$$

$$f_r = 50.3 \times 10^6 \text{ Hz} = 50.3 \text{ MHz}$$

[Now turn to Chapter 10 and study questions A5A06, A5A08 through A5A11 and questions A5C01 through A5C10. Review this section if any of these questions give you trouble.]

Calculating Component Values for Resonance

Using the same technique that we used to derive Equation 5-9, we can easily derive equations to calculate either the inductance or capacitance to resonate with a certain component at a specific frequency:

$$L = \frac{1}{\left(2\pi f_r\right)^2 C}$$ (Equation 5-10)

$$C = \frac{1}{\left(2\pi f_r\right)^2 L}$$ (Equation 5-11)

You should always be sure to change the frequency to hertz, the capacitance to farads and the inductance to henrys when using these equations.

Let's try solving a couple of problems, just to be sure you know how to handle the equations. Suppose you have a 100-pF capacitor, and you want a circuit that is resonant in the 40-meter band. What inductor value should you choose? Well, first of all, you will have to select a frequency in the 40-meter band to use for the calculation. In a practical design problem, you would have a specific frequency in mind, but for a general example, let's pick some frequency in the middle of the band, such as 7.10 MHz. Actually, the required inductance will not change drastically from one end of the band to another. You may want to solve the problem for a resonant frequency of 7.00 MHz and 7.30 MHz just to prove that to yourself.

Equation 5-10 will give us the desired answer, but first we must change the frequency to 7.10×10^6 Hz and the capacitance to 100×10^{-12} F. Then:

$$L = \frac{1}{\left(2\pi\right)^2 \left(7.10 \times 10^6\right)^2 \left(100 \times 10^{-12}\right)}$$

$$L = \frac{1}{\left(39.48\right)\left(5.04 \times 10^{13}\right)\left(100 \times 10^{-12}\right)}$$

$$L = \frac{1}{1.99 \times 10^5} = 5.03 \times 10^{-6} \text{ H} = 5.03 \text{ }\mu\text{H}$$

So you will need a 5-microhenry inductor to build the desired circuit. If you want a circuit that is resonant at 7.00 MHz, you will need a 5.2-μH inductor, and to make it resonant at 7.30 MHz it will take a 4.8-μH coil.

What value capacitor is needed to make a circuit that is resonant in the 20-meter band if you have a 2-μH coil? Choose a frequency in the 20-meter band to work with. Let's pick 14.10 MHz. Then convert to fundamental units: $f_r = 14.10 \times 10^6$ Hz and $L = 2 \times 10^{-6}$ H. This time, select Equation 5-11, since that one is written to find capacitance, the quantity we are looking for:

$$C = \frac{1}{\left(2\pi\right)^2 \left(14.10 \times 10^6\right)^2 \left(2 \times 10^{-6}\right)}$$

$$C = \frac{1}{\left(39.48\right) \left(1.99 \times 10^{14}\right) \left(2 \times 10^{-6}\right)}$$

$$C = \frac{1}{1.57 \times 10^{10}} = 63.7 \times 10^{-12} \text{ F} = 63.7 \text{ pF}$$

You will need a 64-pF capacitor. If you try solving this problem for both ends of the 20-meter band, you will find that you need a 65-pF capacitor at 14.00 MHz and a 62-pF unit at 14.35 MHz. So any capacitor value within this range will resonate in the 20-meter band with the 2-μH inductor.

[For more practice calculating inductance and capacitance values for resonant circuits, study questions A5B11 and A5C11 in Chapter 10. Also study questions A7D02 through A7D06. Review the procedures described in this section if you have difficulty with these questions.]

Q — THE QUALITY FACTOR OF REAL COMPONENTS

We have talked about ideal resistors, capacitors and inductors, and how they behave in ac circuits. We have shown that resistance in a circuit causes some departure from the ideal conditions. But how can we determine how close to the ideal a certain component comes? Or how much of an effect it will have on the designed

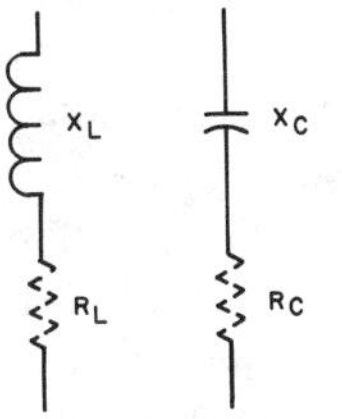

Figure 5-16 — A practical coil can be considered as an ideal inductor in series with a resistor, and a practical capacitor can be considered as an ideal capacitor in series with a resistor.

circuit conditions? We can assign a number to the coil or capacitor that will tell us the relative merits of that component — a quality factor of sorts. We call that number **Q**. We can also assign a Q value to an entire circuit, and that is a measure of how close to the ideal that circuit performs — at least in terms of its resonance properties.

One definition of Q is that it is the ratio of reactance to resistance. Figure 5-16 shows that a capacitor can be thought of as an ideal capacitor in series with a resistor and a coil can be considered as an ideal inductor in series with a resistor. This internal resistance can't actually be separated from the coil or capacitor, of course, but it acts just the same as if it were in series with an ideal, lossless component. The Q of a real inductor, L, is equal to the inductive reactance divided by the resistance and the Q of a real capacitor, C, is equal to the capacitive reactance divided by the resistance:

$$Q = \frac{X}{R}$$

(Equation 5-12)

If you want to know the Q of a circuit containing both internal and external resistance, both resistances must be added together to find the value of R used in the equation. Since added external resistance can only raise the total resistance, the Q always goes down when resistance is added in series. There is no way to reduce the internal resistance of a coil or capacitor and raise the Q, therefore, except by building a better component. The internal resistance of a capacitor is usually much less than that for a coil, so we often ignore the resistance of a capacitor and consider only that associated with the coil. Figure 5-17A shows a series RLC circuit with a Q of 10. To calculate that value, select either value of reactance and divide it by the resistor value.

At Figure 5-17B, we have increased the input frequency 5 times. The reactance of our inductor has increased 5 times, and we have selected a new capacitor to provide a resonant circuit. This time the components are arranged to provide a parallel resonant circuit. The circuit Q is still found using Equation 5-12. You will notice that the Q for this circuit is 50. Increasing the frequency increased the inductive reactance, so as long as the internal resistance stays the same, the Q increases by the same factor.

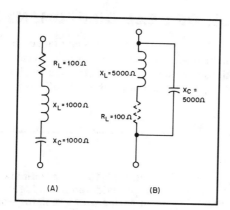

Figure 5-17 — The Q in a series-resonant circuit such as is shown at A and a parallel-resonant circuit such as is shown at B is found by dividing the inductive reactance by the resistance.

Skin Effect Increases Resistance

Unfortunately, the internal resistance of the coil (due mainly to the resistance of the wire used to wind it) increases somewhat as the frequency increases. In fact, the coil Q will increase with increasing frequency up to a point, but then the internal resistance becomes greater and the coil Q degrades. Figure 5-18 illustrates how coil Q changes with increasing frequency.

The major cause of this increased resistance at higher frequencies is something known as **skin effect**. As the frequency increases, the electric and magnetic fields of the signal do not penetrate as deeply into the conductor. At dc, the entire thickness of the wire is used to carry currents but as the frequency increases, the effective area gets smaller and smaller. In the HF range, the current all flows in the outer few thousandths of an inch of the conductor, and at VHF and UHF, the depth is on the range of a few ten thousandths of an inch. This makes the wire less able to carry the electron flow, and increases the effective resistance.

[Before going on to the next section, you should turn to Chapter 10 and study questions A5D01 through A5D05. Review this section if you are uncertain about the importance of skin effect.]

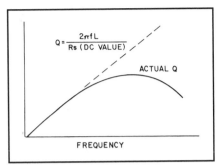

Figure 5-18 — For low frequencies, the Q of an inductor is proportional to frequency. At high frequencies, increased losses in the coil cause the Q to be degraded from the expected value.

Q in Parallel-Resonant Circuits

Often a parallel resistor is added to a parallel-resonant circuit to decrease the Q and increase the **bandwidth** of the circuit, as shown in Figure 5-19. Such a resis-

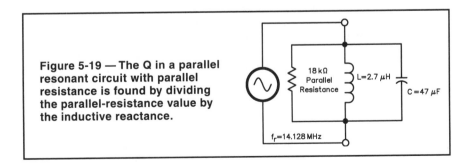

Figure 5-19 — The Q in a parallel resonant circuit with parallel resistance is found by dividing the parallel-resistance value by the inductive reactance.

tor should have a value more than 10 times the reactance of the coil (or capacitor) at resonance or the resonance conditions will change. With the added parallel resistor, circuit Q is found by dividing the resistor value by the reactance value. So for a parallel resonant circuit:

$$Q = \frac{R}{X}$$

(Equation 5-13)

Figure 5-19 shows a parallel circuit made with a 2.7-μH inductor and a 47-pF capacitor. The circuit has an 18-kΩ resistor in parallel with the inductor and capacitor. The resonant frequency of this circuit is 14.128 MHz. What is the circuit Q?

First we must calculate the reactance of either the inductor or capacitor. For this example, we will use the inductor.

$$X_L = 2 \pi f_r L = 6.28 \times 14.128 \times 10^6 \text{ MHz} \times 2.7 \times 10^{-6} \text{ H} = 239.6 \ \Omega$$

Next we will divide this reactance value by the circuit resistance to find the Q.

$$Q = \frac{X}{R} = \frac{18 \times 10^3 \ \Omega}{239.6 \ \Omega} = 75.1$$

As in the case of a series circuit, any added resistance degrades the circuit Q and increases the **bandwidth**. (Bandwidth is discussed in the next section.) When a resistor is added in parallel, the circuit Q is always less than if no resistor were included. If no additional resistor were included, the parallel resistance would be infinitely large. Adding a resistor to the circuit always decreases that value, and so the Q is also reduced. The larger the value of added resistance, the higher the Q.

[Now study questions A5F01 through A5F11 in Chapter 10. Review this section as needed.]

Resonant-Circuit Bandwidth

The higher the circuit Q, the sharper the frequency response of a resonant circuit will be. Figure 5-20 shows the relative **bandwidth** of a circuit with two different Q values. Bandwidth refers to the frequency range over which the cir-

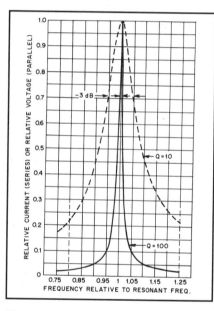

Figure 5-20 — The relative – 3-dB bandwidth of two resonant circuits is shown. The circuit with the higher Q has a steeper response, and a narrower bandwidth. Notice that the vertical scale represents current for a series circuit and voltage for a parallel one. The two circuits are considered to have equal peak response for either case.

cuit response is no more than 3 dB below the peak response. The –3-dB points are shown on Figure 5-20, and the bandwidths are indicated. (The decibel, or dB, represents the logarithm of a ratio of two power levels. If you are not familiar with the use of decibels, see ARRL's *Understanding Basic Electronics*, *The ARRL Handbook*, or a math text on the subject.) Since this 3-dB decrease in signal represents the points where the circuit power is one half of the resonant power, the –3-dB points are also called **half-power points**. The voltage and current have been reduced to 0.707 times their peak values.

The half-power points are called f_1 and f_2; Δf is the difference between these two frequencies, and represents the half-power (or 3-dB) bandwidth. It is possible to calculate the bandwidth of a resonant circuit based on the circuit Q and the resonant frequency.

$$\Delta f = \frac{f_r}{Q} \hspace{4cm} \text{(Equation 5-14)}$$

where:

Δf = the half-power bandwidth.

f_r = the resonant frequency of the circuit.

Q = the circuit Q, as given by Equation 5-12 or 5-13 as appropriate.

Let's calculate the half-power bandwidth of a parallel circuit that has a resonant frequency of 1.8 MHz and a Q of 95. The half-power bandwidth is found by Equation 5-14:

$$\Delta f = \frac{f_r}{Q} = \frac{1.8 \times 10^6 \text{ Hz}}{95} = 18.9 \times 10^3 \text{ Hz} = 18.9 \text{ kHz}$$

Notice that the response of this circuit will be at least half of the peak signal power for signals in the range 1.79055 to 1.80945 MHz. To find the upper and lower frequency limits you have to subtract *half* the total bandwidth from the center frequency for the lower limit and add half the bandwidth to get the upper frequency limit.

[This is a good time to practice the half-power bandwidth calculations from the Advanced class question pool. You can expect to have one of these questions on your Element 4A exam. Study questions A5E01 through A5E11. Review the example in this section if you have difficulty with any of these questions.]

POWER IN REACTIVE CIRCUITS

Earlier in this chapter, we learned that energy is stored in the magnetic field of an inductor when current increases through it, and in the electric field of a capacitor when the voltage across it increases. That energy is returned to the circuit when the current through the inductor decreases or when the voltage across the capacitor decreases. We also learned that the voltages across and currents through these components are 90° out of phase with each other. One way to think of this situation is that in one half of the cycle the power source gives some energy to the inductor or capacitor, only to have the same amount of energy handed back on the next half cycle. A perfect capacitor or coil does not consume any energy, but current does flow in the circuit when a voltage is applied to it.

Power is the time rate of doing work, or the time rate of using energy. Going back to our example of work and energy at the beginning of this chapter, if you did the 10 foot-pounds of work in 5 seconds, then you have developed a power of 2 foot-pounds per second. If you could develop 550 foot-pounds per second of power, then we would say you developed 1 horsepower. So power is a way to express not only how much work you are doing (or how much energy is being stored); it also tells how fast you are doing it. In the metric system of measure, which is used to express all of our common electrical units, power is expressed in terms of the watt, which means energy is being stored at the rate of 1 joule per second, or work is being done at the rate of 1 joule per second. To pass the General class license exam, you learned that electrical power is equal to the current times the voltage:

$$P = I E$$

(Equation 5-15)

But there is one catch. That equation is only true when the current is through a resistor and the voltage is across that resistor. In other words, the current and voltage must be in phase.

Peak and Average Power

In a dc circuit, it is easy to calculate the power. Simply measure or calculate the current and voltage, and multiply the values. But in an ac circuit, you must specify whether you are using the peak or effective (root-mean-square — RMS) values. Peak voltage times peak current will give the **peak power**, and RMS voltage times RMS current gives the **average power** value. Notice that we do not call this RMS power! It is interesting to note that with pure sine waves, the peak power is just two times the average power. With a modulated RF wave, which does not result in a pure sine wave, the relationship between peak and average power is not quite that simple. To avoid any possible confusion, we usually refer to the average power output of an RF amplifier. So when you say the output power from your transmitter is 100 W, you mean that the average output power is 100 W.

Another term that is often used in relation to RF power amplifiers is **peak-envelope power (PEP)**. This is the power specification used by the FCC to determine the maximum permissible output power. PEP is defined as the average power of the RF envelope during a modulation peak. So PEP refers to the *maximum average power* of a modulated wave, and not the *peak power* as defined in the previous paragraph! It is interesting to note that in the case of a CW transmitter with the key held down, the PEP and average power will be the same, and can be measured on a peak-reading wattmeter.

Power Factor

An ammeter and a voltmeter connected in an ac circuit to measure voltage across and current through an inductor or capacitor will read the correct RMS values, but multiplying them together does not give a true indication of the power being dissipated in the component. If you multiply the RMS values of voltage and current read from these meters, you will get a quantity that is referred to as **apparent power**. This term should tell you that the power found in that way is not quite correct! The apparent power should be expressed in units of volt-amperes (VA) rather than watts. The apparent power in an inductor or capacitor is called **reactive power** or nonproductive, wattless power. Reactive power is expressed in volt-amperes-reactive (VARs).

When there are inductors and capacitors in the circuit, the voltage is not in phase with the current. As we have already said, there is no power developed in a circuit containing a pure capacitor or a pure inductor. With ideal inductors and reactors, the power is handed back and forth between the magnetic field in the inductor and the electric field in the capacitor, but no power is dissipated. Only the resistive part of the circuit will dissipate power.

There is something about the phase angle between the voltage and current that must be taken into account. That something is called the **power factor**. Power factor is a quantity that relates the apparent power in a circuit to the real power. You can find the **real power** in a circuit by:

$$P = I^2 R$$ (Equation 5-16)

for a series circuit, or:

$$P = \frac{E^2}{R}$$ (Equation 5-17)

for a parallel circuit. Notice that both of these equations are easily derived by using Ohm's Law to solve for either voltage or current, ($E = I \times R$ and $I = E / R$) and replacing that term with the Ohm's Law equivalent.

Figure 5-21 — Only the resistance actually consumes power. The voltmeter and ammeter read the proper RMS value for the circuit, but their product is apparent power, not real average power.

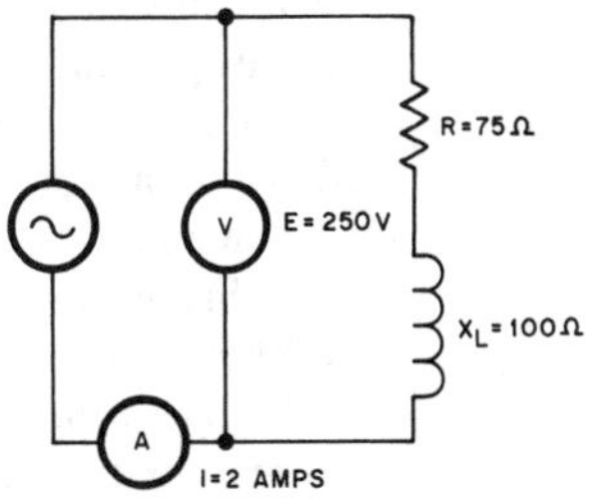

$R = 75\,\Omega$

$E = 250\,V$

$X_L = 100\,\Omega$

$I = 2\ AMPS$

One way to calculate the power factor is to simply divide the real power by the apparent power:

$$\text{Power factor} = \frac{P_{\text{REAL}}}{P_{\text{APPARENT}}}$$

(Equation 5-18)

Figure 5-21 shows a series circuit containing a 75-Ω resistor and a coil with an inductive reactance of 100 Ω at the signal frequency. The voltmeter reads 250-V RMS and the ammeter indicates a current of 2-A RMS. This is an apparent power of 500 VA. Use Equation 5-16 to calculate the power dissipated in the resistor, $P_{\text{REAL}} = (2\text{ A})^2 \times 75\ \Omega = 4\text{ A}^2 \times 75\ \Omega = 300\text{ W}$. Now by using Equation 5-18, we can calculate the power factor:

$$\text{Power factor} = \frac{300\text{ W}}{500\text{ VA}} = 0.6$$

Another way to calculate the real power, if you know the power factor, is given by:

$$P_{\text{REAL}} = P_{\text{APPARENT}} \times \text{Power factor}$$

(Equation 5-19)

In our example,

$$P_{\text{REAL}} = 500\text{ VA} \times 0.6 = 300\text{ W}$$

Of course the value found using Equation 5-19 must agree with the value found by either Equation 5-16 or 5-17, depending on whether the circuit is a series or a parallel one.

What if you don't have the benefit of a voltmeter and an ammeter in a circuit? How can you calculate the real power or power factor? Well, it turns out that the phase angle between the total applied voltage and the circuit current can be used. We already learned how to calculate the phase angle of either a series or a parallel circuit, so that should be no problem. If you don't remember how to calculate the phase angle, go back and review that section of this chapter.

The power factor can be calculated from the phase angle by finding the cosine value of the phase angle:

$$\text{Power factor} = \cos \theta$$

(Equation 5-20)

where θ is the phase angle between voltage and current in the circuit and cosine is the trigonometric function with values that vary between 0 for an angle of $-90°$ (such as for the voltage across a capacitor lagging the current) to 1 for an angle of $0°$ (such as for the voltage and current being in phase for a resistor) and back to 0 for an angle of $90°$ (such as for the voltage across an inductor leading the current).

From this discussion, we can see that for a circuit containing only resistance, where the voltage and current are in phase, the power factor is 1, and the real power is equal to the apparent power. For a circuit containing only pure capacitance or pure inductance, the power factor is 0, so there is no real power! For most practical circuits, which contain resistance, inductance and capacitance, and the phase angle is some value greater than or less than 0°, the power factor will be something less than one. In such a circuit, the real power will always be something less than the apparent power. This is an important point to remember.

Let's try a sample problem, just to be sure you understand all this. We will assume you can calculate the phase angle between voltage and current, given the component values and generator frequency. After that, it doesn't matter if the circuit is a series or parallel one. The procedure is the same. What is the power factor for an R-L circuit having a phase angle of 60°? Use Equation 5-20 to answer this question. Find the cosine value using your calculator or a table of trigonometric values.

Power factor = cos 60° = 0.500

Let's try another example. Suppose you have a circuit that draws 4 amperes of current when 100 V ac is applied. The power factor for this circuit is 0.2. What is the real power (how many watts are consumed) for this circuit? The apparent power is calculated using Equation 5-15:

$P_{APPARENT} = 100 \text{ V} \times 4 \text{ A} = 400 \text{ VA}$

Real power is then found using Equation 5-19:

$P_{REAL} = 400 \text{ VA} \times 0.2 = 80 \text{ W}$

[If you have practiced these example calculations with the text you should have no difficulty with questions A5H01 through A5H11 in Chapter 10. Try those questions now, to be sure you understand the answers and the calculations. Review this section as needed.]

EFFECTIVE RADIATED POWER (ERP)

Knowing the output power from your transmitter is important to ensure that you stay within the limits set by FCC rules for your Amateur Radio Station. Sometimes it is more helpful, in evaluating your total station performance, to know how much power is actually being radiated. This is especially true with regard to repeater systems. The effective power radiated from the antenna helps establish the coverage area of the repeater. In addition, the height of the repeater antenna as compared to buildings and mountains in the surrounding area (height above average terrain, or HAAT) has a large effect on the repeater coverage. In general, for a given coverage area, with a greater antenna HAAT, less **effective radiated power (ERP)** is needed. A frequency coordinator may even specify a maximum ERP for a repeater, to help reduce interference between stations using the same repeater frequencies.

You may be wondering why the transmitter power output is not the same as the power radiated from the antenna. Well, there is always some power lost in the feed line, and often there are other devices inserted in the line, such as a wattmeter, SWR bridge or an impedance-matching network. In the case of a repeater system, there is usually a duplexer so the transmitter and receiver can use the same antenna, and perhaps a circulator to reduce the possibility of intermodulation interference. These devices also introduce some loss to the system. Antennas are compared to a reference, and they may exhibit gain or loss as compared to that reference.

The two types of antennas commonly used for reference are a half-wave dipole and a theoretical isotropic radiator. For our discussion and calculations here, we will assume the antenna gain is with reference to a half-wave dipole. You will learn more about the isotropic radiator when you study for the Amateur Extra class license. A beam antenna will have some gain over a dipole, at least in the desired radiation direction. The exact amount of gain will depend on the design and installation.

These system gains and losses are usually expressed in **decibels (dB)**. The decibel is a logarithm of the ratio of two power levels, and the gain in dB is calculated by:

$$dB = 10 \log \left(\frac{P_2}{P_1} \right)$$

(Equation 5-21)

where:

P_1 is the reference power.
P_2 is the power being compared to the reference.

In the case of calculating ERP, the transmitter output power is considered as the reference, and the power at any other point is P_2.

The main advantage of using decibels is that system gains and losses expressed in these units can simply be added, with losses written as negative values. Suppose we have a repeater station that uses a 50-W transmitter and a feed line with 4 dB of loss. There is a duplexer in the line that exhibits 2 dB of loss and a circulator that adds another 1 dB of loss. This repeater uses an antenna that has a gain of 6-dBd (the antenna gain is compared to a dipole). Our total system gain looks like:

System gain = −4 dB + −2 dB + −1 dB + 6 dBd = −1 dB

Note that this is a loss of 1 dB total for the system. Using Equation 5-21:

$$-1 \text{ dB} = 10 \log \left(\frac{P_2}{50 \text{ W}} \right)$$

$$\frac{-1 \text{ dB}}{10} = \log \left(\frac{P_2}{50 \text{ W}} \right)$$

$$\log^{-1}\left(\frac{-1 \text{ dB}}{10}\right) = \log^{-1}\left(\log\left(\frac{P_2}{50 \text{ W}}\right)\right)$$

$$\left(\log^{-1} \text{ means the antilog, or inverse log function}\right)$$

$$\log^{-1}\left(-0.1 \text{ dB}\right) = \frac{P_2}{50 \text{ W}}$$

$$P_2 = \log^{-1}\left(-0.1\right) \times 50 \text{ W} = 0.79 \times 50 \text{ W} = 39.7 \text{ W}$$

This is consistent with our expectation that with a 1-dB system loss we would have somewhat less ERP than transmitter output power.

As another example, suppose we have a transmitter that feeds a 100-W output signal into a feed line that has 1-dB of loss. The feed line connects to an antenna that has a gain of 6-dBd. What is the effective radiated power from the antenna? To calculate the total system gain (or loss) we add the decibel values given:

System gain = $-$ 1 dB + 6 dBd = 5 dB

Then we can use Equation 5-21 to find the ERP:

$$5 \text{ dB} = 10 \log\left(\frac{P_2}{100 \text{ W}}\right)$$

$$\frac{5 \text{ dB}}{10} = \log\left(\frac{P_2}{100 \text{ W}}\right)$$

$$\log^{-1}\left(\frac{5 \text{ dB}}{10}\right) = \log^{-1}\left(\log\left(\frac{P_2}{100 \text{ W}}\right)\right)$$

$$\log^{-1}\left(0.5 \text{ dB}\right) = \frac{P_2}{100 \text{ W}}$$

$$P_2 = \log^{-1}\left(0.5 \text{ dB}\right) \times 100 \text{ W} = 3.16 \times 100 \text{ W} = 316 \text{ W}$$

The total system has positive gain, so we should have expected a larger value for ERP than the transmitter power. Keep in mind that the gain antenna concentrates more of the signal in a desired direction, with less signal in undesired directions. So the antenna doesn't really increase the total available power.

[Turn to Chapter 10 now and study examination questions A5I01 through A5I11. If you have any difficulty with those questions, review this section.]

THEVENIN'S THEOREM

Thevenin's Theorem is a useful tool for simplifying complex networks of resistors (or other components). Our discussion here is limited to resistors, to make it easier for you to understand. (Besides, you don't have to know how to handle this technique with reactive components for the Element 4A examination!) The theorem states that any two-terminal network of resistors and voltage sources, no matter how complex, can be replaced by a circuit consisting of a single voltage source and a single series resistor. Doing this conversion for a part of a complex circuit greatly simplifies calculations involving other parts of the circuit.

We will use a simple example to illustrate the technique provided by Thevenin's Theorem. While it may seem easier to solve this problem using other methods that you are familiar with, the advantages become obvious with more complex problems. Figure 5-22 shows a simple circuit involving a voltage source, two parallel resistors and a series resistor. We want to know the current through R_L, so that is considered as the load for the Thevenin-equivalent circuit.

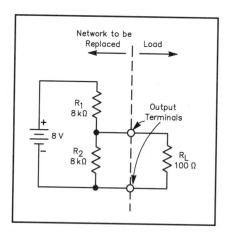

Figure 5-22 — Circuit showing a combination of a voltage source and a resistive network that can be simplified using Thevenin's Theorem to find the load current.

We could solve this problem by calculating the resistance of the parallel-resistor combination, adding the series resistance, then using Ohm's Law to find the total circuit current. By calculating the voltage drops across the series resistor and the parallel pair, we could finally solve for the current through R_L. A lot of work, even for this simple problem.

To apply Thevenin's Theorem, first remove the load from the network to be simplified. The new Thevenin-equivalent voltage source has a voltage equal to the open-circuit (no load) voltage at the output terminals. Figure 5-23A shows the network to be replaced, with the output voltage. You can find that output voltage by using Ohm's Law to calculate the total circuit current and the voltage drop across R_2:

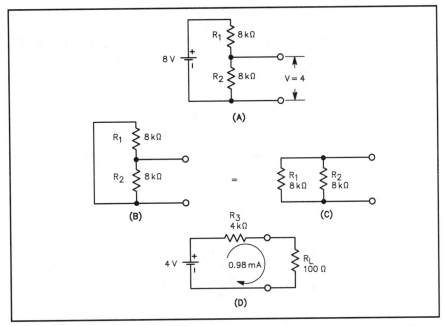

Figure 5-23 — Steps showing how a two-terminal network can be simplified to find a Thevenin-equivalent voltage source (A) and the Thevenin-equivalent resistance (B and C) to calculate the load current (D).

$$V_{OUT} = I \times R_2$$

But the circuit current is equal to the source voltage divided by the total resistance:

$$I = \frac{E}{R_1 + R_2}$$

Then:

$$V_{OUT} = \frac{E \times R_2}{R_1 + R_2} = \frac{8\,V \times 8 \times 10^3\,\Omega}{8 \times 10^3\,\Omega + 8 \times 10^3\,\Omega} = \frac{64 \times 10^3\,V}{16 \times 10^3} = 4\,V$$

Now replace the voltage source in the original network with its "internal resistance." (Most voltage sources can be assumed to have an internal resistance of zero ohms, so we will use a piece of wire in place of the voltage source.) See Figure 5-23B. The Thevenin-equivalent series resistor is equal to that resistance which would be measured by an ohmmeter connected to the load terminals of the circuit with the voltage source shorted out. We can redraw the circuit of Figure 5-23B as shown in part C, which clearly shows that these two resistors are simply in parallel. Resistors in parallel are combined by adding their reciprocals, then taking the reciprocal of that sum. When there are only two resistors to combine, you probably remember the equation from when you studied for the Technician or Technician Plus license exam:

$$R_T = \frac{R_1 \times R_2}{R_1 + R_2} \qquad \text{(Equation 5-22)}$$

$$R_T = \frac{8 \text{ k}\Omega \times 8 \text{ k}\Omega}{8 \text{ k}\Omega + 8 \text{ k}\Omega} = \frac{64 \text{ k}\Omega^2}{16 \text{ k}\Omega} = 4 \text{ k}\Omega$$

Now we can replace the original voltage-divider network with our new Thevenin equivalent circuit, consisting of a 4-V source in series with an 8-kilohm resistor. Adding the load resistor back in, as shown in Figure 5-23D, enables us to calculate the load current easily using Ohm's Law:

$$I = \frac{E}{R_3 + R_L} = \frac{4 \text{ V}}{4000 \text{ W} + 100 \text{ W}} = \frac{4 \text{ V}}{4100 \text{ W}} = 9.8 \times 10^{-4} \text{ A} = 0.98 \text{ mA}$$

The whole procedure is much more difficult to describe than it is to carry out, and after practicing a few problems, you should be ready to tackle problems with much more complicated networks, and maybe even a few inductors and capacitors thrown in with an ac voltage source!

[To get some of the practice that you need, turn to Chapter 10, and solve questions A5J01 through A5J11. Review this section as needed.]

**CHAPTER 6
KEYWORDS
KEYWORDS
KEYWORDS**

Alpha (α) — The ratio of transistor collector current to emitter current. It is between 0.92 and 0.98 for a junction transistor.

Alpha cutoff frequency — A term used to express the useful upper frequency limit of a transistor. The point at which the gain of a common-base amplifier is 0.707 times the gain at 1 kHz.

Anode — The terminal that connects to the positive supply lead for current to flow through a device.

Avalanche point — That point on a diode characteristic curve where the amount of reverse current increases greatly for small increases in reverse bias voltage.

Beta (β) — The ratio of transistor collector current to base current. Betas show wide variations, even between individual devices of the same type.

Beta cutoff frequency — The point at which the gain of a common-emitter amplifier is 0.707 times the gain at 1 kHz.

Bipolar junction transistor — A transistor made of two PN semiconductor junctions using two layers of similar-type material (N or P) with a third layer of the opposite type between them.

Cathode — The terminal that connects to the negative supply lead for current to flow through a device

Crystal-lattice filter — A filter that employs piezoelectric crystals (usually quartz) as the reactive elements. They are most often used in the IF stages of a receiver or transmitter.

Depletion region — An area around the semiconductor junction where the charge density is very small. This creates a potential barrier for current to flow across the junction. In general, the region is thin when the junction is forward biased, and becomes thicker under reverse-bias conditions. Also called the **transition region**.

Doping — The addition of impurities to a semiconductor material, with the intent to provide either excess electrons or positive charge carriers (holes) in the material.

Forward bias — A voltage applied across a semiconductor junction so that it will tend to produce current.

Hot-carrier diode — A type of diode in which a small metal dot is placed on a single semiconductor layer. It is superior to a point-contact diode in most respects.

Light-emitting diode — A device that uses a semiconductor junction to produce light when current flows through it.

Maximum average forward current — The highest average current that can flow through the diode in the forward direction for a specified junction temperature.

Monolithic microwave integrated circuit (MMIC) — A small pill-sized amplifying device that simplifies amplifier designs for microwave-frequency circuits. An MMIC has an input lead, an output lead and two ground leads.

Neon lamp — A cold-cathode (no heater or filament), gas-filled tube used to give a visual indication of voltage in a circuit, or of an RF field.

N-type material — Semiconductor material that has been treated with impurities to give it an excess of electrons. We call this a "donor material."

Peak inverse voltage (PIV) — The maximum instantaneous anode-to-cathode reverse voltage that is to be applied to a diode.

Piezoelectric effect — The physical deformation of a crystal when a voltage is applied across the crystal surfaces.

PIN diode — A diode consisting of a relatively thick layer of nearly pure semiconductor material (intrinsic semiconductor) with a layer of P-type material on one side and a layer of N-type material on the other.

PN junction — The contact area between two layers of opposite-type semiconductor material.

Point-contact diode — A diode that is made by a pressure contact between a semiconductor material and a metal point.

P-type material — A semiconductor material that has been treated with impurities to give it an electron shortage. This creates excess positive charge carriers, or "holes," so it becomes an "acceptor material."

Reverse bias — A voltage applied across a semiconductor junction so that it will tend to prevent current.

Semiconductor material — A material with resistivity between that of metals and insulators. Pure semiconductor materials are usually doped with impurities to control the electrical properties.

Silicon-controlled rectifier (SCR) — A bistable semiconductor device that can be switched between the off and on states by a control voltage.

Thyristor — Another name for a silicon-controlled rectifier.

Toroid — A coil wound on a donut-shaped ferrite or powdered-iron form.

Transition region — An area around the semiconductor junction where the charge density is very small. This creates a potential barrier for current to flow across the junction. In general, the region is thin when the junction is forward biased, and becomes thicker under reverse-bias conditions. Also called the **depletion region**.

Triac — A bidirectional SCR, primarily used to control ac voltages.

Tunnel diode — A diode with an especially thin depletion region, so that it exhibits a negative resistance characteristic.

Unijunction transistor (UJT) — A three-terminal, single-junction device that exhibits negative resistance and switching characteristics unlike bipolar transistors.

Varactor diode — A component whose capacitance varies as the reverse-bias voltage is changed. This diode has a voltage-variable capacitance.

Zener diode — A diode that is designed to be operated in the reverse-breakdown region of its characteristic curve.

Zener voltage — A reverse-bias voltage that produces a sudden change in apparent resistance across the diode junction, from a large value to a small value.

CIRCUIT COMPONENTS

Before we look at some complex electronic circuits, let's discuss some basic information about the parts that make up those circuits. This chapter presents the information about circuit components that you need to know in order to pass your Advanced class Amateur Radio license exam. You will find descriptions of several types of diodes, a variety of transistors, silicon-controlled rectifiers, light-emitting diodes, neon lamps and crystal-lattice filters. You will also find a discussion of toroidal inductors and the core materials used to make these popular components. This chapter also includes information about a type of integrated circuit: the monolithic microwave integrated circuit, or MMIC. These components are combined with other devices to build practical electronic circuits, which are described in the next chapter.

As you study the characteristics of the components described in this chapter, be sure to turn to the Element 4A questions in Chapter 10 when you reach the end of a section. That will provide a review of bite-sized chunks of the material, as it is covered. It will show you where you need to do some extra studying. There are entire books written about nearly every topic in this chapter, so if you want additional information, check out some other sources of information. *The ARRL Handbook* is always a good place to start. If you thoroughly understand how these components work, then you should have no problem learning how they can be connected to make a circuit perform a specific task.

The Circuit Components subelement includes six syllabus groups, and from those groups your exam will have six questions. The syllabus groups for this section are:

A6A   Semiconductor material: Germanium, Silicon, P-type, N-type.
A6B   Diodes: Zener, tunnel, varactor, hot-carrier, junction, point contact, PIN and light emitting.
A6C   Toroids: Permeability, core material, selecting, winding.
A6D   Transistor types: NPN, PNP, junction, unijunction, power.
A6E   Silicon controlled rectifier (SCR); triac; neon lamp.
A6F   Quartz crystal (frequency determining properties as used in oscillators and filters); monolithic amplifiers (MMICs).

SEMICONDUCTOR MATERIALS

Silicon and germanium are the materials normally used to make **semiconductor materials**. (The element silicon is not the same as the household lubricants and rubber-like sealers called silicone.) Silicon has 14 protons and 14 electrons, while germanium has 32 of each. Silicon and germanium each have an electron structure with four electrons in the outer energy layers. The silicon and germanium atoms will share these four electrons with other atoms around them.

Some atoms share their electrons so the atoms arrange themselves into a regular pattern. We say these atoms form crystals. Figure 6-1 shows how silicon and germanium atoms produce crystals. (Different kinds of atoms might arrange themselves into other patterns.) The crystals made by silicon or germanium atoms do not make good electrical conductors. They aren't good insulators either, however. That's why we call them semiconductor materials. Sometimes they act like conductors and sometimes they act like insulators. Semiconductor material exhibits properties of both metallic and nonmetallic substances.

Manufacturers add other atoms to these crystals through a carefully controlled process, called **doping**. The atoms added in this way produce a material that is no longer pure silicon or pure germanium. We call the added atoms impurity atoms.

As an example, the manufacturer might add some atoms of arsenic or antimony to the silicon or germanium while making the crystals. Arsenic and antimony atoms each have five electrons to share.

Figure 6-2 shows how an atom with five electrons in its outer layer fits into the crystal structure. In such a case, there is an extra electron in the crystal. We refer to this as a *free electron*, and we call the semiconductor material made in this way **N-type material**. (This name comes from the extra negative charge in the crystal structure.)

The impurity atoms are electrically neutral, just as the silicon or germanium

Figure 6-1 — Silicon and germanium atoms arrange themselves into a regular pattern, called crystals. Notice that each atom in this crystal structure is sharing four electrons with other nearby atoms.

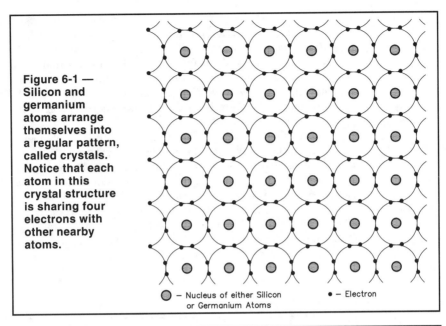

⊛ — Nucleus of either Silicon or Germanium Atoms   • — Electron

Figure 6-2 — Manufacturers can add antimony or arsenic atoms to the silicon or germanium crystals. These impurity atoms add an extra electron to the crystal structure, producing N-type semiconductor material.

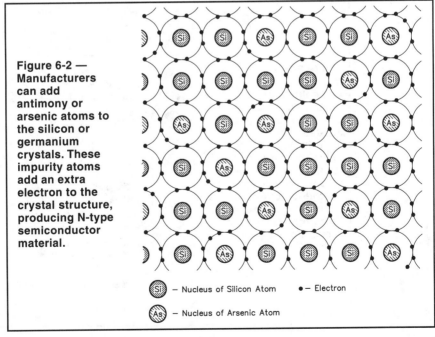

Ⓢ⒤ — Nucleus of Silicon Atom   • — Electron

Ⓐⓢ — Nucleus of Arsenic Atom

atoms are. The free electrons result from the crystal structure itself. These impurity atoms create (donate) free electrons in the crystal structure. That is why we call them *donor* impurity atoms.

Now let's suppose the manufacturer adds some gallium or indium atoms instead of arsenic or antimony. Gallium and indium atoms only have three electrons that they can share with other nearby atoms. When there are gallium or indium atoms in the crystal there is an extra space where an electron could fit into the structure.

Figure 6-3 shows an example of a crystal structure with an extra space where an electron could fit. We call this space for an electron a *hole*. The semiconductor material produced in this way is **P-type material**. These impurity atoms produce holes that will accept extra electrons in the semiconductor material. That is why we call them *acceptor* impurity atoms.

Again, you should realize that the impurity atoms have the same number of electrons as protons. The material is still electrically neutral. The *crystal structure* is missing an electron in P-type material. Similarly, the *crystal structure* has an extra electron in N-type material.

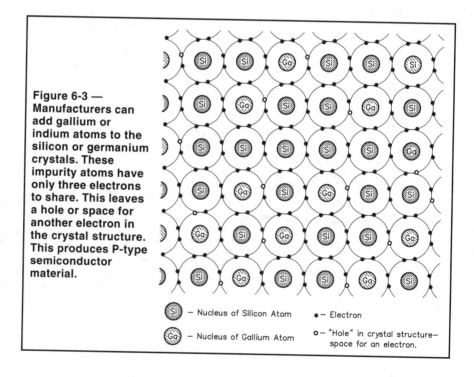

**Figure 6-3 —
Manufacturers can add gallium or indium atoms to the silicon or germanium crystals. These impurity atoms have only three electrons to share. This leaves a hole or space for another electron in the crystal structure. This produces P-type semiconductor material.**

(Si) — Nucleus of Silicon Atom   • — Electron

(Ga) — Nucleus of Gallium Atom   o — "Hole" in crystal structure—
space for an electron.

Suppose we apply a voltage across a crystal of N-type semiconductor. The positive side of the voltage attracts electrons. The free electrons in the structure move through the crystal toward the positive side. Since most of the current through this N-type semiconductor material is produced by these free electrons, we call them the *majority charge carriers*.

Next suppose we apply a voltage to a crystal of P-type material. The negative voltage attracts the holes, and they move through the material toward the negative side. The *majority charge carriers* in P-type semiconductor material are holes.

Free electrons and holes move in opposite directions through a crystal. Manufacturers can control the electrical properties of semiconductor materials by carefully controlling the amount and type of impurities that they add to the silicon or germanium crystals.

Semiconductors are solid crystals. They are strong and not easily damaged by vibration or rough handling. We refer to electronic parts made with semiconductor materials as solid-state devices.

Other materials are used to make semiconductor materials for special-purpose applications. For example, gallium arsenide semiconductor material has performance advantages for use at microwave frequencies, and is often used to make solid-state devices for operation on those frequencies.

[Turn to Chapter 10 and study questions A6A01 through A6A12 before you go on to the next section. Review this material if you have any difficulty with those questions.]

DIODES

To qualify for the Advanced class license, you must be familiar with some specialized types of diodes. Rectifier circuits were covered on the General class examination; for the Advanced class test, you must be familiar with the two main structural categories of diodes: junction and point-contact types. PIN, Zener, light-emitting, tunnel, varactor and hot-carrier diodes are special-purpose devices that come under those main headings.

Junction Diodes

The junction diode, also called the **PN-junction** diode, is made from two layers of semiconductor material joined together. One layer is made from P-type (positive) material. The other layer is made from N-type (negative) material. The name PN junction comes from the way the P and N layers are joined to form a semiconductor diode. Figure 6-4 illustrates the basic concept of a junction diode.

The P-type side of the diode is called the **anode**, which is the lead that normally connects to the positive supply lead. The N-type side is called the **cathode**, and that lead normally connects to the negative supply lead. When voltage is applied to a junction diode as shown at A in Figure 6-5, carriers flow across the barrier and the diode conducts (that is, electrons will flow through it). When the diode anode is

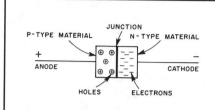

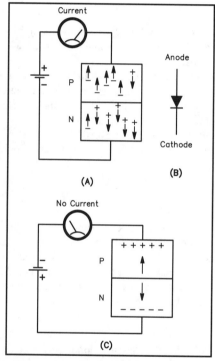

Figure 6-5 — At A, the PN junction is forward biased and conducting. B shows the schematic symbol used to represent a diode, drawn as it would be used instead of the semiconductor blocks of part A. *Conventional current* flows in the direction indicated by the arrow in this symbol. At C, the PN junction is reverse biased, so it does not conduct.

positive with respect to the cathode, electrons are attracted across the junction from the N-type material, through the P-type material and on through the circuit to the positive battery terminal. Holes are attracted in the opposite direction by the negative potential of the battery. When the diode anode is connected in this manner it is said to be **forward biased**.

Conventional current (which flows from positive to negative) in a diode flows from the anode to the cathode. The electrons flow in the opposite direction. Electrons and holes are also called carriers because they are the means by which current is carried from one side of the junction to the other. When no voltage is applied to a diode, the junction between the P-type and N-type material acts as a barrier that prevents carriers from flowing between the layers.

Figure 6-5B shows the schematic symbol for a diode, drawn as it would be used instead of the semiconductor blocks used in part A. Here we can see one of the advantages of considering conventional current instead of electron flow through the diode. The arrow on the schematic symbol points in the direction of conventional current.

If the battery polarity is reversed, as shown at C in Figure 6-5, the excess

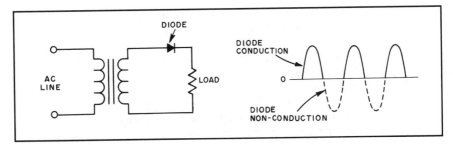

Figure 6-6 — PN-junction diodes are used as rectifiers because they allow current in one direction only.

electrons in the N-type material are attracted away from the junction by the positive battery terminal. Similarly, the holes in the P-type material are attracted away from the junction by the negative battery side. When this happens, the area around the junction has no current carriers; electrons do not flow across the junction to the P-type material, and the diode does not conduct. When the anode is connected to a negative voltage source and the cathode is connected to a positive voltage source, the diode does not conduct, and the device is said to be **reverse biased**.

Junction diodes are used as rectifiers because they allow current in one direction only. See Figure 6-6. When an ac signal is applied to a diode, it will be forward biased during one half of the cycle, so it will conduct, and there will be current to the load. During the other half of the cycle, the diode is reverse biased, and there will be no current. The diode output is pulsed dc, and current always flows in the same direction.

In a junction diode, the P and N layers are sandwiched together, separated by the junction. Although the spacing between the layers is extremely small, there is some capacitance at the junction. The structure can be thought of in much the same way as a simple capacitor: two charged plates separated by a thin dielectric.

Although the internal capacitance of a PN junction diode may be only a few picofarads, this capacitance can cause problems in RF circuits, especially at VHF and above. Junction diodes may be used from dc to the microwave region, but there is a special type of diode with low internal capacitance that is specially designed for RF applications. This device, called the **point-contact diode**, is discussed later in this chapter.

Diode Ratings

Junction diodes have maximum voltage and current ratings that must be observed, or damage to the diode could result. The voltage rating is called **peak inverse voltage (PIV)**, and the rectified current rating is called **maximum average forward current**. With present technology, diodes are commonly available with ratings up to 1000 PIV and 100 A.

Peak inverse voltage is the voltage that a diode must withstand when it isn't conducting. Although a diode is normally used in the forward direction, it will conduct in the reverse direction if enough voltage is applied. A few hole/electron pairs are thermally generated at the junction when a diode is reverse biased. These pairs cause a very small reverse current, called leakage current. Semiconductor diodes can withstand some leakage current. If the inverse voltage reaches a high enough value, however, the leakage current rises abruptly, resulting in a large reverse current. The point where the leakage current rises abruptly is called the **avalanche point**. A large reverse current usually damages or destroys the diode.

The maximum average forward current is the highest average current that can flow through the diode in the forward direction for a specified junction temperature. This specification varies from device to device, and it depends on the maximum allowable junction temperature and on the amount of heat the device can dissipate. As the forward current increases, the junction temperature will increase. If allowed to get too hot, the diode will be destroyed.

Impurities at the PN junction cause some resistance in the diode. This resistance results in a voltage drop across the junction. For silicon diodes, this drop is approximately 0.6 to 0.7 V; it is 0.2 to 0.3 V for germanium diodes. When current flows through the junction, some power is dissipated in the form of heat. The amount of power depends on the current through the diode. For example, it would be approximately 6 W for a silicon rectifier with 10 A flowing through it (P = I E; P = 10 A × 0.6 V). If the junction temperature exceeds the safe level specified by the manufacturer the diode is likely to be damaged or destroyed.

Diodes designed to safely handle forward currents in excess of 6 A generally are packaged so they may be mounted on a heat sink. These diodes are often referred to as stud-mount devices. The heat sink helps the diode package dissipate heat more rapidly, thereby keeping the diode junction temperature at a safe level. The metal case of a stud-mount diode is usually one of the contact points, so it must be insulated from ground.

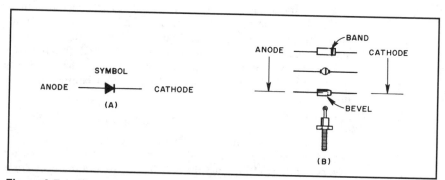

Figure 6-7 — The schematic symbol for a diode is shown at A. Diodes typically are packaged in one of the case styles shown at B.

Figure 6-7 shows some of the more common diode-case styles, as well as the general schematic symbol for a diode. The line, or spot, on a diode case indicates the cathode lead. On a high-power, stud-mount diode, the stud may be either the anode or cathode. Check the case or the manufacturer's data sheet for the correct polarity.

[This is a good time to take a break from the text and turn to Chapter 10 for some review. Study questions A6B09, A6B10, A6B11 and A6B13. Review this section if you have difficulty with any of those questions.]

Varactor and Varicap Diodes

As mentioned before, junction diodes exhibit an appreciable internal capacitance. It is possible to change the internal capacitance of a diode by varying the amount of reverse bias applied to it. Manufacturers have designed certain kinds of diodes, called voltage-variable capacitors or variable-capacitance diodes (Varicaps) and **varactor diodes** (variable reactance diodes) to take advantage of this property.

Varactors are designed to provide various capacitance ranges from a few picofarads to more than 100 pF. Each style has a specific minimum and maximum capacitance, and the higher the maximum amount, the greater the minimum amount. A typical varactor can provide capacitance changes over a 10:1 range with bias voltages in the 0- to 100-V range.

Varactors are similar in appearance to junction diodes. Common schematic symbols for a varactor diode are given in Figure 6-8. These devices are used in frequency multipliers at power levels as great as 25 W, remotely tuned circuits and simple frequency modulators.

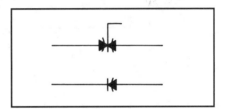

Figure 6-8 — These schematic symbols are commonly used to represent varactor diodes.

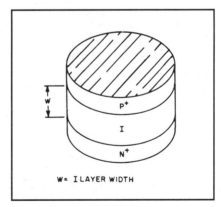

Figure 6-9 — This diagram illustrates the inner structure of a PIN diode.

PIN Diodes

A **PIN** (positive/intrinsic/negative) **diode** is formed by diffusing P-type and N-type layers onto opposite sides of an almost pure silicon layer, called the I region. See Figure 6-9. This layer is not "doped" with P-type or N-type charge carriers, as are the other layers. Any charge carriers found in this layer are a result of the natural properties of the pure semiconductor material. In the case of silicon, there are relatively few free charge carriers. PIN-diode characteristics are determined primarily by the thickness and area of the I region. The outside layers are designated P+ and N+ to indicate heavier than normal doping of these layers.

PIN diodes have a forward resistance that varies inversely with the amount of forward bias applied. When a PIN diode is at zero or reverse bias, there is essentially no charge, and the intrinsic region can be considered as a low-loss dielectric. Under reverse-bias conditions, the charge carriers move very slowly. This slow response time causes the PIN diode to look like a resistor, blocking RF currents.

When forward bias is applied, holes and electrons are injected into the I region from the P and N regions. These charges do not recombine immediately. Rather, a finite quantity of charge always remains stored, resulting in a lower I-region resis-

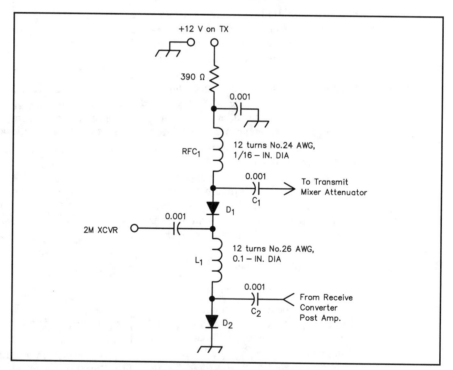

Figure 6-10 — PIN diodes may be used as RF switches. This diagram shows a PIN-diode T/R switch for use between a 2-meter transceiver and a UHF or microwave transverter.

D1, D2 — Phillips BA182, Motorola MPN3401 or equivalent.

tivity. The amount of resistivity that a
PIN diode exhibits to RF can be controlled by changing the amount of forward bias applied.

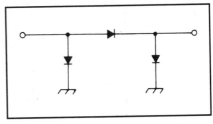

Figure 6-11 — PIN diodes are often used as variable RF attenuators.

PIN diodes are commonly used as
RF switches, variable attenuators and
phase shifters. PIN diodes are faster,
smaller, more rugged and more reliable
than relays or other electromechanical
switching devices.

Figure 6-10 shows how PIN diodes
can be used to build an RF switch. This
diagram shows a transmit/receive switch for use between a 2-meter transceiver and
a UHF or microwave transverter. With no bias, or with reverse bias applied to the
diode, the PIN diode exhibits a high resistance to RF, so no signal will flow from
the generator to the load. When forward bias is applied, the diode resistance will
decrease, allowing the RF signal to pass. The amount of insertion loss (resistance
to RF current) is determined primarily by the amount of forward bias applied; the
greater the forward bias current, the lower the RF resistance.

A PIN-diode attenuator is shown in Figure 6-11. The PIN diodes are connected
as resistors would be in a standard Pi-type resistive pad. The major difference,
however, is that the attenuation of this pad can be varied by changing the value of
forward bias applied to the diodes, changing their RF resistivity.

PIN diodes are packaged in case styles similar to conventional diodes. The
package size depends on the intended application. PIN diodes intended for low-
power UHF and microwave work are packaged in small epoxy or glass cases to
minimize internal capacitance. Others, intended for high-power switching, are of
the stud-mount variety so they can be attached to heat sinks. PIN diodes are shown
by the schematic symbol shown in Figure 6-7.

Zener Diodes

Zener diodes are a special class of junction diode used as voltage references
and as voltage regulators. When they are used as voltage regulators, Zener diodes
provide a nearly constant dc output voltage, even though there may be large changes
in load resistance or input voltage. As voltage references, they provide an extremely
stable voltage that remains constant over a wide temperature range.

As discussed earlier, leakage current rises as reverse (inverse) voltage is applied to a diode. At first, this leakage current is very small and changes very little
with increasing reverse voltage. There is a point, however, where the leakage current rises suddenly. Beyond this point, the current increases very rapidly for a small
increase in voltage; this is called the **avalanche point. Zener voltage** is the voltage necessary to cause avalanche. Normal junction diodes would be destroyed immediately if they were operated in this region, but Zener diodes are specially manufactured to safely withstand the avalanche current.

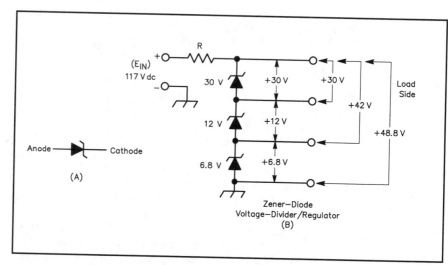

Figure 6-12 — The schematic symbol for a Zener diode is shown at A. B is an example of how Zener diodes are used as voltage regulators.

Since the current in the avalanche region can change over a wide range while the voltage stays practically constant, this kind of diode can be used as a voltage regulator. The voltage at which avalanche occurs can be controlled precisely in the manufacturing process. Zener diodes are calibrated in terms of avalanche voltage.

Zener diodes are currently available with ratings between 1.8 and 200 V. The power ratings range from 250 mW to 50 W. They are packaged in the same case styles as junction diodes. Usually, Zener diodes rated for 10-W dissipation or more are made in stud-mount cases. The schematic symbol for a Zener diode is shown in Figure 6-12, along with an example of how such a device is used as a voltage regulator.

Tunnel Diodes

The **tunnel diode** is a special type of device that has no rectifying properties. When properly biased, it possesses an unusual characteristic: negative resistance. Negative resistance means that when the voltage across the diode increases, the current decreases. This property makes the tunnel diode capable of *amplification* and *oscillation*.

Figure 6-13 — This is the schematic symbol for a tunnel diode.

At one time, tunnel diodes were expected to dominate in microwave applications, but other devices with better performance soon replaced them. The tunnel diode is seldom used today. The schematic symbol for a tunnel diode is shown in Figure 6-13.

Light-Emitting Diodes

Light-emitting diodes (LEDs) are designed to emit light when they are forward biased, and current passes through their PN junctions. The junction of an LED is made from gallium arsenide, gallium phosphide or a combination of these two materials. The color and intensity of the LED depends on the material, or combination of materials, used for the junction. LEDs are available in many colors.

LEDs are packaged in plastic cases, or in metal cases with a transparent end. LEDs are useful as replacements for incandescent panel and indicator lamps. In this application they offer long life, low current drain and small size. One of their most important applications is in numeric displays, in which arrays of tiny LEDs are arranged to provide illuminated segments that form the numbers. Schematic symbols and typical case styles for the LED are shown in Figure 6-14.

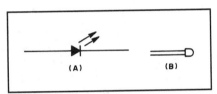

Figure 6-14 — The schematic symbol for an LED is shown at A. At B is a drawing of a typical LED case style.

A typical red LED has a voltage drop of 1.6 V. Yellow and green LEDs have higher voltage drops (2 V for yellow and 4 V for green). The forward-bias current for a typical LED ranges between 10 and 20 mA for maximum brilliance. Bias currents of about 10 mA are recommended for longest device life. As with other diodes, the current through an LED can be varied with series resistors. Varying the current through an LED will affect its intensity; the voltage drop, however, will remain fairly constant.

Point-Contact Diodes

Figure 6-15 illustrates the internal structural differences between a junction diode and a **point-contact diode**. As you can see by this diagram, the point-contact diode has a much smaller surface area at the junction than does a PN-junction diode. When a point-contact diode is manufactured, the main portion of the device is made from N-type material, and a thin aluminum wire, often called a *whisker*, is placed in contact with the semiconductor surface. This forms a metal-semiconductor junction. The result is a diode that exhibits much less internal capacitance than PN-junction diodes, typically 1 pF or less. This means point-contact diodes are better suited for VHF and UHF work than are PN-junction diodes.

Point-contact diodes are packaged in a variety of cases, as are junction diodes.

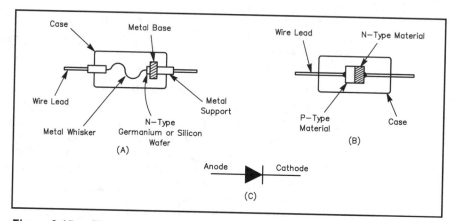

Figure 6-15 — The internal structure of a point-contact diode is shown at A. B shows the internal structure of a PN-junction diode.

The schematic symbol is the same as is used for junction diodes, and is included in Figure 6-15. Point-contact diodes are generally used as UHF mixers and as RF detectors at VHF and below.

Hot-Carrier Diodes

Another type of diode with low internal capacitance and good high-frequency characteristics is the **hot-carrier diode**. This device is very similar in construction to the point-contact diode, but with an important difference. Compare the inner structure of the hot-carrier diode depicted in Figure 6-16 to the point-contact diode shown in Figure 6-15.

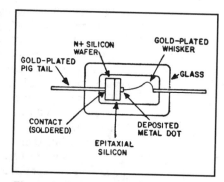

Figure 6-16 — This drawing represents the internal structure of a hot-carrier diode.

The point-contact device relies on the touch of a metal whisker to make contact with the active element. In contrast, the whisker in a hot-carrier diode is physically attached to a metal dot that is deposited on the element. The hot-carrier diode is mechanically and electrically superior to the point-contact diode. Some of the advantages of the hot-carrier type are improved power-handling characteristics, lower contact resistance and improved immunity to burnout caused by transient noise pulses.

Hot-carrier diodes are similar in appearance to point-contact and junction diodes and share the same schematic symbol. They are often used in mixers and detectors at VHF and UHF. In this application, hot-carrier diodes are superior to point-contact diodes because they exhibit greater conversion efficiency and lower noise figure.

[This completes your study of diode types and characteristics for your Advanced class license exam. You should turn to Chapter 10 now and study examination questions A6B01 through A6B08, A6B12 and A6B14 through A6B16. Review this section as needed.]

TRANSISTORS

The **bipolar junction transistor** is a type of three-terminal, PN-junction device that is able to amplify signal energy (current). It is made up of two layers of like semiconductor material with a layer of the opposite-type material sandwiched in between. See Figure 6-17. If the outer layers are P-type material, and the middle layer is N-type material, the device is called a PNP transistor because of the layer arrangement. If the outer layers are N-type material, the device is called an NPN transistor. A transistor is, in effect, two PN-junction diodes back-to-back. Figure 6-18 shows the schematic symbols for PNP and NPN bipolar transistors.

The three layers of the transistor sandwich are called the emitter, base and collector. These are functionally analogous to the cathode, grid and plate of a vacuum tube. A diagram of the construction of a typical PNP transistor is given in Figure 6-17. In an actual bipolar transistor, the center layer (in this case, N-type material) is much thinner than the outer layers. As shown in the diagram, forward-bias voltage across the emitter-base section of the sandwich causes electrons to flow through it from the base to the emitter. As the free electrons from the N-type material flow into the holes of the P-type material, the holes in effect travel into the base. Some

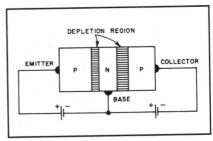

Figure 6-17 — A bipolar junction transistor consists of two layers of like semiconductor material separated by a layer of the opposite material. This drawing represents the internal structure of a PNP transistor.

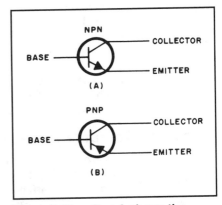

Figure 6-18 — Part A shows the schematic symbol for an NPN bipolar transistor. B is the schematic symbol for a PNP bipolar transistor.

of the holes will be neutralized in the base by free electrons, but because the base layer is so thin, some will move right on through into the P-type material in the collector.

As shown, the collector is connected to a negative voltage with respect to the base. Normally, no current will flow because the base-collector junction is reverse biased. The collector, however, now contains an excess of holes because of those from the emitter that overshot the base. Since the voltage source connected to the collector produces a negative charge, the holes from the emitter will be attracted to the power-supply connection. The amount of emitter-to-collector current is approximately proportional to the base-to-emitter current. Because of the transistor construction, however, the current through the collector will be considerably larger than that flowing through the base.

When a transistor is forward biased, collector current increases in proportion to the amount of bias applied. The transistor is saturated when the collector current reaches its maximum value, and the transistor is said to be fully on. Further increases in bias voltage do not increase the collector current when the transistor is saturated. At the other end of the curve, when the transistor is reverse-biased, the transistor is turned off. There is no current from the emitter to the collector, and the transistor is at cutoff.

A load line is a graphical representation of the range of the transistor resistance for collector-current points between cutoff and saturation. At one end of the load line, the transistor has an infinite resistance to current (cutoff); at the other, it has zero resistance (saturation). Normally, a transistor is operated at some point on the load line between these two extremes. For best efficiency and stability, the transistor in a solid-state power amplifier is operated at a point on the load line that is just below saturation.

Figure 6-17 shows an area around each junction that is called the **depletion region**. The depletion region, sometimes called the **transition region**, is an area near a PN junction that is devoid of holes and excess electrons. This region is caused by the repelling forces of the ions on opposite sides of the junction. When the PN junction is reverse biased, the depletion region becomes larger because the electrons and holes are attracted away from the junction. When the PN junction is forward biased, the depletion region becomes smaller because the electrons and holes move toward each other.

Transistor Characteristics

As mentioned before, the current through the collector of a bipolar transistor is approximately proportional to the current through the base. The ratio of collector current to base current is called the current gain, or **beta**. Beta is expressed by the Greek symbol β. It can be calculated from the equation:

$$\beta = \frac{I_c}{I_b}$$

(Equation 6-1)

where:

   I_c = collector current
   I_b = base current

For example, if a 1-mA base current results in a collector current of 100 mA, the beta is 100. Typical betas for junction transistors range from as low as 10 to as high as several hundred. Manufacturers' data sheets specify a range of values for β. Individual transistors of a given type will have widely varying betas.

Another important transistor characteristic is **alpha**, expressed by the Greek letter α. Alpha is the ratio of collector current to emitter current, given by:

Figure 6-19 — Transistors are packaged in a wide variety of case styles, depending on their intended application.

$$\alpha = \frac{I_c}{I_e} \qquad\qquad \text{(Equation 6-2)}$$

where:

   I_c = collector current
   I_e = emitter current

The smaller the base current, the closer the collector current comes to being equal to that of the emitter, and the closer alpha comes to being 1. For a junction transistor, alpha is usually between 0.92 and 0.98.

Transistors have important frequency characteristics. The **alpha cutoff frequency** is the frequency at which the current gain of a transistor in the common-base configuration decreases to 0.707 times its gain at 1 kHz. Alpha cutoff frequency is considered to be the practical upper frequency limit of a transistor configured as a common-base amplifier.

Beta cutoff frequency is similar to alpha cutoff frequency, but it applies to transistors connected as common-emitter amplifiers. Beta cutoff frequency is the frequency at which the current gain of a transistor in the common-emitter configuration decreases to 0.707 times its gain at 1 kHz. (These amplifier configurations are explained in Chapter 7.)

Bipolar junction transistors are used in a wide variety of applications, including amplifiers (from very low level to very high power), oscillators and power supplies. They are used at all frequency ranges from dc through the UHF and microwave range.

Transistors are packaged in a wide variety of case styles. Some of the more common case styles are depicted in Figure 6-19.

Unijunction Transistors

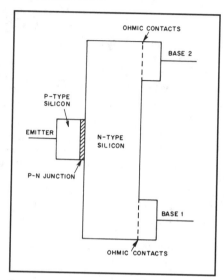

Figure 6-20 — This drawing represents the internal structure of a unijunction transistor.

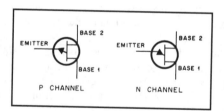

Figure 6-21 — These schematic symbols are used to represent P-channel and N-channel unijunction transistors.

Another three-terminal semiconductor device is the **unijunction transistor (UJT)**, sometimes called a double-base diode. The internal structure of a UJT is shown in Figure 6-20. The elements of a UJT are base 1, base 2 and emitter. There is only one PN junction, and this is between the emitter and the silicon substrate. The base terminals are ohmic contacts; that is, the current is a linear function of the applied voltage. Current flowing between the bases sets up a voltage gradient along the substrate. In operation, the direction of current flow causes the emitter junction to be reverse biased.

The most common application for the UJT is in relaxation oscillator circuits. UJTs are packaged in cases similar to small-signal bipolar transistors. Two schematic symbols for UJTs are given in Figure 6-21. The substrate is made from one type of semiconductor material, and the emitter is made from the other type of material. The large block of material that the two bases connect to in Figure 6-20 is called the substrate. When the substrate is N-type material, we call the UJT an N-channel device. Similarly, the substrate of a P-channel UJT is formed from P-type semiconductor material.

[Proceed to Chapter 10 and study questions A6D01 through A6D13. Review this section as needed.]

SILICON-CONTROLLED RECTIFIERS

Also known as a **thyristor**, a **silicon-controlled rectifier (SCR)** is a three-terminal, solid-state device. The three terminals are called the cathode, anode and gate. An SCR is a diode whose forward conduction from cathode to anode is controlled by a third terminal, the gate. The schematic symbol for the SCR is shown in Figure 6-22, along with an equivalent circuit made from discrete components.

In operation, the SCR will not conduct until the voltage across its cathode and anode terminals exceeds the forward-breakdown voltage. Forward-breakdown voltage is determined by the gate current. Without gate current, the SCR presents an open circuit in both directions. When sufficient gate current is applied, the SCR is "triggered" (begins to conduct) and looks like a closed circuit in the forward direction. Once the SCR is triggered, the gate no longer has any control, and the device behaves like a forward-biased junction diode that is conducting. The SCR will continue to conduct, regardless of gate current, until the anode voltage drops to zero again; when this happens, the forward-breakdown voltage barrier is reestablished, and the gate regains control. SCRs have only two stable operating conditions; conducting and nonconducting. There is no partially conducting condition.

SCRs come in a variety of case styles, depending on intended application. See Figure 6-23. They are often used in power-supply overvoltage-protection circuits (also called crowbars), electronic ignition systems, alarms, and many other applications that require high-speed, unidirectional dc switching.

Triacs

A **triac** is a type of bidirectional SCR. Electrically, a triac is equivalent to two reverse-connected (anode-to-cathode) SCRs wired in parallel, with their gates tied together. Figure 6-24 shows the schematic symbol for the triac. Its three leads are called anode 1, anode 2 and gate. When the gate of a triac is triggered, the device will conduct either polarity of applied voltage, so triacs are used to switch alternating currents.

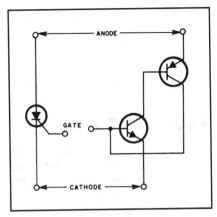

Figure 6-22 — The schematic symbol for an SCR, along with an equivalent circuit made from two transistors, is shown in this diagram.

Figure 6-23 — SCRs are packaged in a wide variety of cases, depending on their intended application.

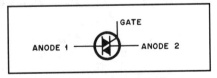

Figure 6-24 — The schematic symbol for a triac is shown in this diagram.

The gate voltage of a triac can be set so that it conducts only during part of an ac waveform cycle. Varying the gate voltage varies the amount of the ac voltage that passes through. Triacs can control the current to ac-operated devices, and find common application as light dimmers and motor speed controls.

[Study examination questions A6E01 through A6E08 in Chapter 10. Review this section as needed.]

NEON LAMPS

Neon lamps are used as indicators, or pilot lamps, in amateur equipment. Neon lamps are often used in the 120-V primary circuit to indicate that the equipment is on. Neon lamps are assigned part numbers with the prefix NE, such as NE-1, NE-2, NE-3 and so on. The NE-2 is the lamp most often found in Amateur Radio equipment.

It takes approximately 67 V dc to light a neon lamp. If the supply voltage is ac, then the peak voltage must be 67 V, so approximately 48 V RMS will fire the lamp. By including about a 150-kΩ resistor in series with one lead, to limit the current through the bulb, you can connect it across the 120 V ac line to use it as a power-on indicator. Another useful feature of the neon lamp is that it will fire in the presence of RF, so these lamps are sometimes used as RF indicators.

Neon lamps are often incorporated into colorful plastic and metal cases to provide a wide variety of panel indicators. Figure 6-25 depicts a "bare bones" NE-2 lamp and the schematic symbol for a neon lamp.

[Now turn to Chapter 10 and study questions A6E09 through A6E11. Review this section as needed.]

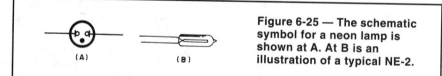

(A)   (B)

Figure 6-25 — The schematic symbol for a neon lamp is shown at A. At B is an illustration of a typical NE-2.

CRYSTAL-LATTICE FILTERS

Crystal-lattice filters are used in SSB transmitters and receivers where high-Q, narrow-bandwidth filtering is required. Such filters typically are used at intermediate frequencies above 500 kHz in receivers. In SSB transmitters, crystal-lattice filters frequently are used after the balanced modulator to attenuate the unwanted sideband. A quartz crystal acts as an extremely high-Q circuit. The equivalent electrical circuit is depicted in Figure 6-26A. Figure 6-26B shows a graph of the reactance versus frequency for the crystal.

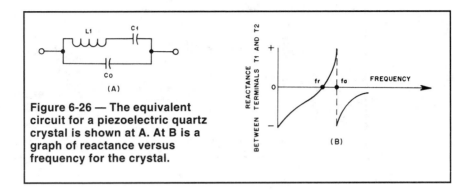

Figure 6-26 — The equivalent circuit for a piezoelectric quartz crystal is shown at A. At B is a graph of reactance versus frequency for the crystal.

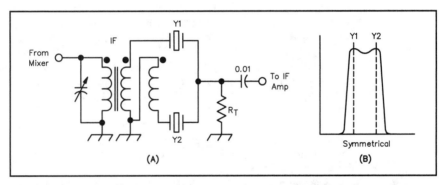

Figure 6-27 — Part A shows a schematic diagram of a half-lattice crystal filter. B shows a typical response curve for this type of filter. Note the steep skirts on the response, representing good rejection of signals outside the passband.

A quartz crystal is known as a *piezoelectric device*. That means if a voltage is applied across a crystal, the crystal will be physically deformed. This physical deformation is known as the **piezoelectric effect**. This physical deformation results in crystal vibrations at a particular frequency, and these vibrations can be used to control the operating frequency of a circuit. This is the operating principle for a crystal oscillator. Crystal filters use the piezoelectric effect to pass desired frequencies while blocking undesired ones.

Although single crystals can be used as filtering devices, the normal practice is to wire two or more together in various configurations to provide a desired response curve. Figure 6-27 depicts a configuration known as the half-lattice filter. In this arrangement, crystals Y1 and Y2 are on different frequencies. The bandwidth and response shape of the filter depends on the relative frequencies of these two crystals. The overall filter bandwidth is equal to approximately 1 to 1.5 times

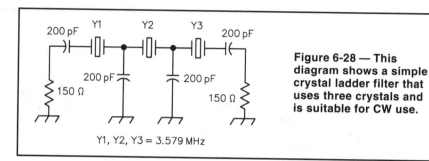

Figure 6-28 — This diagram shows a simple crystal ladder filter that uses three crystals and is suitable for CW use.

Y1, Y2, Y3 = 3.579 MHz

the frequency separation of the crystals. The closer the crystal frequencies, the narrower the bandwidth of the filter. In general, a crystal lattice filter has narrow bandwidth and steep response skirts, as shown in Figure 6-27B.

A good crystal-lattice filter for double-sideband (DSB) voice use would have a bandwidth of approximately 6 kHz at the – 6 dB points on the response curve. A good crystal filter for single-sideband (SSB) phone service is significantly narrower; typical bandwidth is 2.1 kHz at the – 6 dB points. For CW use, crystal filters typically have 250 to 500-Hz bandwidths.

The home construction of crystal filters can be time consuming but can be relatively inexpensive. For home builders, crystal ladder filters may be easier to design than the lattice variety. Figure 6-28 illustrates a simple crystal ladder filter for CW, using three crystals.

Start with a collection of crystals that have approximately the same frequency and characteristics. TV color-burst-oscillator crystals are inexpensive, easy to obtain and work well for such filters. Use an oscillator circuit to measure the actual operating frequency of each of your crystals. Carefully select the crystals to build your filter. The frequency difference between the selected crystals should be less than 10% of the desired filter *bandwidth*. In other words, if you want to build a 500-Hz bandwidth filter, select crystals that are all within 50 Hz of each other.

[To check your understanding of this section, turn to Chapter 10 and study questions A6F01 through A6F06. Review this section as needed.]

MONOLITHIC MICROWAVE INTEGRATED CIRCUITS

The **monolithic microwave integrated circuit (MMIC)** is unlike most other ICs that you may be familiar with. These ICs are quite small, often classified as "pill sized" devices, perhaps because they look like a small pill, with four leads coming out of the device at 90° to each other.

MMIC devices typically have predefined operating characteristics, and require only a few external components for proper operation. This can greatly simplify

Figure 6-29 — This simple utility amplifier is suitable for transmitting and receiving on 903, 1296, 2304, 3456 and 5760 MHz. It consists of three MMICs, three ordinary resistors and four chip (surface mount) capacitors. In addition, a feed-through capacitor is used to bring the bias-supply voltage into the project case box.

Figure 6-30 — This is the schematic diagram of the utility amplifier pictured in Figure 6-29. Note that the operating bias voltage is applied to the MMICs through a resistor connected to the device output leads. An RF choke is also often included in series with the supply lead.

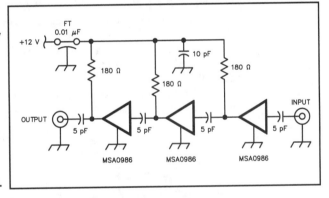

an amplifier design for circuits at UHF and microwave frequencies. For example, an MSA-0135 MMIC would be an excellent choice to build a receive preamplifier for a 1296-MHz receiver. This device provides 18.5 dB of gain for signals up to 1300 MHz, with a noise figure of 5.5 dB.

An MSA-0735 MMIC might be used to construct a 3456-MHz receiver amplifier. This device can be expected to provide 13 dB of gain for signals up to about 2500 MHz, and slightly less gain for higher-frequency signals. This amplifier would have a noise figure of about 4.5 dB.

Circuits built using MMICs generally employ *microstrip* construction techniques. Double-sided circuit board material is used, and one side serves to form a ground plane for the circuit. Circuit traces connecting the MMICs or other active devices to the signal input and output connectors form sections of feed line. The line widths, along with the circuit-board thickness and dielectric constant of the insulating material form sections of feed line with the desired characteristic impedance. Components are soldered to these feed line sections. Figure 6-29 is a photo of a general-purpose amplifier constructed using this technique with three MMICs, three resistors, four chip (surface-mount) capacitors and a feed-through capacitor to bring the supply voltage into the box.

The operating bias voltage is supplied to an MMIC through a resistor connected to the device output lead. Figure 6-30 is the schematic diagram of the general-purpose amplifier shown in the photo of Figure 6-29. Although this example does not include one, an RF choke is also often included in series with the bias voltage supply.

[Before you go on to the next section, you should turn to Chapter 10 and study questions A6F07 through A6F11. Review this section if you have difficulty with any of these questions.]

TOROIDS

A donut-shaped coil form is called a **toroid**. When a wire is wound on such a coil form, a *toroidal inductor* is produced. Toroidal inductors are one of the most popular inductor types in RF circuits. See Figure 6-31 for a photo of a variety of toroidal inductors. Toroidal inductors are also called ferromagnetic inductors, because the coil forms are made with ferrite and powdered-iron materials. The chemical names for iron compounds are based on the Latin word for iron, *ferrum*, so this is how these materials get the name ferrite.

Figure 6-31 — This photo shows a variety of inductors wound on powdered-iron and ferrite toroids.

A primary advantage of using a toroidal core to wind an inductor rather than a linear core is that nearly all the magnetic field is contained within the core of a toroid. With a linear core, the magnetic field extends through the space surrounding the inductor. The magnetic field of one linear-core inductor will interact with other nearby inductors, so external shields or other isolation methods must be used. Toroidal inductors can be located close to each other on the circuit board and there will be almost no interaction, however.

Manufacturers offer a wide variety of materials, or mixes, to provide cores that will perform well over a desired frequency range. By careful selection of core material, it is possible to produce toroidal inductors that can be used from dc to at least 1000 MHz. Cores made by mixing various amounts of powdered iron with binder materials are called *powdered-iron* toroids. If other materials, such as nickel-zinc and manganese-zinc compounds are mixed with the iron, *ferrite* toroids are produced.

The inductance of a toroidal core is determined by the number of turns of wire on the core, and on the core *permeability*. Permeability refers to the strength of a magnetic field in the core as compared to the strength of the field if no core were used. Cores with higher values of permeability will produce larger inductance values for the same number of turns on the coil. In other words, if you make two inductors with 10 turns on the coil forms, the core with a higher permeability will have more inductance.

The choice of core materials for a particular inductor presents a compromise of features. The powdered-iron cores generally have better temperature stability. Ferrite toroids generally have higher permeability values, however, so coils made with ferrite toroids usually require fewer turns to produce a given inductance value.

Calculating the inductance of a particular toroidal coil is simple. First you must know the *inductance index* value for the particular core you will use. This value, known as A_L, is found in the manufacturer's data. For powdered-iron toroids, A_L values are given in microhenrys per 100 turns. Table 6-1 gives an example

Table 6-1

A_L Values for Selected Powdered-Iron and Ferrite Toroids

A_L Values for Powered-Iron Cores (µH per 100 turns)

Size	Mix				
	2	3	6	10	12
T-12	20	60	17	12	7.5
T-20	27	76	22	16	10.0
T-30	43	140	36	25	16.0
T-50	49	175	40	31	18.0
T-200	120	425	100	na	na

A_L Values for Ferrite Cores (mH per 1000 turns)

Size	Mix			
	43	61	63	77
FT-23	188.0	24.8	7.9	396
FT-37	420.0	55.3	19.7	884
FT-50	523.0	68.0	22.0	1100
FT-114	603.0	79.3	25.4	1270

Data from *The ARRL Handbook*, 1995, courtesy of Amidon Associates and Micrometals.

of the data for several core types. The information for this table is taken from the 1995 edition of The ARRL Handbook, and is courtesy of Amidon Associates (one of the major distributors of toroidal cores in small quantities to the amateur market) and Micrometals (the manufacturer of the cores distributed by Amidon). See The ARRL Handbook for more complete information about these cores and their applications.

To calculate the inductance of a powdered-iron toroidal coil, when the number of turns and the core material are known, use Equation 6-3.

$$L = \frac{A_L \times N^2}{10,000} \qquad \text{(Equation 6-3)}$$

where:

L = inductance in μH.
A_L = inductance index, in μH per 100 turns.
N = number of turns.

For example, suppose you have a T-50 sized core made from the number 6 mix, which is good for inductors from about 10 to 50 MHz. From Table 6-1 we find that this core has an A_L value of 40. What is the inductance of a coil that has 10 turns on this core?

$$L = \frac{A_L \times N^2}{10,000} = \frac{40 \times 10^2}{10,000} = \frac{40 \times 100}{10,000} = \frac{4000}{10,000} = 0.4 \ \mu H$$

Often you want to know how many turns to wind on the core to produce an inductor with a specific value. In that case, you simply solve Equation 6-3 for N.

$$N = 100 \sqrt{\frac{L}{A_L}} \qquad \text{(Equation 6-4)}$$

Suppose you want to know how many turns to wind on the T-50-6 core used in the previous example to produce a 5-μH inductor? (The A_L value = 40.)

$$N = 100 \sqrt{\frac{L}{A_L}} = 100 \sqrt{\frac{5}{40}} = 100\sqrt{0.125} = 100 \times 0.35 = 35$$

So we will have to wind 35 turns of wire on this core to produce a 5-μH inductor. Perhaps the most common error made when winding an inductor involves counting the correct number of turns. Keep in mind that if the wire simply passes through the center of the core, you have a 1-turn inductor. Each time the wire passes through the center of the core, it counts as another turn. The common error is to count one complete wrap around the core ring as one turn. That produces a 2-turn inductor, however. See Figure 6-32.

The calculations with ferrite toroids are nearly identical, but the A_L values are given in millihenrys per 1000 turns instead of microhenrys per 100 turns. This requires a change of the constant in Equation 6-3 from 10,000 to 1,000,000. Use

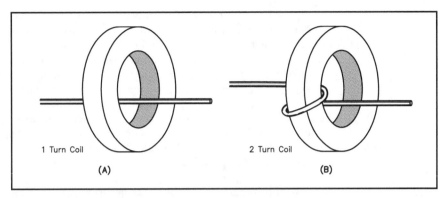

Figure 6-32 — Proper turns counting is important when you wind a toroidal inductor. Each pass through the center of the core must be counted. Part A shows a 1-turn inductor and Part B shows a 2-turn inductor.

Equation 6-5 to calculate the inductance of a ferrite toroidal inductor.

$$L = \frac{A_L \times N^2}{1,000,000}$$

(Equation 6-5)

where:
  L = inductance in mH.
  A_L = inductance index, in mH per 1000 turns.
  N = number of turns.

  Suppose we have an FT-50-sized core, made from 43-mix material. What is the inductance of a 10-turn coil? Table 6-1 shows that the $A_L = 523$ for this core.

$$L = \frac{A_L \times N^2}{1,000,000} = \frac{523 \times 10^2}{1,000,000} = \frac{523 \times 100}{1,000,000} \quad \frac{52300}{1,000,000} = 0.0523 \text{ mH} = 52.3 \ \mu\text{H}$$

  Again, it is a simple matter to solve this equation for N, so you can calculate the number of turns required to produce a specific inductance value for a particular ferrite core.

$$N = 1000 \sqrt{\frac{L}{A_L}}$$

(Equation 6-6)

  How many turns must we wind on a T-50-43 core to produce a 1-mH inductor? ($A_L = 523$.)

$$N = 1000 \sqrt{\frac{L}{A_L}} = 1000 \sqrt{\frac{1}{523}} = 1000 \sqrt{1.91 \times 10^{-3}} = 1000 \times 0.0437 = 43.7$$

Winding 43 or 44 turns on this core will produce an inductor of about 1 mH.

Toroidal cores are available in a wide variety of sizes. It is important to select a core size large enough to be able to hold the required number of turns to produce a particular inductance value. For a high-current application you will have to use a large wire size, so a larger core size is required. To wind an inductor for use in a high-power antenna tuner, for example, you may want to use number 10 or 12 wire. If your inductor requires 30 turns of this wire, you would probably select a 200-size core, which has an outside diameter of 2 inches and an inside diameter of 1¼ inches. You might even want to select a larger core for this application. A *ferrite bead* is a very small core with a hole designed to slip over a component lead. These are often used as parasitic suppressors at the input and output terminals of VHF and UHF amplifiers.

The use of ferrite beads as parasitic suppressors points out another interesting property of these core materials. While we normally want to select a core material that will have low loss at a particular frequency or over a certain range for winding our inductors, at times we want to select a core material that will have high loss. Toroid cores are very useful for solving a variety of radio-frequency interference (RFI) problems. For example, you might select a *type 43 mix* ferrite core and wind several turns of a telephone wire or speaker leads through the core to produce a *common-mode choke*. Such a choke is designed to suppress any RF energy flowing on these wires. So the audio signals flow through the choke unimpeded, but the RF signals are blocked.

RF transformers are often wound on toroidal cores. If two wires are twisted together and wound on the core as a pair to place two windings on the core, we say it is a *bifilar winding*. It is also possible to wind three, four or more wires on the core simultaneously, but the bifilar winding is the most common.

You may wonder how you can tell the difference between, say a T-50-6 core and a T-50-10 core if you found them both in a piece of surplus equipment or in a grab bag of parts. For that matter, how can you tell if a particular core is a powdered-iron or a ferrite core. Unfortunately, the answer is, you can't! There is no standard way of marking or coding these cores for later identification. So it is important to purchase your cores from a reliable source, and store them separate in marked containers.

[Turn to Chapter 10 now and study questions A6C01 through A6C11. Review this section if you have any difficulty with these questions about toroids. By now, you should have a basic understanding of how all of the circuit components required for your Advanced class license work. If you had trouble with any of the related questions on the Element 4A question pool, review those sections before proceeding to the next chapter.]

Amplifier transfer function — A graph or equation that relates the input and output of an amplifier under various conditions.

Automatic gain control — An amplifier circuit designed to provide a relatively constant output amplitude over a wide range of input values.

Balanced modulator — A circuit used in a single-sideband suppressed-carrier transmitter to combine a voice signal and an RF signal. The balanced modulator isolates the input signals from each other and the output, so that only the difference of the two input signals reaches the output.

Band-pass filter — A circuit that allows signals to go through it only if they are within a certain range of frequencies. It attenuates signals above and below this range.

Butterworth filter — A filter whose passband frequency response is as flat as possible. The design is based on a Butterworth polynomial to calculate the input/output characteristics.

Chebyshev filter — A filter whose passband and stopband frequency response has an equal-amplitude ripple, and a sharper transition to the stop band than does a Butterworth filter. The design is based on a Chebyshev polynomial to calculate the input/output characteristics.

Detector — A circuit used in a receiver to recover the modulation (voice or other information) signal from the RF signal.

Double-balanced mixer (DBM) — A mixer circuit that is balanced for both inputs, so that only the sum and the difference frequencies, but neither of the input frequencies, appear at the output. There will be no output unless both input signals are present.

Elliptical filter — A filter with equal-amplitude passband ripple and points of infinite attenuation in the stop band. The design is based on an elliptical function to calculate the input/output characteristics.

Frequency discriminator — A circuit used to recover the audio from an FM signal. The output amplitude depends on the deviation of the received signal from a center (carrier) frequency.

Half section — A basic L-section building block of image-parameter filters.

High-pass filter — A filter that allows signals above the cutoff frequency to pass through. It attenuates signals below the cutoff frequency.

Image rejection — The ability of a receiver (or receiver stage) to prevent unwanted signals from mixing with the local oscillator signal and producing a signal at the **intermediate frequency (IF)**.

Intermediate frequency (IF) — The output frequency of a mixing stage in a superheterodyne receiver. Signal processing (filter and amplification) stages tuned to this frequency allow efficient signal processing.

Linear electronic voltage regulator — A type of voltage-regulator circuit that varies either the current through a fixed dropping resistor or the resistance of the dropping element itself. The conduction of the control element varies in direct proportion to the line voltage or load current.

L network — A combination of a capacitor and an inductor, one of which is connected in series with the signal lead while the other is shunted to ground.

Low-pass filter — A filter that allows signals below the cutoff frequency to pass through. It attenuates signals above the cutoff frequency.

Mixer — A circuit that takes two or more input signals, and produces an output that includes the sum and difference of those signal frequencies.

Modulator — A circuit designed to superimpose an information signal on an RF carrier wave.

Neutralization — Feeding part of the output signal from an amplifier back to the input so it arrives out of phase with the input signal. This negative feedback neutralizes the effect of positive feedback caused by coupling between the input and output circuits in the amplifier. The negative-feedback signal is usually supplied by connecting a capacitor from the output to the input circuit.

Oscillator — A circuit built by adding positive feedback to an amplifier. It produces an alternating current signal with no input signal except the dc operating voltages.

Parasitics — Undesired oscillations or other responses in an amplifier.

Phase modulator — A device capable of modulating an ac signal by varying the reactance of an amplifier circuit in response to the modulating signal. (The modulating signal may be voice, data, video or some other kind.) The circuit capacitance or inductance changes in response to an audio input signal. Used in PM (or FM) systems, this circuit acts as a variable reactance in an amplifier tank circuit.

Pi network output-coupling circuits — A combination of two like reactances (coil or capacitor) and one of the opposite type. The single component is connected in series with the signal lead and the two others are shunted to ground, one on either side of the series element.

Product detector — A detector circuit whose output is equal to the product of a beat-frequency oscillator (BFO) and the modulated RF signal applied to it.

Ratio detector — A circuit used to demodulate FM signals. The output is the ratio of voltages from either side of a discriminator-transformer secondary.

Reactance modulator — A device capable of modulating an ac signal by varying the reactance of an oscillator circuit in response to the modulating signal. (The modulating signal may be voice, data, video or some other kind.) The circuit capacitance or inductance changes in response to an audio input signal. Used in FM systems, this circuit acts as a variable reactance in an oscillator tank circuit.

Slope detection — A method for using an AM receiver to demodulate an FM signal. The signal is tuned to be part way down the slope of the receiver IF filter curve.

Switching regulator — A voltage-regulator circuit in which the output voltage is controlled by turning the pass element on and off at a high rate, often several kilohertz. The control-element duty cycle is proportional to the line or load conditions.

Zener diode — A diode that is designed to be operated in the reverse-breakdown region of its characteristic curve.

PRACTICAL CIRCUITS

Now that you have studied some intermediate-level electrical principles, and have learned about the basic properties of some simple components, you are ready to learn how to apply those ideas to practical Amateur Radio circuits. This chapter will lead you through examples and explanations to help you gain that knowledge. In addition to reading the text, you should be prepared to perform all of the calculations shown in the sample problems.

This chapter covers the syllabus points that summarize the topics covered by the 10 groups of questions for this portion of the Advanced class question pool. Your Element 4A exam will include 10 of these questions. The topics covered here are:

A7A   Amplifier circuits: Class A, Class AB, Class B, Class C, amplifier operating efficiency (ie, dc input versus PEP), transmitter final amplifiers.
A7B   Amplifier circuits: tube, bipolar transistor, FET.
A7C   Impedance-matching networks: Pi, L, Pi-L.
A7D   Filter circuits: constant K, M-derived, band-stop, notch, crystal lattice, pi-section, T-section, L-section, Butterworth, Chebyshev, elliptical.
A7E   Voltage-regulator circuits: discrete, integrated and switched mode.
A7F   Oscillators: types, applications, stability.
A7G   Modulators: reactance, phase, balanced.

A7H   Detectors; filter applications (audio, IF, digital signal processing {DSP}).
A7I   Mixer stages; frequency synthesizers.
A7J   Amplifier applications: AF, IF, RF.

Several of these syllabus points cover related topics, so we've organized this chapter to cover the related material in the same major sections. For example, when we discuss amplifier circuits, we also cover amplifier applications before moving on to the next topic. At appropriate places throughout the chapter, we will instruct you to turn to Chapter 10 and to use the examination questions as a study aid to review your understanding of the material.

You should keep in mind that there have been entire books written on every topic covered in this chapter. So if you do not understand some of the circuits from our brief discussion, it would be a good idea to consult some other reference books. *The ARRL Handbook* is a good starting point, but even that won't tell you everything about each topic. Our discussion in this chapter should help you understand the circuits well enough to pass your Advanced class exam, however.

AMPLIFIERS

The energy picked up by a receiving antenna is extremely small. Therefore, the strength of the signals must be increased by a large factor in a receiver — often a million or more times. You might think that this amplification process could be carried on indefinitely, so that even the weakest radio signals could be brought up to usable strength. Unfortunately, there are limitations on how much you can amplify a signal.

Noise across the radio spectrum limits the amount of useful amplification. Unlimited amplification of a particular frequency cannot reduce the noise mask covering a desired signal, since both are RF emissions. Where does this noise come from? Electrical currents generated in nature (static) and in electrical circuits and devices all produce noise. Occurring at all frequencies, this noise is inescapable.

Amateurs don't give up in their battle against noise, though, because there are means to make it less bothersome. One of the big objectives in designing good receivers is to improve the signal-to-noise ratio. This means amplifying the signal as much as possible while amplifying the noise as little as possible.

Amplifier Operating Class

The basic operation of an amplifier is selected by choosing a bias or dc input when there is no signal applied to the amplifier. In the most common amplifier configuration, tubes and PNP transistors are operated with a slightly negative bias voltage applied to the grid or base, and NPN transistors with a slightly positive bias voltage applied to the base. The actual bias voltage depends on the amplifying device used and the type of amplification desired.

The amplifier operating characteristics define several classes of operation. The three basic operating classes are Class A, Class B and Class C. Another class of operation is sometimes added between the A and B categories, and this is called Class-AB operation.

To understand what we mean by operating class, let's consider something called an **amplifier transfer function**. This is simply a way of expressing the amplifier output in terms of the input. All amplifiers have their own transfer function, but they all have certain characteristics in common.

There is usually a point at which a bigger input signal will not produce a bigger output signal. This is called the *saturation region*. There is also a point at which the output will not decrease if a weaker (or no) input signal is supplied. This is called the *cutoff region*.

Between saturation and cutoff is an area where the input and output signals vary in a linear fashion. Of course, this is called the *linear region*. In practice, the saturation region is considered to be that area where the output signal no longer increases linearly with increasing input signals. Likewise, the cutoff region is the area below the linear-response region. Figure 7-1 illustrates a typical transfer function, showing the important regions on the curve.

Two factors about the input are extremely important for establishing the type of operation required of a specific amplifier. The first, called the bias, is the average signal input level, or the dc input level. The second is the size of the input signal. These levels must be set carefully to produce the desired type of operation.

With a Class-A amplifier, the bias level and input-signal amplitude are set so that all of the input signals appear between the saturation and cutoff regions. This means the amplifier is operating in the linear region of its transfer function, and the output is a linear (but larger) reproduction of the input signal. There is an output signal for the full 360° of input signal. At its weakest point, the input signal is large enough to produce an output signal, and at its strongest point the input signal does not drive the output signal into the saturation region.

That portion of the input cycle for which there is an output is called the conduction angle. For a Class-A amplifier, the conduction angle is 360°. Figure 7-2A is a graph of the output current from a Class-A amplifier. The saturation and cutoff levels are indicated. The *quiescent point* sets the operating bias point, and represents the amplifier voltage or current conditions with no input signal.

The efficiency of a Class-A amplifier is low, because there is always a significant amount of current drawn from the power supply — even with no input signal. This no-signal current is called the quiescent current of the amplifier. The

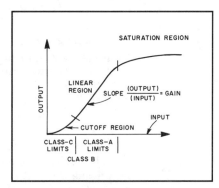

Figure 7-1 — A typical transfer-function graph displaying output versus input for different amplifier classes.

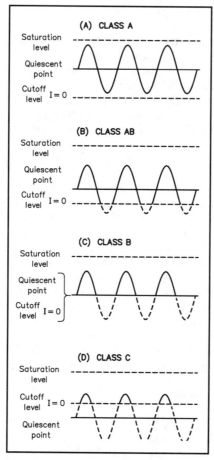

Saturation level

(A) CLASS A

Quiescent point

Cutoff level I = 0

(B) CLASS AB

Saturation level

Quiescent point

Cutoff level I = 0

(C) CLASS B

Saturation level

Quiescent point

Cutoff level I = 0

(D) CLASS C

Saturation level

Cutoff level I = 0

Quiescent point

Figure 7-2 — Amplifying-device output current for various classes of operation. All graphs assume a sine-wave input signal.

maximum theoretical efficiency of a Class-A amplifier is 50%, but in practice it is more like 25 to 30%.

For a Class-AB amplifier, the drive level and dc bias are adjusted so output current flows for more than half the input cycle, but not for the entire cycle. The conduction angle is between 180° and 360°, and the operating efficiency is often more than 50%. Figure 7-2B shows the output signal for a Class-AB amplifier.

Class-B operation sets the bias right at the cutoff level. In this case, there is output current only during half the input sine wave. This represents a conduction angle of 180°. See Figure 7-2C The output is not as linear as with a Class-A amplifier, but is still acceptable for many applications. The advantage is increased efficiency; up to 65% efficiency is theoretically possible with a Class-B amplifier, and practical amplifiers often attain 60% efficiency.

Class-C amplification requires that the bias be well below cutoff on the transfer-characteristic curve, and that the signal be large enough to bring some part of the top half into the conduction region.

A Class-C amplifier has a conduction angle of less than 180°. The output current will just be pulses at the signal frequency, as shown in Figure 7-2D. The amplifier is cut off for considerably more than half the cycle, so the operating efficiency can be quite high — up to 80% with proper design. Linearity is very poor, however.

The linearity of the amplifier stage is important, because it describes how faithfully the input signal will be reproduced at the output. Any nonlinearity results in a distorted output. So you can see that a Class-A amplifier will have the least amount of distortion, while a Class-C amplifier produces a severely distorted output. A side effect of this distortion or nonlinearity is that the output will contain harmonics of the input signal. Odd-order intermodulation products are formed, which are close to the desired frequency, and so will not be filtered out by a resonant tank circuit. So a pure sine wave input signal becomes a complex combination of sine waves at the output.

By now you are probably asking, "But why would anyone want to have an amplifier that generates a distorted output signal?" That certainly is a good question. At first thought, it sure doesn't seem like a very good idea. You must remember, however, that every circuit design consists of compromises between fundamentally opposing ideals. We would like to have perfect linearity for our amplifiers, but we would also like them to have 100% efficiency. You have just learned that those two ideals are exclusive. The closer you get to one of them, the further you get from the other. So you will have to compromise the ideals a bit to achieve a workable design. Your particular application and circuit conditions help you determine which compromises to make.

An important point to keep in mind is that most RF amplifiers will have a tuned-output tank circuit. That tuned circuit stores electrical energy like a mechanical flywheel, which is used to store mechanical energy. A heavy wheel, with its mass concentrated as close to the outer rim as possible, will have a large moment of inertia. That means that it has a lot of angular momentum when it is spinning, so it will continue to spin unless there is something to make it stop. If you turn off the motor that was used to set the flywheel in motion, the wheel will continue to spin for a long time. (In the ideal condition, where there is no friction, and the wheel is spinning on perfect bearings, it would never stop.) Your car engine uses a flywheel to store some of the energy from the gasoline exploding in the engine, and keep the system spinning smoothly. Without a flywheel, there would be a noticeable "bump" every time a cylinder fires.

A parallel tuned circuit is the electrical equivalent of a flywheel. The electrons that make up the current in the tank circuit oscillate, or circulate, back and forth in the inductance (L) and capacitance (C) of the parallel-resonant circuit. (This type of circuit is often called a tank circuit because you can also think of it as a storage tank for electrical energy.) By placing such a circuit in series with the amplifier output, you can use it to smooth out the "bumps" that will occur when there is no output because the amplifier is turned off. This is especially useful if you are amplifying a pure sine-wave signal, such as for CW, and want to take advantage of the increased efficiency offered by a Class-C amplifier.

The tank circuit is a parallel-resonant circuit, as discussed in Chapter 5. You should remember that the electric current flowing in the circuit is passing the energy back and forth between the electric field of the capacitor and the magnetic field of the inductor. The signal source only has to supply small amounts of energy to keep the current flowing. The tuned tank circuit will filter out the unwanted harmonics generated by a nonlinear amplifier stage. Usually, the tank circuit should have a Q of at least 12 to reduce these harmonics to an acceptable level.

If you are amplifying an audio signal, linearity may be the most important consideration. Use a Class-A amplifier for audio stages. For an AM or SSB signal, which is an RF signal envelope that varies at an audio rate, you would want to use

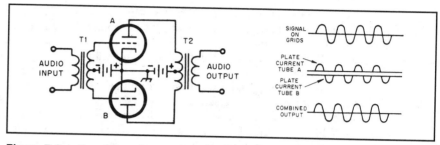

Figure 7-3 — Two Class-B amplifiers connected to operate as a push-pull amplifier. The graphs show current waveforms at the points indicated.

a linear amplifier. You may be willing to accept a bit of nonlinearity to obtain increased efficiency, in which case Class-AB operation would be indicated. In some instances, you may even be willing to use a Class-B amplifier, and let the flywheel effect of a tank circuit fill in the waveform voids for you.

You can take advantage of the nonlinearity of a Class-C amplifier by using it as a frequency-multiplier stage. We mentioned earlier that one consequence of a nonlinear amplifier is that it would generate harmonic signals. If you want to multiply the frequency for operation on another band, sometimes you can do that by using a tank circuit tuned to a harmonic of the input frequency, and selecting the appropriate harmonic from the output of a Class-C amplifier. This is especially useful for generating an FM signal at VHF or UHF, where you may start with a signal in the HF range, apply modulation through a reactance modulator, and then multiply the signal frequency to the desired range on 2 meters or higher. The exact bias point for the Class-C amplifier will determine which harmonic frequency will be the strongest in the output, so careful selection of bias point is important.

Class-B amplifiers are often used for audio frequencies by connecting two of them back to back in push-pull fashion. Figure 7-3 illustrates a simple triode-tube push-pull amplifier and the waveforms associated with this type of operation. A push-pull amplifier can also be used at RF. While one tube is cut off, the other is conducting, so both halves of the signal waveform are present in the output. This reduces the amount of distortion in the output, and will reduce or eliminate even-order harmonics.

Voltage, Current and Power Amplification

When we talk about amplifiers, it is common to think of a *power* amplifier. An amateur amplifier is said to be "1 W in, 10 W out," for example. We can also build circuits to amplify voltage or current, however. An instrumentation amplifier may be designed to amplify a very small voltage so that it can be measured with a voltmeter; in a multistage amplifier, we may wish to amplify the output current from one stage to drive the next stage. *Stage*, by the way, is the name we give to one of a number of signal-handling sections used one after another in an electronic device. The instrumentation amplifier, for example, may increase the voltage level but not the available current, so the available power is not increased significantly.

The input and output impedance of an amplifier circuit will vary, depending on what type of amplification the circuit is designed to provide. Generally speaking, voltage amplifiers have a very low input impedance and a high output impedance. If too much current is drawn from the output of a voltage amplifier, its operation can be affected. Current and power amplifiers have a lower output impedance, and can be used to supply both voltage and current to a load.

Amplifier Gain

The *gain* of an amplifier is the ratio of the output signal to the input signal. Voltage amplifier gain is based on the ratio of output and input voltages, current amplifier gain on the ratio of output and input current levels and power amplifier gain is determined by the ratio of output and input power levels.

This ratio can get to be a very large number when several stages are combined or when the gain is very large. The decibel expresses the ratio in terms of a logarithm, making the number smaller and easier to work with. We often state the gain of a stage as a voltage gain of 16 or a power gain of 250, which are ratios, or refer to a gain of 24 dB, for example.

[Turn to Chapter 10 and study questions A7A01 through A7A06, questions A7A11 and A7A12, and questions A7B02 and A7B03. Review this section as needed.]

Transistor Amplifiers

We are limiting our discussion here to the operation of bipolar-transistor-amplifier circuits, but many of the techniques and general circuit configurations also apply to tube-type amplifiers. One major difference between tube-type amplifiers and transistor amplifiers is that tubes are voltage-operated devices, but transistors are current operated. A transistor amplifier is essentially a current amplifier. To use it as a voltage amplifier, the current is drawn through a resistor, and the resulting voltage drop provides the amplifier output. FET amplifiers operate as voltage amplifiers. To learn more about how tube-type amplifiers function, we recommend that you turn to appropriate sections of *The ARRL Handbook*. At some point in your Amateur Radio career, you will probably need to learn about tube amplifiers, but for the purpose of helping you pass your Advanced class exam, we will concentrate on transistor circuits.

Bipolar-transistor diode junctions must be forward biased in order to conduct significant current. If your circuit includes an NPN transistor, the collector and base must be positive with respect to the emitter, and the collector must be more positive than the base. When working with a PNP transistor, the base and collector must be negative with respect to the emitter and the collector more negative than the base. The required bias is provided by the collector-to-emitter voltage, and by the emitter-to-base voltage. These bias voltages cause two currents to flow: Emitter-to-collector current and emitter-to-base current. Either type of transistor, PNP or NPN, can be used with a negative- or positive-ground power supply. Forward bias must still be maintained, however. Remember that the amount of bias current sets the class of amplifier operation.

The lower the forward bias, the less collector current will flow. As the forward bias is increased, the collector current rises and the junction temperature rises. If the bias is continuously increased, the transistor eventually overloads and burns out. This condition is called thermal runaway. To prevent damage to the transistor, some form of bias stabilization should be included in a transistor amplifier design. Even if the bias is not increased, however, thermal runaway can occur. As the transistor heats up, its beta increases, causing more collector current to flow. This causes more heating, even higher beta and even more current, until eventually the transistor burns out.

Amplifier circuits used with bipolar transistors fall into one of three types, known as the common-base, common-emitter and common-collector circuits. These are shown in Figure 7-4 in elementary form. The three circuits correspond approximately to the grounded-grid, grounded-cathode and cathode-follower vacuum-tube circuits, respectively.

Common-Emitter Circuit

The common-emitter circuit is shown in Figure 7-4A. The base current is small and the input impedance is fairly high — several thousand ohms on average. The collector resistance is some tens of thousands of ohms, depending on the signal-circuit source impedance. The common-emitter circuit has a lower cutoff frequency than does the common-base-circuit, but it gives the highest power gain of the three configurations.

In this circuit, the output (collector) current phase is opposite to that of the input (base) current. So any feedback through the small emitter resistance is negative and that will stabilize the amplifier, as we will show by an example.

Common-emitter amplifiers are probably the most common bipolar-transistor-amplifier type, so we will use this circuit to illustrate some of the design procedures. R_1 and R_2 make up a divider network to provide base bias. These resistors provide a fixed, stable operating point, and tend to prevent thermal runaway. R_1 and R_2 are sometimes called *fixed-bias* resistors. R_3 provides the proper value of bias voltage, when normal dc emitter current flows through it. This biasing technique is often referred to as *self bias*. C_3 is used to bypass the ac signal current around the emitter resistor. This *emitter-bypass* capacitor should have low reactance at the signal frequency. C_1 and C_2 are coupling capacitors, used to allow the desired signals to pass into and out of the amplifier, while blocking the dc bias voltages. Their values should be chosen to provide a low reactance at the signal frequency.

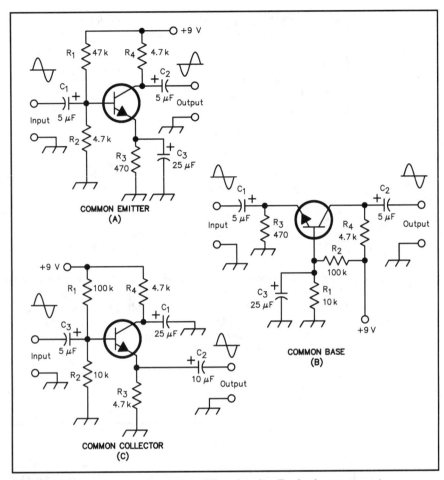

Figure 7-4 — Basic transistor-amplifier circuits. Typical component values are given for use at audio frequencies where these circuits are often used. The input and output phase relationship of each amplifier is shown.

The emitter-to-base resistance is approximately:

$$R_{e-b} = \frac{26}{I_e}$$

(Equation 7-1)

where I_e is the emitter current in milliamperes. The voltage gain is the ratio of collector load resistance to (internal) emitter resistance. The letter A represents amplifier gain, which you can think of as *amplification factor*, to help you identify the meaning of the symbol in the equation.

$$A_V = \frac{R_L}{R_{e-b}}$$

(Equation 7-2)

For the example of Figure 7-4A, if the emitter current is 1.6 mA, then R_{e-b} is 16.25 Ω, and

$$A_V = \frac{R_L}{R_{e-b}} = \frac{4.7 \text{ k}\Omega}{16.25 \text{ }\Omega} = \frac{4.7 \times 10^3 \text{ }\Omega}{16.25 \text{ }\Omega} = 289$$

If you would like to express this voltage-gain ratio in decibels, just take the logarithm of 289, and multiply by 20:

Gain in dB = 20 × log (289) = 20 × 2.46 = 49 dB.

The base input impedance is given by:

$$R_b = \beta \, R_{e-b}$$

(Equation 7-3)

For our example, R_b = 1625 Ω, assuming beta = 100. The actual amplifier input impedance is found by considering R_1 and R_2 to be in parallel with this resistance, so the input impedance for our amplifier is about 1177 Ω.

If you omit the emitter bypass capacitor, then all of the emitter signal current must flow through R_3. This resistor dominates the emitter impedance in the voltage gain equation, so we have:

$$A_V = \frac{R_L}{R_E}$$

(Equation 7-4)

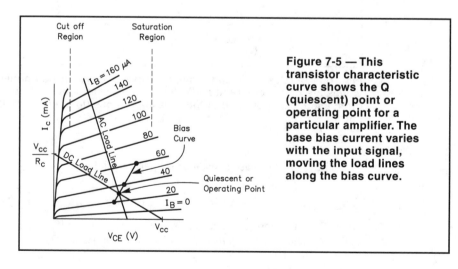

Figure 7-5 — This transistor characteristic curve shows the Q (quiescent) point or operating point for a particular amplifier. The base bias current varies with the input signal, moving the load lines along the bias curve.

Now we obtain a gain of 10 from our amplifier (20 dB). The base impedance of the unbypassed-emitter circuit becomes $\beta R_E = 47$ kΩ, which when combined with the bias resistors, gives an amplifier input impedance of 3.92 kΩ. Notice that the unbypassed emitter resistor introduced 29 dB of negative feedback. This has the effect of stabilizing the gain and impedance values over a wide frequency range. A frequently used amplifier design splits the emitter resistor into two series resistors. The first, connected to the emitter, is unbypassed, and the second, which connects to ground, is bypassed with a capacitor.

Figure 7-5 shows a set of characteristic curves for a typical NPN transistor amplifier. The lines on this graph show how the collector current changes as the collector to emitter voltage varies. Since the base current controls the collector current for this amplifier, there are individual lines to represent various values of base current. The collector supply voltage and bias resistors establish the quiescent point, or operating point for the amplifier. The amplifier load lines represent the amplifier operation. The changing conditions of an ac input signal move the load line along the bias curve. For best efficiency and stability, you usually want to operate the amplifier at a point on the load line that is just below saturation.

Common-Base Circuit

The input circuit of a common-base amplifier must be designed for low impedance. Equation 7-1 can be used to calculate the base-emitter junction resistance. The optimum output load impedance, R_L, may range from a few thousand ohms to 100,000 Ω, depending on the circuit. In the common-base circuit (Figure 7-4B), the phase of the output (collector) current is the same as that of the input (emitter) current. The parts of these currents that flow through the base resistance are likewise in phase, so the circuit tends to be regenerative and will oscillate if the current amplification factor is greater than one. You will notice that the bias resistors have much the same configuration as with the common-emitter circuit, and that input and output-coupling capacitors are used. C_3 bypasses the base bias resistor, placing the base at ac-ground potential.

Common-Collector Circuit

The common-collector transistor amplifier, sometimes called an emitter-follower amplifier, has high input impedance and low output impedance. The input resistance depends on the load resistance, being approximately equal to the load resistance divided by $(1 - \alpha)$. The fact that input resistance is directly related to the load resistance is a disadvantage of this type of amplifier, especially if the load is one whose resistance or impedance varies with frequency.

The cutoff frequency of the common-collector circuit is the same as in the common-emitter amplifier. The input and output currents are in phase, as shown in Figure 7-4C. C_1 is a collector bypass capacitor. This amplifier also uses input- and output-coupling capacitors, C_2 and C_3. R_3 is a feedback resistor, and is also used to develop an output voltage for the next stage. Because of this use to develop the output voltage, R_3 is often called the *emitter load* resistor for the common collector circuit.

[Study the examination question A7A07 and questions A7B04 through A7B12 in Chapter 10 now. Review this section as needed.]

RF Amplifiers

Receiver RF amplifiers are useful primarily to improve the receiver noise figure at frequencies around 28 MHz and above. The idea is to increase the level of weak RF signals above the internal noise of the receiver. An RF amplifier can degrade the receiver dynamic range, however, because the mixer will be overloaded more easily. If the mixer overloads, it is more likely to generate spurious mixing products, degenerating performance. When atmospheric and man-made noise levels exceed the noise generated in the mixer, an RF amplifier may not provide such a big advantage. Actually, it should be possible to realize better dynamic range by not using an RF amplifier. The gain of the RF stage, when one is used, should be set for the minimum level needed to override the mixer noise. This helps reduce the generation of spurious (unwanted) mixer products. Sometimes the necessary gain is only a few decibels. A good low-noise device should be employed as the RF amplifier in such instances.

All of the FET amplifiers in Figure 7-6 are capable of providing low-noise operation and good dynamic range. The common-source circuits at parts B and C can provide up to 25 dB of gain. They are more prone to instability than is the circuit at A, however. Therefore, the gates are shown tapped down on the gate tank, placing the input at a low impedance point on the tuned circuit to discourage self-oscillation. The same is true of the drain tap. JFETs are able to withstand higher RF voltages without damage than can MOSFET devices.

A broadband bipolar-transistor RF amplifier is shown in Figure 7-6D. This type of amplifier will yield approximately 16 dB of gain up to 148 MHz, and it will be unconditionally stable because of the negative feedback in the emitter and base circuits. A broadband 4:1 transformer is used in the collector to provide a 200-Ω load on the transistor, and step the impedance down to approximately 50 Ω at the amplifier output. The input impedance to the 2N5179 is also about 50 Ω. A band-pass filter should be used at the amplifier input and output to provide selectivity.

RF Power Amplifiers

Most of our discussion here has centered around amplifiers used in receiver circuits. Of primary importance to most Amateur Radio operators is the goal of getting some RF energy to leave the station and travel to another location where it can be received — in other words, communication! Achieving that goal requires that the weak signals generated in an RF oscillator be amplified to some higher power level. Either tube or transistor amplifier circuits are used for that purpose.

Of course, we are always concerned about the efficiency of our circuits, but since a power amplifier is a larger drain on the power supply, it is a good idea to

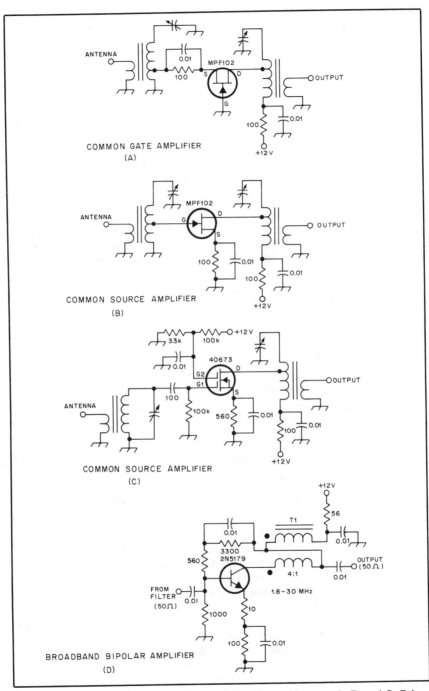

Figure 7-6 — Narrowband FET RF amplifiers are shown at A, B and C. D is a broadband bipolar-transistor RF amplifier.

pay a little extra attention to that factor with a high-power amplifier. The goal is to transfer as much power to the load as possible. The total power generated by the amplifier is given by:

$$P_{IN} = P_{OUT} + P_D$$ (Equation 7-5)

where:
  P_{IN} = dc- and drive-power input.
  P_{OUT} = power delivered to the load.
  P_D = power dissipated in the amplifier resistances.

The efficiency is calculated from the input and output power:

$$\text{Efficiency} = \frac{P_{OUT}}{P_{IN}} \times 100\% = \frac{P_{OUT}}{P_{OUT} + P_D} \times 100\%$$ (Equation 7-6)

where the efficiency is expressed as a percentage. This is sometimes simplified to ignore the drive power input to the amplifier, which is reasonable if the drive power is a small percentage of the total power. In that case, the amplifier efficiency is expressed as:

$$\text{Efficiency} = \frac{\text{RF power out}}{\text{DC power in}} \times 100\%$$ (Equation 7-7)

As an example, suppose the output from an amplifier is 1500 W and the dc input power is 1950 W with 50 W of drive power. This means the total input power is 2000 W, with 500 W dissipated in the amplifier itself. Then the efficiency of that amplifier is:

$$\text{Efficiency} = \frac{P_{OUT}}{P_{OUT} + P_D} \times 100\% = \frac{1500\,W}{1500\,W + 500\,W} \times 100\% = \frac{1500\,W}{2000\,W} \times 100\%$$

$$\text{Efficiency} = 75\%$$

Let's also calculate the efficiency using Equation 7-7, ignoring the drive power.

$$\text{Efficiency} = \frac{\text{RF power out}}{\text{DC power in}} \times 100\% = \frac{1500\,W}{1950\,W} \times 100\% = 0.77 \times 100\% = 77\%$$

The two calculations give the same results, to within a few percent, proving that this simplification is valid.

You can easily solve Equation 7-7 for either RF power input or dc power input, if you know the amplifier efficiency. Further, if you know the operating class of an amplifier, you can estimate the efficiency and calculate a reasonable estimate of the dc input power to your amplifier. Earlier in this chapter you learned that a Class-A amplifier has a maximum efficiency of 50%, and practical amplifiers often have efficiencies more in the range of 25% to 30%. A Class-AB amplifier usually has an efficiency around 50% and a Class-B amplifier is about 60% efficient. Class-C amplifiers are often as much as 80% efficient.

Suppose you have a Class B RF power amplifier that produces 1500 W PEP output. What is the approximate dc input power to this amplifier? (Such an amplifier might be used as the power amplifier stage in an FM phone station.) To answer this question, we will assume an efficiency of 60% for our amplifier, and solve Equation 7-7 for dc input power.

$$\text{DC power in} = \frac{\text{RF power out}}{\text{Efficiency}} \times 100\% = \frac{1500 \text{ W}}{60\%} \times 100\% = 25 \text{ W} \times 100 = 2500 \text{ W}$$

[This is a good time to check on your understanding of the amplifier topics we've been discussing. Turn to Chapter 10 and study question A7A08 and questions A8F09 through A8F11. Also study questions A7J09 through A7J11. Review this section if you have difficulty with any of these questions.]

IF Amplifiers

The main reason for using an **intermediate frequency (IF)** in a transmitter or receiver is that the circuits can process signals more efficiently if they are all converted to a single frequency first. Tuned circuits at the IF will not have to be readjusted for every small frequency change, and amplifiers can be designed for optimum operation at a single frequency. These amplifiers are designed to pass a small band of frequencies, instead of a wide range, and that means the circuit will have improved selectivity to block signals that are slightly off frequency.

An IF amplifier, then, is a fixed-tuned, band-pass amplifier. One important purpose of the first IF amplifier is to provide selectivity.

The amount of amplification used in a receiver IF stage will depend on the signal level available at the input to the IF strip. Sufficient gain is needed to ensure ample audio output to drive headphones or a speaker. Another consideration is the amount of IF-gain change initiated by the **automatic gain control (AGC)**. The more IF stages used (two is typical), the greater the gain change caused by AGC action. AGC refers to a method of using a voltage derived from the input signal to control the stage gain. Stronger signals require less gain, and weaker ones require more. The problem is that if you have the audio gain turned all the way up to hear a weak signal and then a strong signal suddenly comes through, it will blast your eardrums. A good AGC system will quickly reduce the amplifier gain so the strong signal will not be amplified as much. The range is on the order of 80 to 120 dB of gain variation with AGC applied to a pair of typical IC amplifiers.

Nearly all modern receiver circuits use IC amplifiers in the IF section. Numerous types of ICs are available to provide linear RF and IF amplification at low cost. With careful layout techniques, IC amplifiers will be very stable. Bypass capacitors should be placed as close to the IC pins as possible. Input and output circuit elements must be separated to prevent mutual coupling, which can cause unstable operation. If IC sockets are used, they should be of the low-profile variety, with short socket conductors. Figure 7-7 contains examples of bipolar transistor, FET, and IC IF amplifiers. Typical component values are given.

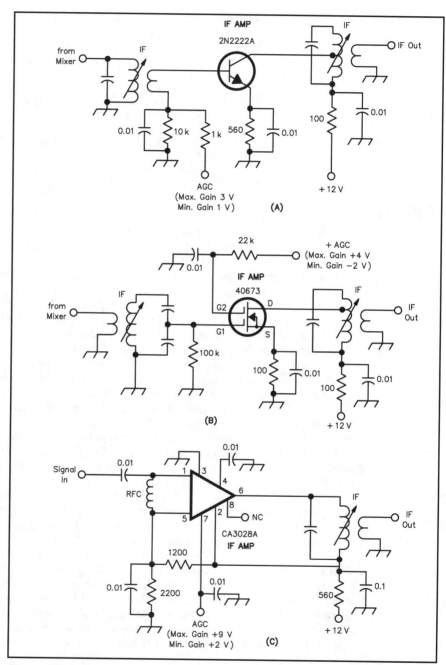

Figure 7-7 — Typical IF amplifier stages are shown, including input for an AGC voltage. At A is a bipolar-transistor amplifier and B is a dual-gate MOSFET stage. An IC amplifier is shown at C.

Intermediate Frequency Choice

Selection of an intermediate frequency is a compromise between two (or more) conflicting factors. With a lower IF, the amplifier gain will be higher, and the selectivity will be improved. But this means the image frequency is closer to the desired receive frequency, which reduces the **image rejection**. (There is more information about image response and image rejection in Chapter 4 of this book.) A higher IF improves the image-rejection ratio, but the gain and selectivity are reduced. Many modern commercially built receivers convert the incoming signals to a frequency in the VHF range for a first IF. This moves any image response far away from the filter passband.

It is a good idea to avoid selecting an IF on which there are strong signals, such as the broadcast band, because these signals may get into the IF amplifier and cause interference or other problems. To select an intermediate frequency you must consider selectivity and image rejection.

In a typical receiver IF system, the first amplifier stage is primarily responsible for providing selectivity. You want to limit the signals going through the IF section to only the ones you want to receive at that instant! Succeeding stages provide more gain and filtering. The final stage must provide an impedance match to the detector, and supply any final gain needed.

AF Amplifiers

After the detector, the received signal has been converted to an audio-frequency (AF) signal. For most amateur phone communications, the audio amplifiers should amplify signals from about 300 Hz to 3000 Hz. This range includes those frequencies necessary to understand speech, and any signals outside this frequency range represent wasted energy (and bandwidth) for communications.

Most of our discussion about RF and IF amplifiers revolved around receiver discussions. Most of those principles apply to transmitters (or the shared circuits in transceivers) as well. The same is true of AF amplifiers. For example, an AF amplifier is used in an amateur phone transmitter to raise the level of the microphone audio output to the level required by the modulator. This microphone amplifier should restrict the signals it amplifies to the same 300 Hz to 3000 Hz range as the AF amplifier in a receiver.

A common figure of merit used to evaluate RF and IF amplifiers is the *signal-to-noise ratio*. We are interested in the amount of noise generated within the amplifier itself, as compared to the signal levels being amplified. Obviously, if an amplifier generates a lot of noise, then that noise may cover a weak desired signal.

A perfect amplifier would only increase the strength of the input signal, and the output would be a perfect enlargement of the input. Any stage that includes a nonlinear device, such as a diode, transistor, tube or magnetic materials will introduce some distortion, however. That distortion can be described in terms of signals at other frequencies, usually harmonics of the input signals.

Total harmonic distortion (THD) is a common figure of merit for audio ampli-

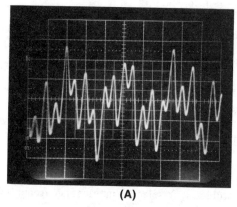 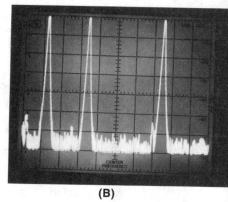

(A) **(B)**

Figure 7-8 — Part A is an oscilloscope display of a complex signal waveform. The spectrum analyzer display at B shows the individual signal frequencies and their relative strengths from the same waveform.

fiers. We must be able to describe the output signal in terms of the input-signal frequency and harmonics of that frequency to calculate THD. The mathematical term for an analysis that takes a complex waveform and breaks it into a series of signals at specific frequencies is *Fourier analysis,* named for the French mathematician who developed the technique. A spectrum analyzer performs Fourier analysis on a signal, and the display shows individual lines at distinct frequencies. The height of each line represents the relative strengths of each signal. See Figure 7-8.

Once the output signal has been broken into individual frequency components, the total harmonic distortion is easily calculated. THD is the ratio of the total RMS voltage for all the harmonic signals in the output to the total RMS voltage of the output signal at the input frequency, when the input signal is a pure sine wave. THD is often expressed as a percentage. We can express THD as an equation:

$$THD = \frac{V_H}{V_F} \times 100\%$$

where:

THD = Total harmonic distortion.

V_H = RMS voltage of all harmonics in the output signal.

V_F = RMS voltage of fundamental frequency in the output signal.

A pair of transistors connected as a *Darlington pair* are often used to build an audio amplifier. Such an amplifier offers several advantages over a single-transistor amplifier. A Darlington pair audio amplifier has high gain, a high input impedance and low output impedance. Figure 7-9 is an example of such an amplifier.

[It's time for another question pool break. Turn to Chapter 10 and study questions A7J01 through A7J08. Review this section if you have difficulty with any of these questions.]

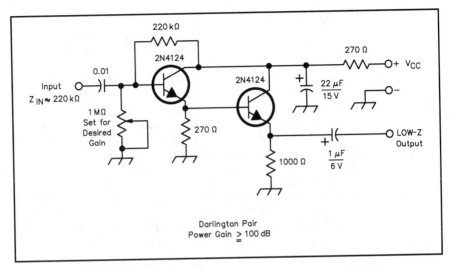

Figure 7-9 — The transistors in this audio amplifier are connected as a Darlington pair.

Amplifier Stability

Excessive gain or undesired feedback may cause amplifier instability. Oscillation may occur in unstable amplifiers under certain conditions. Damage to the active device from over-dissipation is only the most obvious effect of oscillation. Deterioration of noise figure, spurious signals generated by the oscillation and reradiation of the oscillation through the antenna, causing RFI to other services, can also occur from amplifier instability. Negative feedback will stabilize an RF amplifier. Care in terminating both the amplifier input and output can produce stable results from an otherwise unstable amplifier. Attention to proper grounding and proper isolation of the input from the output by means of shielding can also yield stable operating conditions.

Neutralization

A certain amount of capacitance exists between the input and output circuits in any active device. In the bipolar transistor it is the capacitance between the collector and the base. In an FET it's the capacitance between the drain and gate. In vacuum tubes it is the capacitance between the plate and grid circuit. So far we have simply ignored the effect that this capacitance has on the amplifier operation. In fact, it doesn't have much effect at the lower frequencies. Above 10 MHz or so, however, the capacitive reactance may be low enough to cause complications. Oscillations can occur when some of the output signal is fed back in phase, so that it adds to the input (positive feedback). As the output voltage increases so will the feedback voltage: The circuit adds fuel to its own fire and the amplifier is now an oscillator. The output signal is no longer dependent on the input signal, and the circuit is useless as an amplifier.

In order to rid the amplifier of this positive feedback, it is necessary to provide a second feedback path, which will supply a signal that is 180° out of phase with the positive feedback voltage. This is called *negative feedback*. This path should supply a voltage that is equal to that causing the oscillation, but of opposite polarity.

One **neutralization** technique for vacuum-tube amplifiers is shown in Figure 7-10. In this circuit, the neutralization capacitor is adjusted to have the same value as the interelectrode capacitance that is causing the oscillation.

With solid-state amplifiers, a similar technique could be used, although the interelement capacitances tend to be much smaller. It is more common to include a small value of resistance in either the base or collector lead of a low-power amplifier. Values between 10 and 20 Ω are typical. For higher power levels (above about 0.5 W), one or two ferrite beads are often used on the base or collector leads.

Parasitic Oscillations

Oscillations can occur in an amplifier on frequencies that have no relation to those intended to be amplified. Oscillations of this sort are called **parasitics** mainly because they absorb power from the circuits in which they occur. Parasitics are brought on by resonances that exist in either the input or output circuits. They can also occur below the operating frequency, which is usually the result of an improper choice of RF chokes and bypass capacitors. High-Q RF chokes should be avoided, because they are most likely to cause a problem.

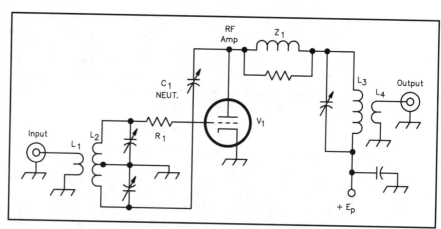

Figure 7-10 — An example of a neutralization technique used with a tube-type RF amplifier.

Parasitics are more likely to occur above the operating frequency as a result of stray capacitance and lead inductance along with interelectrode capacitances. In some cases it is possible to eliminate such oscillations by changing lead lengths or the position of leads so as to change the capacitance and inductance values. An effective method with vacuum tubes is to insert a parallel combination of a small coil and a resistor in series with the grid or plate lead. The coil serves to couple the VHF energy to the resistor, and the resistor value is chosen so that it loads the VHF circuit so heavily that the oscillation is prevented. Values for the coil and resistor have to be found experimentally as each different layout will probably require different suppressor networks.

With transistor circuits, ferrite beads are often used on the device leads, or on a short connecting wire placed near the transistor. These beads act as a high impedance to the VHF or UHF oscillation, and block the parasitic current flow. In general, proper neutralization will help prevent parasitic oscillations.

Transmitter Tuning Procedure

Most of the newer solid-state transceivers have broadband, no-tune final amplifiers. Many older rigs that are still popular include a tube-type final amplifier, however. Also, most external power amplifiers used to increase your signal beyond the typical 100 W level, rely on vacuum tubes. These usually have a pi-network output circuit that must be adjusted carefully for proper operation. You should be familiar with the manufacturer's operating instructions, or the function of each control on a piece of homemade gear. It is a good idea to know the general tune-up procedure for such equipment, however. The outline given here is a general one at best, but will give you an idea of how it's done. This description refers to Figure 7-11.

It is a good idea to operate the transmitter into a dummy load while learning the ropes of transmitter tuning. The best dummy load is a high-power noninductive resistor submerged in an oil bath. The value of resistance used is normally 50 ohms, the same as the output impedance of most transmitters. Turn on the POWER and HEATER switches and allow a warm-up period of a minute or so for the tube filaments to reach operating temperature. The BAND switch must be placed in the proper position for the intended band of operation.

Set the meter switch to the GRID or RF OUTPUT position, turn the FUNCTION switch to TUNE and key the transmitter. With the DRIVE LEVEL advanced to the approximate mid-rotation position, adjust the DRIVE TUNE control for a maximum reading. Let up on the key and switch the meter to the CURRENT or PLATE position, and turn the FUNCTION switch to the OPERATE position. With the LOAD control set for maximum capacitance, again key the transmitter and quickly adjust the TUNE control for a minimum (dip) plate current. Adjust the final-amplifier LOAD control for a peak current reading, then re-dip the plate current with the TUNE control. Continue this procedure through several iterations, until the tube is drawing the specified amount of current. It is a good idea to check the grid current after this procedure, since it

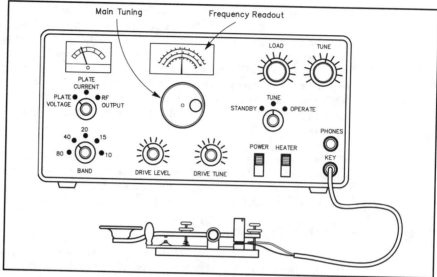

Figure 7-11 — Typical controls found on an amateur transceiver with a tube-type final amplifier.

has likely changed during tuning. It should be readjusted to the manufacturer's specification by means of the DRIVE TUNE control. Never run the transmitter outside the ratings specified by the manufacturer.

To summarize this tuning procedure, you should alternately increase the plate current with the loading capacitor and dip the plate current with the tuning capacitor. After several iterations you should reach a point where you can no longer increase the plate current as you adjust the loading capacitor. At this point the transmitter final amplifier should be properly tuned. This should coincide with the point of maximum power out of the transmitter.

[Before proceeding, turn to Chapter 10 and study examination questions A7A09 and A7A10. Also study question A7B01. Review this section as needed.]

IMPEDANCE MATCHING NETWORKS

When most hams talk about impedance matching networks, they probably think of a circuit used between a transmitter or transceiver and an antenna system. Such an *antenna-coupling circuit* has two basic purposes: (1) to match the output impedance of a power-amplifier tube or transistor to the input impedance of the antenna feed line, so the amplifier has a proper resistive load, and (2) to reduce unwanted emissions (mainly harmonics) to a very low value.

Most tube-type transmitters or amplifiers use **pi-network output-coupling cir-**

cuits (Figure 7-12). A pi-network consists of one inductor and two capacitors or two inductors and one capacitor. As used to match a transmitter output impedance with an antenna system impedance, a pi-network consists of one capacitor in parallel with the input and another capacitor in parallel with the output. An inductor is in series between the two capacitors. The circuit is called a pi network because it resembles the Greek letter pi (π) — if you use your imagination a bit

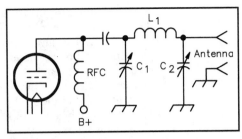

Figure 7-12 — A pi-network output-coupling circuit. C_2 is the coupling (loading) control and C_1 adjusts the tuning.

while you look at the two capacitors drawn down from the ends of the horizontally drawn inductor. Because of the series coil and parallel capacitors, this circuit acts as a low-pass filter to reduce harmonics, as well as acting as an impedance-matching device. (A pi network with two coils shunted to ground and a series capacitor would make a high-pass filter, and is virtually never used as an amateur output-coupling circuit.) The circuit Q will be equal to the plate load impedance divided by the reactance of C_1. Coupling is adjusted by varying C_2, which generally has a reactance somewhat less than the load resistance (usually 50 Ω). Circuit design information for pi networks appears in *The ARRL Handbook*.

Harmonic radiation can be reduced to any desired level by sufficient shielding of the transmitter, filtering of all external power and control leads, and inclusion of

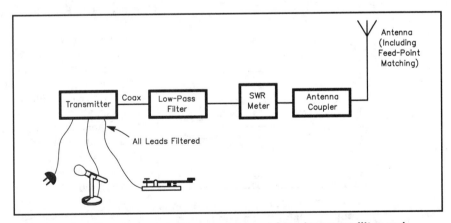

Figure 7-13 — Proper connecting arrangement for a low-pass filter and antenna coupler. The SWR meter is included for use as a tuning indicator for the antenna coupler.

Practical Circuits   7-23

a low-pass filter (of the proper cutoff frequency) connected with shielded cable to the transmitter antenna terminals (see Figure 7-13). Unfortunately, low-pass filters must be operated into a load of close to their design impedance or their filtering properties will be impaired, and damage may occur to the filter if high power is used. For this reason, if the filter load impedance is not within limits, a device must be used to transform the load impedance of the antenna system (as seen at the transmitter end of the feed line) into the proper value. For this discussion we will assume this impedance to be 50 Ω. This device is usually an external antenna-matching network.

Impedance-matching devices are variously referred to as Transmatches, matchboxes or antenna couplers. Whatever you call it, an impedance-matching unit transforms one impedance to be equivalent to another. To accomplish this it must be able to cancel reactances (provide an equal-magnitude reactance of the opposite type) and change the value of the resistive part of a complex impedance.

The **L network**, which consists of an inductor and a capacitor, (Figure 7-14A) will match any unbalanced load with a series resistance higher than 50 Ω. (At least it will if you have an unlimited choice of values for L and C.) Most unbalanced antennas will have an impedance that can be matched with an L network. To adjust this L network for a proper match, the coil tap is moved one turn at a time, each time adjusting C for lowest SWR. Eventually a combination should be found that will give an acceptable SWR value. If, however, no combination of L and C is available to perform the proper impedance transformation, the network may be reversed input-to-output by moving the capacitor to the transmitter side of the coil.

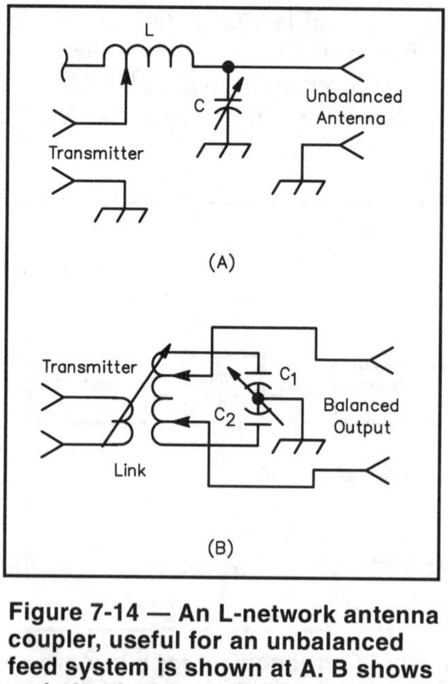

The major limitation of an L network is that a combination of inductor and capacitor is normally chosen to operate on only one frequency band because a given LC combination has a relatively small impedance-matching range. If the operating frequency varies too greatly, a different set of components will be needed.

If you are using balanced feed lines to your antenna, they may be tuned by means of the circuit shown in Figure 7-14B. A capacitor may be added in series with the input to tune out link inductance. As with the L network, the coil taps and tuning-capacitor settings

Figure 7-14 — An L-network antenna coupler, useful for an unbalanced feed system is shown at A. B shows an inductively coupled circuit for use with a balanced feed line.

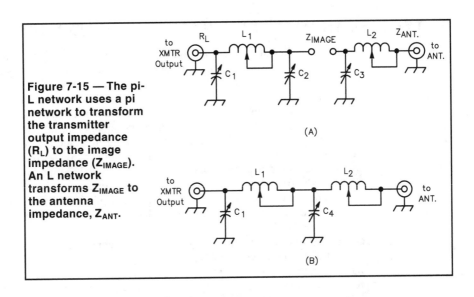

Figure 7-15 — The pi-L network uses a pi network to transform the transmitter output impedance (R_L) to the image impedance (Z_{IMAGE}). An L network transforms Z_{IMAGE} to the antenna impedance, Z_{ANT}.

are adjusted for lowest SWR, with higher impedance loads being tapped farther out from the coil center. For very low load impedances, it may be necessary to put C_1 and C_2 in series with the antenna leads (with the coil taps at the extreme ends of the coil).

You can convert an L network into a pi network by adding a variable capacitor to the transmitter side of the coil. Using this circuit, any value of load impedance (greater or less than 50 Ω) can be matched using some values of inductance and capacitance, so it provides a greater impedance-transformation range. Harmonic suppression with a pi network depends on the impedance-transformation ratio and the circuit Q.

If you need more attenuation of the harmonics from your transmitter, you can add an L network in series with a pi network, to build a pi-L network. Figure 7-15A shows a pi network and an L network connected in series. It is common to include the value of C_2 and C_3 in one variable capacitor, as shown at Figure 7-15B, so the pi-L network consists of two inductors and two capacitors. The pi-L network provides the greatest harmonic attenuation of the three most-used matching networks — the L, pi and pi-L networks.

A T-network configuration consists of two capacitors in series with the signal lead and a parallel, or shunt-connected inductor between them to ground. While this type of T-network will transform impedances, it also acts as a high-pass filter, and so it is virtually never used in an amateur antenna-matching network.

[Turn to Chapter 10 and study the questions A7C01 through A7C11. Review this section as needed.]

FILTERS: HIGH PASS, LOW PASS AND BAND PASS

The function of a filter is to transmit a desired band of frequencies without attenuation and to block all other frequencies. The resonant circuits discussed in previous sections all do this. The term *filter* is frequently reserved for those networks that transmit a desired band with little variation in output, and in which the transition from the "pass" band to the "stop" band is very sharp, rather than being a gradual change, as is the case with simple resonant circuits.

In this section you will learn about *passive filters* and *active filters*. Passive filters always result in some loss of signal strength at the desired signal frequencies. This is called *insertion loss*. Active filters include an amplifying device, to overcome the filter insertion loss, and sometimes even provide signal gain.

The passive filters discussed in this section use inductors and capacitors to set the filter frequency response. There are other types of passive filters, however. For example, *mechanical filters* used to be very popular as receiver IF filters, although they are seldom used in modern receivers. A mechanical filter uses mechanically resonant disks at the design frequency, and a pair of electromagnetic transducers to change the electrical signal into a mechanical wave and back again. *Cavity filters* are used to build a duplexer for a 2-meter repeater. The cavity filters in a duplexer isolate the transmitter and receiver so they can both use a single antenna. We won't go into the construction or operational details of mechanical filters or cavity filters, but you should be aware of their existence.

Filter Classification

Filters are classified into two general groups. A **low-pass filter** is one in which all frequencies below a specified frequency (called the cutoff frequency) are passed without attenuation. Above the cutoff frequency, the attenuation changes with frequency in a way that is determined by the network design. Usually you want this transition region to be as sharp as possible. A **high-pass filter** is just the opposite; there is no attenuation above the cutoff frequency, but attenuation does occur below that point.

High- and low-pass filters can be combined to make a third filter type, the **band-pass filter**. With these filters there are two cutoff frequencies, an upper and a lower one. If a pair of filters, one high-pass and one low-pass type, are tuned to have overlapping passbands, frequencies on both sides of the passband are attenuated. This makes a simple band-pass circuit. If the cutoff frequencies of the high- and low-pass filters are brought close together, but not overlapping, you have a *notch filter*. In this case, signals on either side of a certain band are passed, while signals in the middle are attenuated. Figure 7-16 illustrates idealized response curves for the four types of filters.

Pi- and T-Network Filter Sections

A filter section can be either a π- or T-type network in which the series and shunt reactances are of opposite types, as shown in Figure 7-17. Unlike the pi

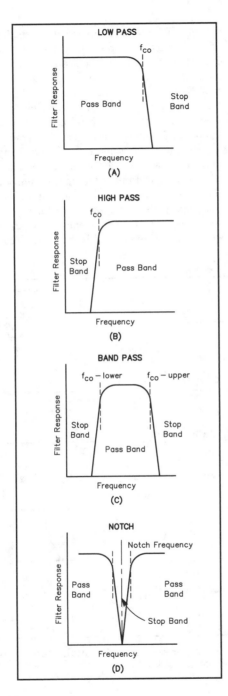

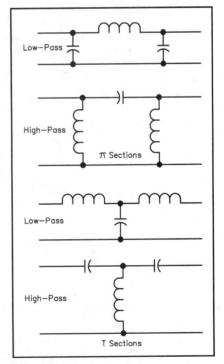

← **Figure 7-16 — Ideal filter-response curves for low-pass, high-pass, band-pass and notch filters.**

Figure 7-17 — High- and low-pass pi and T filter networks are shown.

networks described in the section on antenna couplers, these sections are not designed for transforming a given value of load resistance into a different value that will be suitable for the source of power. Instead, the design is such that the output impedances will be reflected to the input, for frequencies in the passband.

For this to occur, the values of L and C in the section must have a characteristic, or *image*, impedance (in the passband) that equals the load resistance. Then power is transferred from the source to the load without attenuation. Another consequence of having the input and output impedances the same is that sections may be cascaded, or strung together one after another, with no change in the input impedance shown to the source of power, at least at the desired pass frequencies.

The advantage to cascading filter sections is that in the stopband — the range of frequencies in which attenuation occurs — the attenuation for signals increases with the number of filter sections, while remaining zero for signals in the passband. If the attenuation of a section at a given frequency is expressed in decibels, each similar section that is added to a filter also adds the same number of decibels

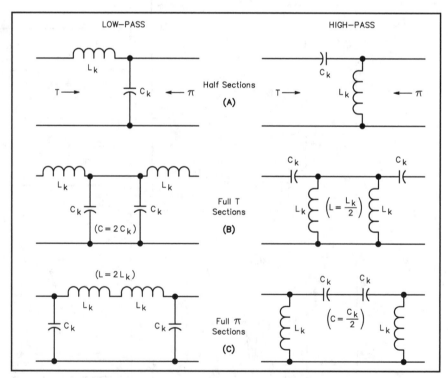

Figure 7-18 — Diagram showing how filter half sections can be combined to form full pi and T filter sections.

to the overall attenuation at the same frequency.

The basis for filter design is the **half section**, shown in Figure 7-18A. The inductance and capacitance, usually designated L_k and C_k, in the half section are related to a desired value of resistance, R, which is the termination or load impedance for the filter. Looking into the half section from the series end shows the beginning of a T section, and looking into it from the opposite side, the shunt end, shows the beginning of a pi section. The full sections are formed by connecting two half sections together as shown in Figure 7-18, parts B and C. When two like reactances are in parallel, as in the shunt arm of the full T sections, they are usually combined into a single physical element. Similarly, where two are in series they may be combined into a single element.

Pay special attention to the way the elements are arranged in these sections. A low-pass filter uses an inductor in series with the signal and a capacitor shunted to ground. The inductor has little reactance at low frequencies, and increasing reactance at higher frequencies, so it will tend to block the high-frequency signals more. The capacitor will shunt any high-frequency energy to ground, but will look like a high impedance to any low-frequency energy. Likewise, a high-pass filter uses a capacitor in series with the signal to pass high frequencies and block low frequencies. An inductor shunted to ground will take any remaining low-frequency energy to ground, but will act as a high impedance to high-frequency signals. You should be able to determine if a filter section is a high-pass or a low-pass one by looking at the element arrangement.

Image Impedance

Filter components may be selected to build a filter of specified design by using the *image-parameter technique*. This is a filter-design method that uses image impedance and other fundamental network functions to approximate the desired characteristics. This method requires values to be calculated using rather complicated equations, which you do not need to know for your exam. The technique is based on the fact that the image impedance (the load impedance reflected back to the input) of a given filter section is not constant at all frequencies in the passband, nor are the image impedances alike for pi and T sections using the same L_k and C_k. The terminating impedance, R, is therefore a "nominal" value.

Because the filter input impedance has some variation, there will be an impedance mismatch at the input, and this results in a signal loss at some frequencies. This mismatch loss is entirely between the source and the filter input. There is no loss of power in the filter itself. Since the filter contains only pure reactances, any power that enters it is either delivered to the terminating resistance R or returned to the source. If there is an adjustable impedance-matching network between the power source and the filter, the mismatch loss can be overcome. This is usually the case when the filter follows a transmitting power amplifier.

Attenuation in the Stopband

In the stopband, the impedances are deliberately mismatched to prevent transfer of power to the load. In this band, with the type of filter sections considered so far, the attenuation increases progressively as the applied frequency is moved away from the cutoff frequency.

Note that the transition from the passband to the stopband is rather abrupt. With an actual filter section there will be variations in attenuation in the stopband near the cutoff frequency, and the sharp transition will be smoothed off. At frequencies somewhat removed from the cutoff point, the real curve becomes free from these variations and approaches the theoretical curve. Beyond about twice the cutoff frequency in the low-pass case, and about half the cutoff frequency in the high-pass case, the attenuation increases uniformly at 12 dB per octave. (An octave is equal to a frequency ratio of 2 to 1.)

Modern Network Design

Image-parameter filter designs are inexact approximations for actual filter frequency responses. So aside from the limitations of impedance variation and steepness of the cutoff-response curve, the calculations will give results that have limited accuracy. These designs have been largely surpassed by newer techniques that are based on exact mathematical equations that can be applied to filter characteristics. Some examples are filters based on equations called Butterworth and Chebyshev polynomials and elliptical functions. With these mathematical techniques, it is possible to build a catalog of filter characteristics, with appropriate component values, and to select a design from the tabulated data. Tables summarizing these computations can be found in *The ARRL Handbook,* and other reference books.

There are many kinds of so-called "modern filters," and they are usually referred to by the name of the mathematical function used to calculate the design. **Butterworth**, **Chebyshev** and **elliptical filters** are three kinds that have many applications in Amateur Radio. A Butterworth filter is used when you want a response that is as flat as possible in the passband, with no ripple. (Ripple is a variation of attenuation, and you could get these "ups and downs" inside the passband and/or outside the passband.) Unfortunately, the transition from passband to stopband is not very sharp with a Butterworth filter. The Chebyshev filter has a sharp cutoff, but with some ripple in the passband. Higher SWR increases the ripple. The elliptical filter has the sharpest cutoff, but it has ripple in the passband and stopband. The elliptical filter also has one or more infinitely deep notches in the stopband, which can be positioned at specific frequencies that you want to attenuate. Figure 7-19 illustrates the filter-response curves for these three types.

Crystal Filters

The filters we have been talking about so far all use electronic circuits to provide the filtering action. The IF section of a radio requires very good band-pass filters to provide the narrow bandwidth needed for a top-performance rig. Filters that do not use inductors and capacitors as the primary circuit elements are often used in the IF section of a radio. In Chapter 6, we described crystal filters and their operation.

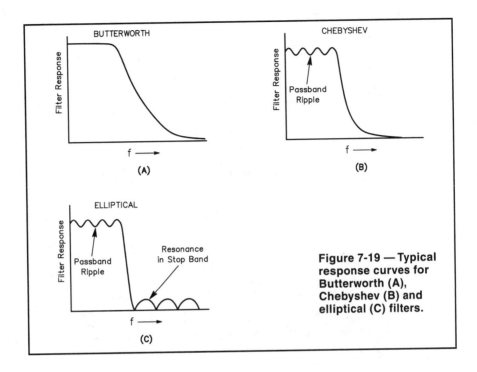

Figure 7-19 — Typical response curves for Butterworth (A), Chebyshev (B) and elliptical (C) filters.

Filters of this type, with piezoelectric quartz crystals to provide high-Q, narrow-bandwidth characteristics are often used in modern receivers and transmitters.

Active Filters

Operational amplifiers (op amps) are often used to build *active filters* for the audio range because the gain and frequency response of the filter are controlled by a few resistors (R) and capacitors (C) connected to the op amp. Such filters are often called RC active audio filters. Figure 7-20 is a simple RC active band-pass filter suitable for CW use.

You can also build an RC active notch filter using an op amp. A notch filter is used to reject a narrow band of frequencies. This type of filter is particularly useful to attenuate an interfering carrier signal while receiving an SSB transmission.

Active filters have a number of advantages over passive (LC) filters at audio frequencies. They provide gain and good frequency-selection characteristics. They do not require the use of inductors, and they can be accurately tuned to a specific design frequency using a potentiometer.

There are a few disadvantages to using active filters as well. The useful upper

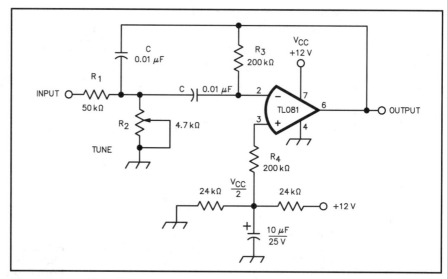

Figure 7-20 — This RC active band-pass audio filter has a center frequency of 800 Hz, and would be suitable for receiving CW signals. R₂ is used to adjust the filter operating frequency. This adjustment compensates for the tolerance variations of the other resistors and capacitors in the circuit.

frequency is limited to a few hundred kilohertz with low-cost op amps. The output voltage swing must be less than the dc supply voltage. Strong out-of-band input signals may overload the active device and distort the output signal. The op amp may generate some noise and add that to the signals, resulting in a lower signal-to-noise ratio than you would have with an LC filter.

Digital Signal Processing (DSP) Filters

Digital signal processing (DSP) is one of the great technological innovations of recent time in electronics. The basic idea of DSP is to represent a signal waveform by a series of numbers, perform some manipulation on those numbers (usually with a computer) and then convert the new series of numbers to a modified signal waveform.

We won't go into detail here about how DSP works. If you want to learn more about this fascinating new technology, start with the Digital Signal Processing chapter of the 1995 or later edition of *The ARRL Handbook*. From there you may want to check out some of the other books and articles referenced in that chapter.

There are numerous advantages to processing signals digitally. Since the processing takes place in a computer, the "circuit" never needs tuning because the computer program doesn't change characteristics with age or temperature. Since the signal processing is controlled by the computer, it is simply a matter of changing the program to process the signal in a new way. In fact, this leads to some interesting applications, because the computer can be programmed to respond differently to different types of signals or conditions. This is called *adaptive processing*.

An adaptive DSP filter can be useful for removing unwanted noise from a received SSB signal, for example. A notch filter might automatically identify an interfering heterodyne, lock onto that signal and remove it from the received audio. Such a filter can even track the interfering signal as it moves through the receiver passband!

Digital communications modes require a filter response that is the same to all signals in the passband. This means the filter must have a linear phase response. Typical SSB IF filters introduce phase distortion to signals because the inductive and capacitive reactance of the elements changes with frequency. A DSP finite impulse response (FIR) filter provides a constant delay to signals regardless of their frequency, so this type of filter has a linear phase response.

A Hilbert-transform filter introduces a 90° phase shift to all frequency components of a signal. This phase shift can be used to generate a single sideband (SSB) signal by the phasing method. (You will learn more about the phasing method of generating SSB signals in Chapter 8.)

[Before proceeding to the next section, turn to Chapter 10 and study examination questions A7D01 and questions A7D07 through A7D11. In addition, study questions A7H06 through A7H11. Review this section as needed.]

ELECTRONIC VOLTAGE-REGULATION PRINCIPLES

Almost every electronic device requires some type of power supply. The power supply must provide the required voltages when the device is operating and drawing a certain current. The output voltage of most power supplies varies inversely with the load current. If the device starts to draw more current, the applied voltage will be pulled down. In addition, the operation of most circuits will change as the power-supply voltage changes. Modern solid-state devices are more sensitive to slight voltage changes than many tube circuits are. For this reason, there is a voltage-regulator circuit included in the power supply of almost every electronic device that uses transistors or integrated circuits. The purpose of this circuit is to stabilize the power-supply output voltage and/or current under changing load conditions.

Linear electronic voltage regulators make up one major category of regulator. With these, the regulation is accomplished by varying the conduction of a control element in direct proportion to the line voltage or load current. The control element can be a fixed dropping resistance through which the current changes as input voltage or load currents change, or it can be the resistance of the dropping element that changes.

Zener-diode regulator circuits and gaseous-regulator-tube circuits control the current through a fixed dropping resistance. Electronic regulators use a tube or transistor as the voltage-dropping rather than a resistor. By varying the dc voltage at the grid or the current at the base of these elements, the conductivity of the device may be varied as necessary to hold the output voltage constant. In solid-state regulators, the series dropping element is called a pass transistor. Power transistors are available which will handle several amperes of current at several hundred volts, but solid-state regulators of this type are usually operated at potentials below 100 V.

The second major regulator category is the **switching regulator**, where the dc source voltage is switched on and off electronically, with the duty cycle proportional to the line or load conditions. The average dc voltage available from the regulator is proportional to the duty cycle of the switching waveform, or the ratio of the on time to the total switching-cycle period. Switching frequencies of several kilohertz are normally used, to avoid the need for extensive filtering to smooth the switching frequency from the dc output. We won't go into the operating details of switching regulators in this manual, but you should at least know they exist.

Discrete-Component Regulators

Zener-Diode Shunt Regulators

A **Zener diode** can be used to stabilize a voltage source, as shown in Figure 7-21. Note that the cathode side of the diode is connected to the positive supply side. This places a reverse bias voltage across the diode. A Zener diode limits the voltage drop across its junction when a specified current passes through it in the reverse-breakdown direction. The diode is connected in parallel with, or shunted across, the load. This type of linear voltage regulator is often referred to as a shunt regulator.

Zener diodes are available in a wide variety of voltage and power ratings. Voltage ratings range from less than two to a few hundred volts. Power ratings specify the power the diode can dissipate, and run from less than 0.25 W to 50 W. The ability of the Zener diode to stabilize a voltage depends on the diode conducting impedance, which can be as low as one ohm or less in a low-voltage, high-power diode, to as high as a thousand ohms in a low-power, high-voltage diode.

Zener diodes of a particular voltage rating have varied maximum current capabilities, depending on the diode power ratings. The Ohm's Law relationships you

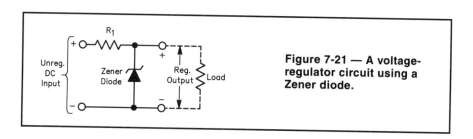

Figure 7-21 — A voltage-regulator circuit using a Zener diode.

are familiar with can be used to calculate power dissipation, current rating and conducting impedance of a Zener diode.

$$P = I \times E \qquad \qquad \text{(Equation 7-8)}$$

$$I = \frac{P}{E} \qquad \qquad \text{(Equation 7-9)}$$

and

$$Z = \frac{E}{I} \qquad \qquad \text{(Equation 7-10)}$$

where:
   P = diode maximum-safe-power-dissipation rating.
   E = Zener voltage.
   I = maximum current that can safely flow through the diode.
   Z = diode conducting impedance.

The power-handling capability of most Zener diodes is rated at 25°C; approximately room temperature. If the diode is operated at a higher ambient temperature, its power-handling capability must be derated. A typical 1-W diode can safely dissipate only $\frac{1}{2}$ W at 100°C. The breakdown voltage of Zener diodes also varies with temperature. Those rated for operation at 5 or 6 V have the smallest variation with temperature changes, and so are most often used as a voltage reference where temperature stability is a consideration.

Obtaining Other Voltages

Figure 7-22 shows how two Zener diodes may be connected in series to obtain regulated voltages that you could not achieve otherwise. This is especially useful when you want to use a 6-V Zener diode for maximum temperature stability, but you need a reference voltage other than 6 V. Two 6-V diodes in series, for example, will provide a 12-V reference. Another advantage with a circuit of this type is that you have two regulated output voltages. The diodes need not have equal breakdown voltages. You must pay attention to the current-handling capability of each diode, however.

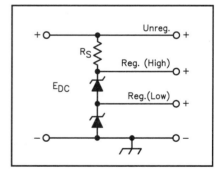

Figure 7-22 — Zener diodes can be connected in series to obtain multiple output voltages from one regulator circuit.

Series Regulators

The previous section outlines some of the limitations when Zener diodes are used as regulators. Greater currents can be accommodated if the Zener diode is

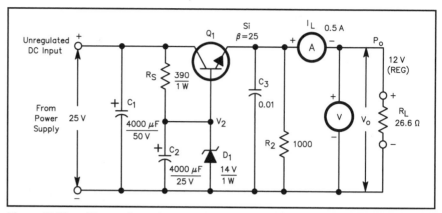

Figure 7-23 — Illustration of a power-supply regular including a pass transistor to provide more current than is available from a circuit using only a Zener diode.

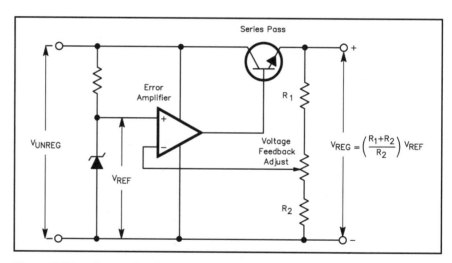

$$V_{REG} = \left(\frac{R_1 + R_2}{R_2}\right) V_{REF}$$

Figure 7-24 — A sample of the output voltage can be fed back to an error amplifier to obtain better regulation.

used as a voltage reference at low current, permitting the bulk of the load current to flow through a series pass transistor (Q_1 of Figure 7-23). Q_1, then, increases the current-handling capability of this linear voltage regulator circuit. An added benefit in using a pass transistor is that ripple on the output waveform is reduced. This technique is commonly referred to as "electronic filtering."

The pass transistor serves as a simple emitter-follower dc amplifier. It increases the load resistance seen by the Zener diode by a factor of beta (β). In this circuit arrangement, the Zener diode, D_1, provides a voltage reference. It is only required to supply the base current for Q_1. The net result is that the load regulation and ripple characteristics are improved by a factor of beta. C_1 charges to the peak value of the input voltage ripple, smoothing, or filtering, the dc input supply voltage to the regulator. R_1 supplies operating current to D_1. The addition of C_2 bypasses any remaining hum or ripple around the reference element, D_1, reducing the output ripple even more. Many simple supplies do not make use of a capacitor in this part of the circuit. C_3 provides some final filtering and helps to stabilize the transistor circuit, preventing it from oscillating. R_2 provides a constant minimum load for the pass transistor, Q_1.

The greater the value of transformer secondary voltage, the higher the power dissipation in Q_1. This not only reduces the overall power-supply efficiency, but requires stringent heat sinking at Q_1 if the dissipation will be more than a small percentage of the transistor rating.

It is possible to obtain better regulation by adding a few components to monitor the output voltage from your regulated supply. This technique, illustrated in Figure 7-24, is called feedback regulation. You will notice that an error amplifier detects the difference between the reference voltage and a feedback voltage. This amplified signal is then applied to the base of the pass transistor. The basic circuit operation is the same as for Figure 7-23. The voltage-divider circuit, connected at the supply output, is adjusted to set a feedback voltage that matches the Zener-diode reference voltage. In this way, you can use any convenient reference diode. It does not have to be the same as the full regulated output voltage.

If the load is connected at the end of a long supply cable, there may be a significant voltage drop when a large current is drawn. This would be the case where the supply is being used to power a solid-state 100-W HF transceiver, for example. By monitoring the voltage at the load rather than at the regulated-supply output, the regulating element can respond to the voltage drop in the cable. This technique is called remote-sensed feedback regulation. The feedback connection to the error amplifier is made directly to the load to provide remote sensing.

Discrete-Component-Regulator Limitations

Shunt-regulator circuits are inherently inefficient because the regulating element draws maximum current when the load is drawing none. (The Zener diode regulators shown in Figures 7-21 and 7-22 are examples of shunt-regulator circuits.) This type of circuit is most useful if the load current will be fairly constant, or if you want to maintain a nearly constant load on the unregulated supply. The current drawn by the shunt element represents wasted power, however.

A series-regulator circuit (such as the one shown in Figure 7-23) will make more efficient use of the unregulated supply, because it draws a minimum current when the load current is zero. The primary limitation of using a series pass transistor is that it can be destroyed almost immediately if a severe overload occurs at R_L. A fuse cannot blow fast enough to protect Q_1 in case of an accidental short circuit

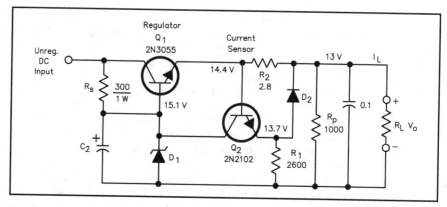

Figure 7-25 — Overload protection for a regulated supply can be effected by addition of a current-sense transistor to turn off the regulator in case of a short circuit on the output.

at the output, so a current-limiting circuit is required. An example of a suitable circuit is shown in Figure 7-25.

All of the load current is routed through R_2. There will be a voltage difference across R_2 that depends on the exact load current at a given instant. When the load current exceeds a predetermined safe value, the voltage drop across R_2 will forward bias Q_2, causing it to conduct. If you select a silicon transistor for Q_2 and a silicon diode for D_2, the combined voltage drops through them (roughly 0.6 V each) will be 1.2 V. Therefore, the voltage drop across R_2 must exceed 1.2 V before Q_2 can turn on. Choose a value for R_2 that provides a drop of 1.2 V when the maximum safe load current is drawn. In this example, there will be a 1.2-V drop across R_2 when I_L reaches 0.43 A.

When Q_2 turns on, some of the current through R_S flows through Q_2, thereby depriving Q_1 of some of its base current. Depending on the amount of Q_1 base current at a precise moment, this action cuts off Q_1 conduction to some degree, thus limiting the current through it.

If the collector-emitter junction of the pass transistor becomes shorted, the full unregulated rectifier voltage can be applied to the load. This can lead to a disaster, especially if the load is an expensive piece of electronic equipment! So you see, the load can draw too much current and damage the supply, or the supply can provide too high a voltage, and damage the load. The current-sense transistor protects the supply, and a "crowbar circuit" is often used to protect the load.

Figure 7-26 shows a simple crowbar circuit. A silicon-controlled rectifier (SCR) is selected that will turn on when a set voltage is applied to its gate terminal. Zener diode D_2, and resistors R_1 and R_2, are used to sense a voltage for the SCR. When the set-point voltage is exceeded, the SCR turns on, creating a short circuit across

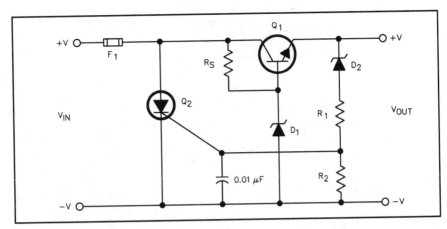

Figure 7-26 — An SCR provides a "crowbar" feature to short the unregulated supply and blow the fuse if the pass transistor fails.

the power-supply output terminals. This sustained short circuit will blow a fuse, turning the supply off. The key is that the SCR "fires" quickly, shorting the output voltage before it can damage the load equipment.

Just as it is possible to add several Zener diodes in series to increase the regulated output voltage, you can connect pass transistors in parallel to increase the current handling capability of the regulator. You can include current limiting, overvoltage protection, remote-sensed feedback regulation and other features in a single supply.

[We've covered quite a bit of information about voltage-regulator circuits, so this is a good time to take a break from learning new material. Turn to Chapter 10 and study questions A7B13 through A7B20. Also study questions A7E01 through A7E07 and question A7E10. Review this section about discrete-component regulators if you have difficulty with any of these questions.]

IC Regulators

The modern trend in voltage regulators is toward the use of integrated-circuit devices known as *three-terminal regulators*. Inside these tiny packages is a voltage reference, a high-gain error amplifier, current-sense resistors and transistors, and a series pass element. Some of the more sophisticated units have thermal shutdown, overvoltage protection and foldback current limiting.

Three-terminal regulators have a connection for unregulated dc input, one for regulated dc output and one for ground. They are available in a wide range of voltage and current ratings. It is easy to see why regulators of this sort are so popular when you consider the low price and the number of individual components they

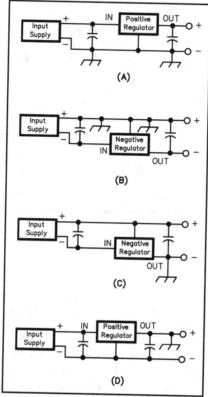

Figure 7-27 — Parts A and B illustrate the conventional manner of connecting three-terminal regulators. Parts C and D show how one regulator polarity can be used to provide an output voltage of the opposite polarity.

can replace. The regulators are available in several different package styles. The package and mounting methods you choose will depend on the amount of current required from your supply. The larger metal TO-3 package, mounted on a heat sink, for example, will handle quite a bit more current than a plastic DIP IC.

Three-terminal regulators are available as positive or negative types. In most cases, a positive regulator is used to regulate a positive voltage and a negative regulator for a negative voltage (with respect to ground). Depending on the system ground requirements, however, each regulator type may be used to regulate the "opposite" voltage.

Parts A and B of Figure 7-27 illustrate the conventional method of connecting an IC regulator. Several regulators can be used with a common input supply to deliver a variety of voltages with a common ground. Negative regulators may be used in the same manner, to provide several negative voltages, or with positive regulators to provide supplies with both positive and negative polarities.

Parts C and D of Figure 7-27 show how a regulator can be connected to provide opposite-polarity voltages, as long as no other supplies operate from the unregulated input source. In these configurations the input supply is floated; neither side of the input is tied to the system ground.

When choosing a three-terminal regulator for a given application, there are several important specifications to look for. Be sure the maximum and minimum input voltage and the maximum output current and voltage ratings aren't exceeded. In addition, you should consider the line regulation, load regulation and power dissipation of the device you choose, to be sure the device will handle your circuit requirements.

In use, most of these regulators require an adequate heat sink, since they may be called on to dissipate a fair amount of power. Also, since the chip contains a high-gain error amplifier, bypass capacitors on the input and output leads are essential for stable operation. Most manufacturers recommend bypassing the input and output directly at the IC leads. Tantalum capacitors are usually recommended because of their excellent bypass capabilities up into the VHF range.

Adjustable-Voltage IC Regulators

Adjustable-voltage regulators are similar to the fixed-voltage devices just described. The main difference is the lead provided to connect the error-amplifier sense voltage to a voltage divider. See Figure 7-28. This allows you to select the portion of the output voltage that is fed back to the amplifier, thus providing an output voltage to suit your needs, at least within certain limits. The ratio of reference voltage to total output voltage will be the same as the ratio $R_1 / (R_2 + R_1)$. For example, the regulator shown in Figure 7-28 (the LM-317 is a 1.2 to 37-V adjustable regulator)

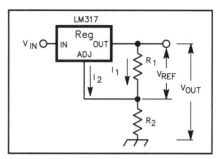

Figure 7-28 — By varying the ratio of R_2 to R_1 in this simple regulator circuit, a wide range of output voltages is possible.

requires a reference voltage of 1.2 V. To use this regulator to build a 12-V supply, the voltage ratio is 1.2 V / 12 V = 0.1. If we select $R_1 = 1$ kΩ, then:

$R_2 + R_1 = 1000 \ \Omega \ /0.1 = 10,000 \ \Omega$,

so

$R_2 = 10,000 \ \Omega - 1000 \ \Omega = 9000 \ \Omega$.

Select a 9.1-kW standard-value component for R_2.

IC regulators are readily available to provide an output voltage of from 5 to 24 V at up to 5 A. The same precautions should be taken with these types of regulators as with the fixed-voltage units. Proper heat sinking and lead bypassing is essential for proper circuit operation.

It is a rather simple task to design a regulated power supply around an IC regulator. Multiple voltages can be obtained by simply adding other ICs in parallel. The main consideration is to provide an input voltage within the range specified for the device you select, and to be sure your load will not draw more current than the IC can safely supply.

[Now turn to Chapter 10 and study the examination questions A7E08, A7E09 and A7E11. Review this section as needed.]

OSCILLATORS

When we discussed amplifiers earlier in this chapter, we described an amplifier-circuit problem that occurs if some of the output signal makes its way back to the input in a manner that creates positive feedback. This circuit instability turns the amplifier into an **oscillator**. When we want the circuit to behave as an amplifier, this is a problem. But sometimes we want a circuit that will generate a signal (often at radio frequencies) without any input signal, and in that case we can take advantage of the instability of an amplifier.

To start a circuit oscillating, we need to feed power from the plate back to the grid of a tube or from the collector to the base of a transistor. This is called feedback. Feedback is what happens when you are using a public address system and you get the microphone too close to the loudspeaker. When you speak, your voice is amplified in the public address system and comes out over the loudspeaker. If the sound that leaves the speaker enters the microphone and goes through the whole process again, the amplifier begins to squeal, and the sound keeps getting louder until the "circuit" is broken, even though you are not talking into the microphone anymore. Usually, the microphone has to be moved away from the speaker or the volume must be turned down. This is a good example of an amplifier becoming unstable through positive feedback, and breaking into oscillation.

When the output signal applied to the input reinforces the signal at the input, we call it positive feedback. Negative feedback opposes the regular input signal.

Positive feedback can be created in many ways — so many ways, in fact, that we can't possibly cover them all here. There are three major oscillator circuits used in Amateur Radio. They are shown in Figure 7-29. These oscillators can be built using vacuum tubes or FETs, and the feedback circuits will be much the same as the ones shown here for bipolar transistor circuits. The amount of feedback required to make the circuit oscillate is determined by the circuit losses. There must be at least as much energy fed back to the input as is lost in heating the components, or the oscillations will die out.

The Hartley oscillator uses inductive feedback. Alternating current flowing through the lower part of the tapped coil induces a voltage in the upper part, which is connected to the circuit input. A Hartley oscillator is the least stable of the three major oscillator types. (You might find it helpful to remember the H for henrys and Hartley as a mnemonic to remind you that this oscillator uses a tapped coil.)

The second general type is the Colpitts oscillator circuit. It uses capacitive feedback. (The Cs give a good mnemonic clue to remind you that the Colpitts oscillator gets the feedback energy from the capacitors.) The collector-circuit energy is fed back by introducing it across a capacitive voltage divider, which is part of the tuning-circuit capacitance. This coupling sets up an RF voltage across the whole circuit, and is consequently applied to the transistor base. The large values of capacitance used in the tuning circuit tend to stabilize the circuit, but the oscillation frequency still will not be as stable as a crystal oscillator.

The most stable oscillator circuit is the Pierce crystal oscillator. The Pierce circuit uses capacitive feedback, with the necessary capacitances supplied by C_1 and C_2. For a tube-type Pierce oscillator the feedback capacitance is supplied by the tube interelectrode capacitances. Besides its stability, another reason why the

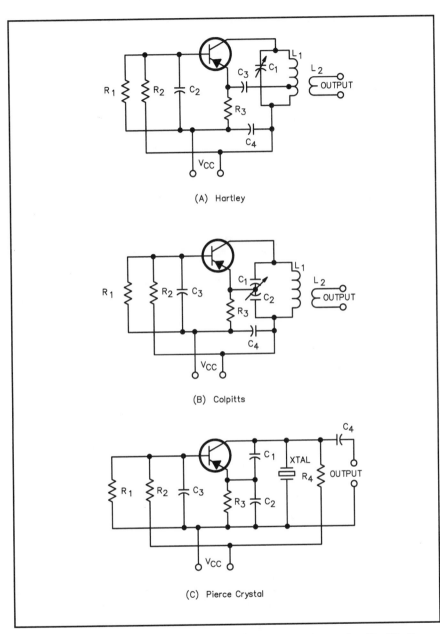

(A) Hartley

(B) Colpitts

(C) Pierce Crystal

Figure 7-29 — Three common types of transistor oscillator circuits. Similar circuits can be built using tubes or FETs.

Pierce circuit is popular is that you do not have to build and tune an LC tank circuit. Adjusting the tank circuit for the exact resonant frequency can be touchy, but it is a simple matter to plug a crystal into a circuit and know it will oscillate at the desired frequency!

Quartz Crystals

A number of crystalline substances can be found in nature. Some have the ability to change mechanical energy into an electrical potential and vice versa. This property is known as the *piezoelectric effect*. A small plate or bar properly cut from a quartz crystal and placed between two conducting electrodes will be mechanically strained when the electrodes are connected to a voltage source. The opposite can happen, too. If the crystal is squeezed, a voltage will develop between the electrodes.

Crystals are used in microphones and phonograph pickups, where mechanical vibrations are transformed into alternating voltages of corresponding frequency. They are also used in headphones to change electrical energy into mechanical vibration.

Crystalline plates have natural frequencies of vibration ranging from a few thousand hertz to tens of megahertz. The vibration frequency depends on the kind of crystal, and the dimensions of the plate. What makes the crystal resonator (vibrator) valuable is that it has an extremely high Q, ranging from a minimum of about 20,000 to as high as 1,000,000.

The mechanical properties of a crystal are very similar to the electrical properties of a tuned circuit. We therefore have an "equivalent circuit" for the crystal. The electrical coupling to the crystal is through the holder plates, which "sandwich" the crystal. These plates form, with the crystal as the dielectric, a small capacitor constructed of two plates with a dielectric between them. The crystal itself is equivalent to a series-resonant circuit and, together with the capacitance of the holder, forms the equivalent circuit shown in Figure 7-30.

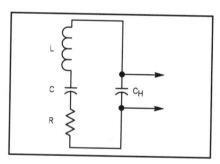

Figure 7-30 — The electrical equivalent circuit of a quartz crystal. L, C and R are the electrical equivalents of the crystal mechanical properties and C$_H$ is the capacitance of the holder plates, with the crystal serving as the dielectric.

Can we change the crystal in some way so that it will resonate at a different frequency? Sure. If we cut a new crystal, and make it longer or thicker, the resonant frequency will go down. On the other hand, if we want the crystal to vibrate at a higher frequency, we would make it thinner and shorter.

There are two major limitations to the use of crystals. First, we can't have more than two terminals to the circuit, since there are only two crystal electrodes. In other words, a crystal can't be tapped as we might tap a coil in a circuit. Second, the crystal is an open circuit for direct current, so you can't feed operating voltages through the crystal to the circuit.

The major advantage of a crystal used in an oscillator circuit is its frequency stability, especially with mechanical vibration. The spacing of coil turns can change with vibration, and the plates of a variable capacitor can move. On the other hand, the frequency of a crystal is much less apt to change when the equipment is bounced around (short of actually cracking the crystal, anyway).

One disadvantage of using crystals to control an oscillator frequency is that they can be easily affected by temperature changes. Manufacturers can cut a crystal at various angles across the plane of the quartz structure, and are thus able to control the temperature coefficient and other parameters, however. If better frequency stability is required, the crystal can also be placed in an "oven" to maintain a constant temperature over varying external environmental conditions.

Variable Frequency Oscillators

The major advantage of a variable-frequency oscillator (VFO) is that a single oscillator can be tuned over a frequency range. Since the frequency is determined by the coil and capacitors in the tuned circuit, the frequency is often not as stable as in a crystal-controlled circuit. The Colpitts oscillator shown in Figure 7-31 is one of the more common types of VFO because it is a stable oscillator. Hartley oscillators are also often used in VFO circuits. The FET in the diagram is connected like a source follower (similar to the bipolar-transistor emitter follower) with the input (gate) connected to the top of the tuned circuit and the output (source) tapped down on the tuned circuit with a capacitive divider. Although the voltage gain of a source follower can never be greater than one, the voltage step-up that results from tapping down on the tuned circuit ensures that the signal at the gate will be large enough to sustain oscillation. Although the FET voltage gain is less than one, the power gain must be great enough to overcome losses in the tank circuit and the load.

The amplitude and frequency stability of a variable-frequency Colpitts oscillator is quite good, although it doesn't approach that of a crystal-controlled oscillator. The large capacitances used in the tank circuit minimize

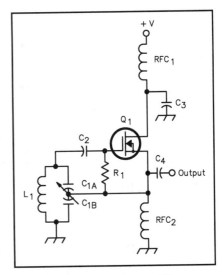

Figure 7-31 — A Colpitts VFO circuit.

frequency shifts that result from small capacitance variations caused by vibration or variations in tube or transistor characteristics.

Most oscillators provide an output rich in harmonic content. The relative amplitude of any given harmonic can be enhanced by choosing the optimum bias voltage. A Colpitts oscillator that has a second tank circuit tuned to the desired harmonic and connected to the output makes an excellent harmonic generator.

Variable capacitors can be large and relatively expensive. Mounting the capacitor so the control shaft reaches through the chassis front panel is sometimes inconvenient. In addition, providing a smoothly operating mechanism with fine tuning adjustments may require the use of an expensive vernier drive. There is an alternative to these problems, however. Varactor diodes have a junction capacitance that is controlled by the reverse bias voltage applied to the diode.

Figure 7-32 is a diagram of a Colpitts oscillator that uses a varactor diode to change the oscillator frequency. The variable resistor, R_2, adjusts the reverse bias voltage applied to the varactor diode. This resistor can be a multiturn potentiometer for fine tuning control. It can be mounted away from the actual VFO circuit, at a convenient location on the front panel. The varactor diode and variable resistor have the added advantage of small size, making this the tuning method of choice for small, battery-operated low power (QRP) equipment.

[Turn to Chapter 10 and study questions A7F01 through A7F10. Review this section as needed.]

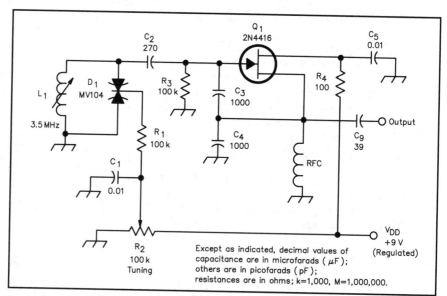

Figure 7-32 — This schematic diagram shows a Colpitts oscillator circuit with varactor diode tuning.

FREQUENCY SYNTHESIZERS

Frequency synthesizers serve much the same purpose as VFOs. That is, they are used to provide a stable, variable tuning range, generally to control the operating frequency of a radio. They are much more stable with changes in temperature and vibration than are Hartley or Colpitts VFOs. There are two major methods of frequency synthesis commonly used in modern commercially built HF radios: phase-locked loop (PLL) synthesis and direct digital synthesis (DDS).

Phase-locked loop synthesizers are the type most commonly used in modern equipment, although more and more radios are beginning to include some form of direct digital synthesizer. The phase-locked-loop (PLL) configuration uses some specialized integrated-circuit chips. Figure 7-33 shows the major blocks of a PLL synthesizer. The signal frequency from a voltage-controlled oscillator (VCO) is divided by some integer value in a programmable-divider IC. The divider output is compared with a stable reference frequency in a phase-detector circuit. The phase detector produces an output that indicates the phase difference between the VCO and the reference signal. This signal is then fed back to the VCO through a filter. The loop provides a feedback circuit that tends to adjust the phase-detector output to zero. That means the divider output is always on the same frequency as the reference oscillator. By changing the division factor, you can change the synthesizer output. Mathematically:

$$F = N f_r$$ (Equation 7-16)

where:
 F = synthesizer output frequency.
 N = the division factor.
 f_r = reference-oscillator frequency.

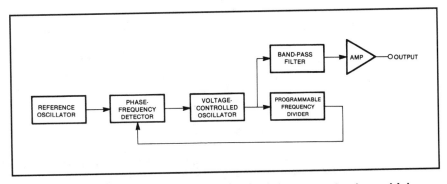

Figure 7-33 — Block diagram of an indirect frequency synthesizer, which uses a phase-locked loop and a variable-ratio divider.

A PLL synthesizer feedback loop is constantly "correcting" the output signal frequency. Any variations in the frequency of the reference oscillator will cause variations in the phase of the output signal, from one cycle to the next. These unwanted variations in the oscillator signal phase are called phase noise. This is a broadband noise around the desired output frequency, and is the major spectral impurity component of a phase-locked loop synthesizer. A very stable reference oscillator must be used for a phase-locked loop frequency synthesizer, to minimize this phase noise on the output signal.

Figure 7-34A is a block diagram of a direct digital synthesizer. This type of synthesizer is based on the concept that we can define a sine wave of any frequency by specifying a series of values (sine or cosine) taken at equal time intervals (or phase angles). The crystal oscillator sets the *sampling rate* for these values. The

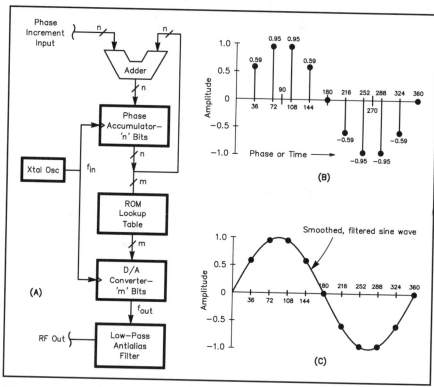

Figure 7-34 — Part A shows a direct digital synthesizer (DDS) block diagram. Part B shows the amplitude values found in the ROM lookup table for a particular sine wave being generated. Part C shows the smoothed output signal from the DDS, after it goes through the low-pass antialias filter.

phase increment input to the adder block sets the number of samples for one cycle. The oscillator *clock* signal tells the phase accumulator to sample the data input from the adder, and then increment the adder value by the phase increment. The phase accumulator value varies between 0 and 360. The ROM lookup table contains the amplitude values for the sine (or cosine) of each angle represented by the phase accumulator.

For example, suppose our synthesizer uses a 10-kHz crystal oscillator. This means there will be one sample every 0.1 ms. If the phase increment is set to 36°, there will be ten samples in each cycle: 0°, 36°, 72°, 108°, 144°, 180°, 216°, 252°, 288° and 324°. The next sample, at 360° starts the second cycle. The total time for these ten samples is 1 ms, which means the sine wave defined by these samples has a frequency of 1 kHz. Figure 7-34B shows a representation of the sine values found in the lookup table for these phase angles.

The sine values are fed to a digital to analog converter (DAC), and the analog output signal goes through the low-pass antialias filter. A smoothed sine-wave signal results. (See Figure 7-34C.)

We can change the frequency of the signal produced by our direct digital synthesizer by changing the value of the phase increment input to the adder. For example, if the new phase increment is 72° there will be five samples per cycle. Each cycle will take 0.5 ms, so the frequency of this new signal is 2 kHz.

DDS has the disadvantage of requiring a more sophisticated control circuit to set the proper phase increment to achieve a desired output frequency. A computer or microprocessor controller circuit is normally used. Phase noise is not a problem with DDS circuits, however. The major spectral impurity components produced by a direct digital synthesizer are spurious (unwanted) signals — *spurs* — at specific discrete frequencies.

Careful design can place those spurs outside of the amateur bands. Commercially built HF radios are beginning to use hybrid synthesizers, which combine some of the best features of PLL and DDS circuits.

[Study question A7F11 and questions A7I05 through A7I11 in Chapter 10. Review this section as needed.]

MIXERS, DETECTORS AND MODULATORS

Mixer circuits are used to change the frequency of a desired signal. In a receiver, this means converting all received signals to a single frequency (called the intermediate frequency or IF) so they can be processed more efficiently. In this way the amplifier chain can be adjusted for maximum efficiency and the best signal-handling characteristics without the need to retune many circuit elements every time you change bands or turn the radio dial. By converting all received signals to a single IF for processing, the selectivity of the receiver is improved greatly.

Mixers are also used to change the frequency of a signal as it progresses through a transmitter. In fact, a mixer circuit is even used in most transmitters to produce the modulated radio wave being sent out across the airwaves. If the circuit introduces a signal loss during processing, it is called a passive mixer. If some gain is provided (or at least no signal loss) because an amplifier is included, then the circuit is called an active mixer,

A **detector** circuit is used to "reclaim" the information that has been superimposed on a radio wave at a transmitter. That information may be the operator's voice, or it may be Morse code or RTTY signals. It could even be the information that will allow you to reproduce a slow-scan TV or facsimile picture that has been transmitted across the miles. There are a variety of detector methods described in this section, but they are all associated with receivers. Like mixer circuits, detectors can be classified as either passive or active.

A **modulator** is the circuit that combines an information signal with a radio-wave signal. The type of modulator circuit used at the transmitter will determine the type of signal that is transmitted, and the particular detection methods that can be used in a receiver for that signal. For example, different modulator circuits are used to produce AM phone, single-sideband (SSB) phone and FM phone signals.

The principles of operation are much the same for mixers, detectors and modulators, so when you become familiar with the operation of one type, you will find the others easier to understand.

Mixers

When two sine waves are combined in a nonlinear circuit, the result is a complex waveform that includes the two original signals, a sine wave that is the sum of the two frequencies and a sine wave that is the difference between the two. Also included are combinations of the harmonics from the input signals, although these are usually weak enough to be ignored. One of the product signals can be selected at the output by using a filter. Of course, the better the filter, the less of the unwanted products or the two input signals that will appear in the final output signal. By using balanced-mixer techniques, the mixer circuit provides isolation of the various ports so the signals at the radio frequency (RF), local oscillator (LO) and intermediate frequency (IF) ports will not appear at any other. This prevents the two input signals from reaching the output. In that case the filter needs to remove only the unwanted mixer products.

In a typical amateur receiving application, you want to mix the RF energy of a desired signal with the output from an oscillator in the receiver so you can produce a specific output frequency (IF) for further processing in the radio. The local oscillator (LO) frequency determines the frequency that the input signal is converted to. By using this frequency-conversion technique, the IF stages can be designed to operate over a relatively narrow frequency range, and filters can be designed to provide a high degree of selectivity in the IF stages.

The mixer stages in a high-performance receiver must be given careful consideration. These stages will have a great impact on the dynamic range of the receiver. The RF signal should be amplified only enough to overcome mixer noise.

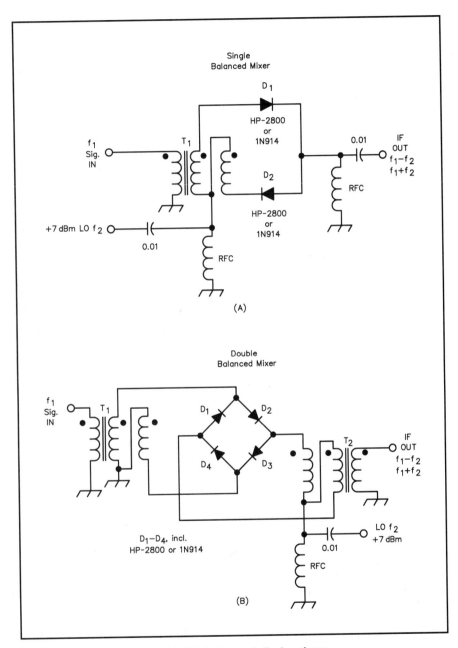

Figure 7-35 — Single and double balanced diode mixers.

Otherwise, strong signals will cause desensitization, cross-modulation and IMD products in the mixer. Spurious mixer products will be produced if an excessive amount of input-signal energy reaches the mixer circuit. The level of these spurious mixer products may be increased to the point that they appear in the output. One result of these effects is that the receiver may be useless in the presence of extremely strong signals. A mixer should be able to handle strong signals without being affected adversely.

Passive Mixers

One simple mixer that has good strong-signal characteristics is the diode mixer shown in Figure 7-35A. This circuit makes use of a trifilar-wound broadband toroidal transformer to balance the mixer, effectively canceling the RF and LO input signals, so only the sum and difference frequencies appear at the IF output terminal. Sometimes this circuit is referred to as a single balanced mixer. Figure 7-35B shows an improved circuit that includes two balance transformers and four diodes arranged in a circular, or ring, pattern. This circuit is usually called a **double balanced mixer (DBM)** or a diode-ring mixer. Note especially that the diodes are not connected as they would be in a bridge rectifier circuit.

The double balanced mixer is the most common. Commercial modules offer electrical balance at the ports that would not be easy to achieve with homemade transformers. They also use diodes whose characteristics have been carefully matched. Typical loss through a DBM is 6 to 9 dB. The port-to-port isolation is usually on the order of 40 dB.

Active Mixers

While passive mixers have good strong-signal-handling ability, they also have some drawbacks. They require a relatively strong LO signal, and they generate a fair amount of noise. Active mixers can be used to advantage if you require less conversion loss, weaker LO signals and less noise. Just be aware that the strong-signal-handling capabilities are not generally as good.

A JFET or dual-gate MOSFET can be used as a mixer, and will provide some gain as well as mixing the signals for you. Bipolar transistors could be used, but seldom are. Figure 7-36 shows an active mixer circuit. Many variations are possible, and this diagram just shows one arrangement.

Integrated-circuit mixers are available in both single and double balanced types. These devices provide at least several decibels of conversion gain, low noise and good port-to-port isolation.

[Before going on to the next section, turn to Chapter 10 and study questions A7I01 through A7I04. Review this section as needed.]

Detectors

The simplest type of **detector**, used in the very first radio receivers, is the diode detector. A complete, simple receiver is shown in Figure 7-37. This circuit would only work for strong AM signals, so it is not used very much today, except

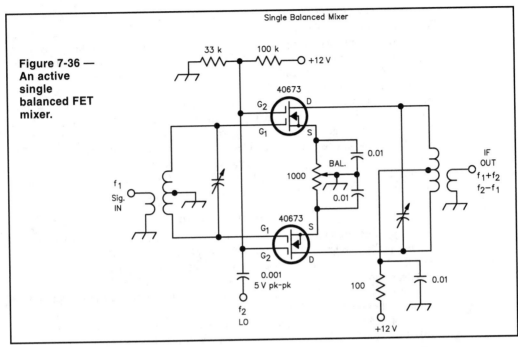

Figure 7-36 — An active single balanced FET mixer.

Single Balanced Mixer

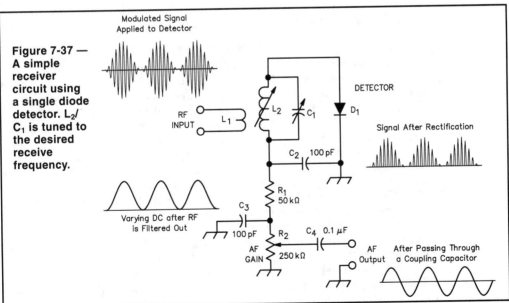

Figure 7-37 — A simple receiver circuit using a single diode detector. L_2/C_1 is tuned to the desired receive frequency.

for experimentation. It does serve as a good starting point to understand detector operation, however. The waveforms shown on the diagram illustrate the changes made to the signal as it progresses through the circuit. L_1 couples the received RF signal to the tuning circuit of L_2-C_1. The diode rectifies the RF waveform, passing only the positive half cycles. C_2 charges to the peak voltage of each pulse, producing a smoothed dc waveform, which then goes through the voltage divider R_1 and R_2. R_2 selects a portion of the signal voltage to be applied to the headphones, providing a volume-control feature. C_4 is a coupling capacitor that serves to remove the dc offset voltage, leaving an ac audio signal.

One drawback of the diode detector is that there is some signal loss in the circuit. An FET can be used as a detector, and the transistor will provide some amplification. The disadvantage of an active detector is that it may be overloaded by strong signals more easily than the passive diode detector.

Product Detectors

A **product detector** is similar to a balanced mixer. It is a detector whose output is equal to the product of a beat-frequency oscillator (BFO) and the RF signal. The BFO signal is like a locally generated RF carrier signal. The BFO frequency is chosen to provide detector output at audio frequencies. A double balanced diode-ring product detector looks very much like the double balanced diode-ring mixer shown in Figure 7-35B. FETs can be used for product detectors, and those circuits also look much the same as the mixer circuits. Special IC packages are also available for use as product detectors, and they will provide the advantages of diode detectors in addition to several decibels of gain. Product detectors are used for SSB, CW and RTTY reception.

Detecting FM Signals

There are three common ways to recover the audio information from a frequency-modulated signal. A **frequency-discriminator** circuit uses a transformer with a center-tapped secondary. The primary signal is introduced to the secondary-side center tap through a capacitor. With an unmodulated input signal, the secondary voltages on either side of the center tap will cancel. But when the signal frequency changes, there is a phase shift in the two output voltages. These two voltages are rectified by a pair of diodes, and the resulting signal varies at an audio rate. Figure 7-38 shows the schematic diagram of a simple frequency discriminator. Crystal discriminators use a quartz-crystal resonator instead of the LC tuned circuit in the frequency discriminator, which is often difficult to adjust properly.

A **ratio detector** can also be used to receive FM signals. The basic operation depends on the rectified output from a transformer similar to the one used in a frequency discriminator being split into two parts by a divider circuit. Then it is the ratio between these two voltages that is used to produce the audio signal.

The last method of receiving an FM signal can be used if you have an AM or SSB receiver capable of tuning the desired frequency range. If you tune the receiver slightly off the center frequency, so the varying signal frequency moves up

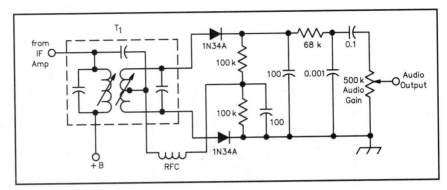

Figure 7-38 — A typical frequency-discriminator circuit used for FM detection. T₁ is a Miller 12-C45 disciminator transformer.

and down the slope of the selectivity curve, an AM signal will be produced. (This technique is also known as **slope detection**.) This AM signal then proceeds through the receiver in the normal fashion, and you can hear the audio in the speaker. Careful tuning will make the FM signal perfectly understandable, although you might have to keep one hand on the tuning knob, and there may be some noise.

[Turn to Chapter 10 now and study questions A7H01 through A7H05. Review this section if you are uncertain of the answers to any of these questions.]

Modulator Circuits

Modulation is really a mixing process whereby information is imposed upon a carrier; any mixer or converter circuit could be used for generating a modulated signal. Instead of introducing two radio frequencies into a mixer circuit, we simply introduce one radio frequency (the carrier frequency) and the voice-band audio frequencies.

Mixer circuits used in receivers are designed to handle a small signal and a large local-oscillator voltage. This means that the percentage of modulation of the IF signal is low. In a transmitter we want to get as close as possible to 100% modulation, and we also want more power output. For these reasons **modulator** circuits differ in detail from receiving mixers, although they are much the same in principle.

Amplitude Modulation

Although double-sideband, full-carrier, amplitude modulation (DSB AM) is seldom used on the amateur bands anymore, we will discuss a simple system just to help you understand how a single-sideband, suppressed-carrier (SSB) signal is generated. When RF and AF signals are combined in a DSB AM transmitter, four principle output signals are generated. First, there is the original audio signal, which is easily rejected by the tuned RF circuits in the stages following the modulator. Then

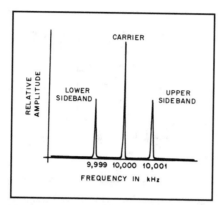

Figure 7-39 — A 10-MHz carrier, modulated by a 1-kHz sine wave, produces an output as shown.

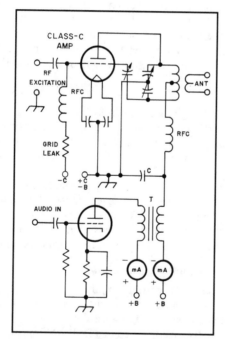

Figure 7-40 — Plate modulation of a Class-C RF amplifier.

there is the RF carrier, also unchanged from its original form. Of primary importance are the other two signals: one being the sum of the carrier frequency and the audio, the other being the difference between the original signals. These two new signals are called sideband signals. The amplitude of these signals at any given instant depends on the amplitude of the original audio signal at that instant. The greater the audio-signal strength, the greater the amplitude of the sideband signals.

The sum component is called the upper sideband. As the audio-signal frequency increases, so does the frequency of the upper-sideband signal. The difference component is called the lower sideband. This sideband is inverted, which means that as the audio-signal frequency increases, the sideband frequency decreases. Figure 7-39 illustrates this principle.

The result of all this is that the RF envelope, as viewed on an oscilloscope, has the general shape of the modulating waveform. The envelope varies in amplitude because it is the vector sum of the carrier and the sidebands. Note that the *carrier* amplitude is not changed in amplitude modulation. This is a common misunderstanding; one that often leads to confusion. It is the RF *envelope* that varies in amplitude.

You can produce amplitude modulation by applying the AF signal to the plate or collector of an RF amplifier stage. Figure 7-40 shows how the audio signal is used to modulate the plate voltage of a Class-C amplifier. You can also modulate the control element of the amplifier (grid, base or gate). A wide variety of modulator circuits have been used over the years.

SSB: The Filter Method

One way to generate an SSB signal is to remove the carrier and one of the sidebands from an ordinary DSB AM signal. The block diagram shown in Figure 7-41 shows how this is done. The RF oscillator generates a carrier wave that is injected into a **balanced modulator**. With this system, the oscillator frequency is changed, depending on which sideband you want to use. Another system main-

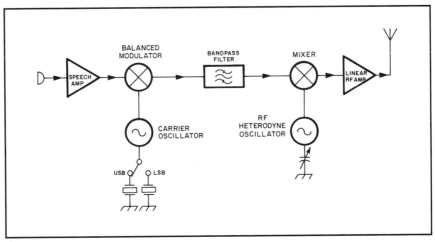

Figure 7-41 — A block diagram showing the filter method of generating an SSB signal.

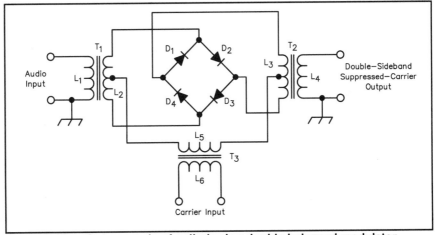

Figure 7-42 — One example of a diode-ring double balanced modulator.

tains the same oscillator frequency, but switches filters to remove the opposite sideband. The audio information is amplified by the speech amplifier and is then applied to the modulator. A balanced modulator takes these two inputs and supplies, as its output, both sidebands without the carrier. This meets the first requirement for the generation of an SSB signal — removal of the carrier.

Let's see how the balanced modulator accomplished this. There are many different types of balanced modulators and it would be impossible to show them all here. One of the more popular types is illustrated in Figure 7-42. This particular circuit is called a diode-ring balanced modulator. If you get the feeling that you have seen all this before, don't worry. The diode-ring balanced modulator is very similar to the double balanced diode-ring mixer. Actually, you could redraw Figure 7-42 to look almost identical to Figure 7-35B.

Audio information is coupled into the circuit through transformer T_1. The carrier is injected through coils L_5 and L_6 and the double-sideband, suppressed-carrier output is taken through L_3 and L_4. To better understand the circuit operation, let's first analyze the circuit with only the carrier applied. See Figure 7-43A. The polarity of voltage shown across L_5 will cause electrons to flow in the direction indicated by the arrows. D_1 and D_3 will conduct. The current that flows through each half of L_3 is equal and opposite, causing a canceling effect. Output at L_4 will be zero. During the next half cycle, the polarity of voltage across L_5 will reverse, D_2 and D_4 will conduct, and again the output at L_4 will be zero.

Now for a moment, let's remove the carrier signal and connect an audio source to the audio-input terminals. Figure 7-43B illustrates this condition. During one half of the audio cycle, D_2 and D_3 will conduct and the output will be zero. On the other half cycle of the audio signal, D_1 and D_4 will conduct and again the output at L_4 will be zero. Notice that in this case, there is no signal through L_3 at all. We can see that if either the carrier or the audio is applied without the other, there will be no output.

Both signals must be applied if the circuit is to work as intended. In practice, the carrier level is made much larger than the audio input. Diode conduction is, therefore, determined by the carrier. There are four conditions that can exist as far as the voltage polarity of the carrier and audio signals are concerned. Both waves can be positive, both can be negative or they can be opposites. We must keep in mind that the carrier is going through many cycles while the audio sine wave goes through only one.

Let's look again at the drawing at Figure 7-43C. With the polarities indicated, the major electron flow will be through D_3 since the audio and carrier voltages are aiding (adding together) in this path. Since the balance through L_3 has been upset (the bottom half of L_3 has more current flowing through it than does the top half), there will be an output present at L_4. Next, the carrier polarity will reverse while the audio polarity remains the same. This is the order in which the polarities would change, since the carrier is reversing polarity at a much faster rate than the audio signal. Later, the carrier polarity will be back to what it was at C; however, now the audio will be on the negative portion of its sine wave and so its polarity will be

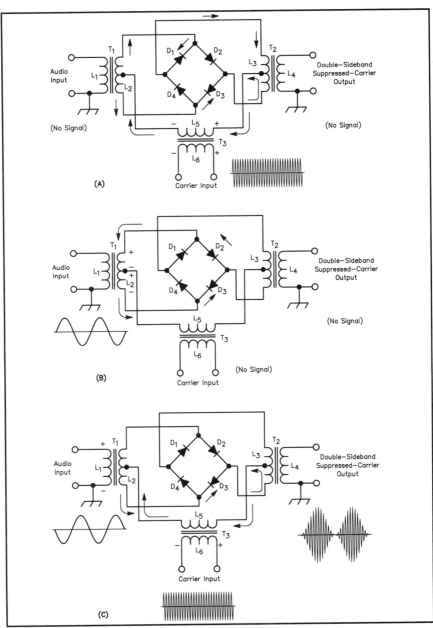

Figure 7-43 — Arrows indicate the direction of electron flow in a diode-ring mixer circuit. Part A shows the condition when only a carrier signal is applied. B shows the conditions with audio applied but no carrier, and C shows the conditions for one polarity relationship of carrier- and audio-input signals. See the text for more information about the other three possible polarity relationships.

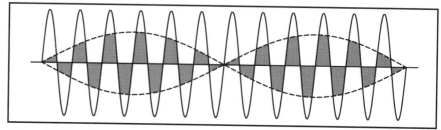

Figure 7-44 — Superimposed audio and RF waveforms. The shaded area represents the double-sideband, suppressed-carrier output from a balance modulator.

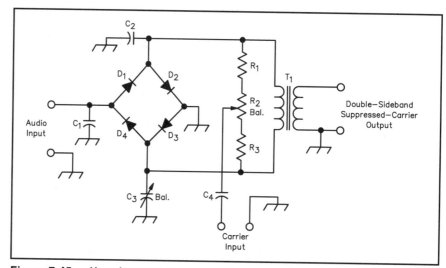

Figure 7-45 — Here is another common form of balanced modulator. C_3 and R_2 are adjusted for maximum carrier suppression.

reversed. Maximum current under these conditions is through D_1 and again an output signal appears at L_4. The fourth condition that will exist is shown when the audio-signal polarity is the same as in C, but the carrier is reversed. This time, D_4 will be the main current path. As you have probably guessed, there is output at L_4 under these conditions.

Figure 7-44 shows a composite drawing of the audio and carrier waveforms. The shaded areas represent the double-sideband, suppressed-carrier output from the balanced modulator.

Another type of diode-ring balanced modulator is shown in Figure 7-45. Two balance controls are provided so that the circuit can be adjusted for optimum carrier

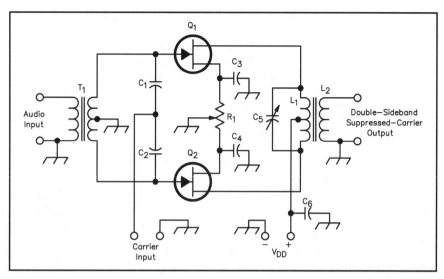

Figure 7-46 — This circuit is an active balanced modulator using two FET devices. R_1 is adjusted for maximum carrier suppression, and C_5 is adjusted for maximum double-sideband suppressed-carrier output.

suppression (50 dB is a practical amount). This circuit operates basically the same way the modulator shown in Figures 7-42 and 7-43 works. With either the audio or carrier applied separately, the circuit is balanced and there will be no output. With both the audio and carrier applied, the balance is upset and output will be present at T_1. We won't go into the detail that we did in the previous circuit. You should be able to determine which diodes are conducting for the different voltage polarities presented by the audio and carrier signals.

The two circuits we have examined can be classified as passive balanced modulators That is, they do not provide gain (amplify) but actually cause a small amount of signal loss (insertion loss). Balanced modulators can also be built using active devices.

One such modulator is shown in Figure 7-46. This circuit makes use of two FETs as the active devices. As was true with the two other modulators, the circuit is in a balanced condition if either the audio or carrier signals are applied separately. In this circuit the tuned output network (C_5/L_1) is adjusted for resonance at the RF carrier frequency. At audio frequencies, this circuit represents a very low impedance, allowing no audio to appear at the output.

Consider the carrier input for a moment. Injection voltage is supplied to each gate in a parallel fashion. The input to each gate is of equal amplitude and of the same phase, and the output circuit is connected for push-pull operation. Currents through each half of the tank are equal and opposite. The signals will effectively cancel and the output will be zero.

Let's analyze the circuit with both the audio and carrier energy applied. Since we have a push-pull input arrangement for the audio information, the bias for the FETs varies at an audio rate. The audio signal applied to one gate is 180° out of phase with the other. While one of the devices is forward biased, the other is reversed biased. The input to each device is the audio signal plus the carrier signal. Sum and difference frequencies are developed at the output to produce a double-sideband, suppressed-carrier signal.

Removing the Unwanted Sideband

We now have a signal that contains both the upper and lower sidebands of what was an amplitude-modulated signal with carrier. The next step in generating our SSB signal is to remove one of the sidebands. Looking back at Figure 7-41, we see that the next stage after the balanced modulator is the filter. This circuit does just as its name suggests — it filters out one of the sidebands.

An example of a simple crystal filter is shown in Figure 7-47. The two crystals

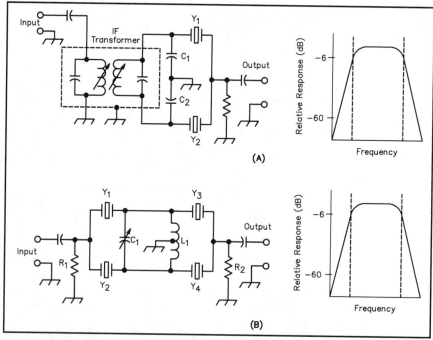

Figure 7-47—A half-lattice band-pass filter is shown at A. B shows two half-lattice filters in cascade. The filter response is shown to the right of each filter.

would be separated by approximately 2 kHz to provide a bandwidth suitable for passing one sideband and not the other. The curve to the right of the filter shows the shape of the filter response. More-elaborate filters using four and six crystals will give steeper slopes on the response curve, without affecting the bandwidth near the top of the curve. As shown at B, a filter with more crystals has a narrower response at the − 60 dB point. The filter at B has better "skirt selectivity" than the one at A. Two half-lattice filters of the type shown at A are connected back to back to form the filter at B. Crystal-lattice filters of this type are available commercially for frequencies up to 40 MHz or so.

Amplification for the Modulator

The last stage shown in the block diagram in Figure 7-41 is the linear amplifier. Since the modulation process occurs at low power levels in a conventional transmitter, it is necessary that any amplifiers following the balanced modulator be linear. In the low-power stages of a transmitter, high voltage gain and maximum linearity are quite a bit more important than efficiency.

Generating an FM Signal

Any type of modulation that changes the phase angle of the transmitted sine wave is called *angle modulation*. Frequency modulation and phase modulation are the two common types of angle modulation. Most methods of producing FM will fall into two general categories. They are direct FM and indirect FM. As you might expect, each has its advantages and disadvantages. Let's look at the direct-FM method first.

Direct FM

A **reactance modulator** is a simple and satisfactory device for producing FM in an amateur transmitter. This is a vacuum tube or transistor connected to the RF tank circuit of an oscillator so that it acts as a variable inductance or capacitance. The only way to produce a true emission F3E signal is with a reactance modulator on the transmitter oscillator.

Figure 7-48 is a representative circuit. Gate 1 of the modulator MOSFET is connected across the oscillator tank circuit (C_1 and L_1) through resistor R_1 and blocking capacitor C_2. C_3 represents the input capacitance of the modulator transistor.

R_1 is made large compared to the reactance of C_3, so the RF current through R_1 and C_3 will be practically in phase with the RF voltage appearing at the terminals of the tank circuit. The voltage across C_3 will lag the current by 90°, however. The RF current in the drain circuit of the modulator will be in phase with the gate voltage, and consequently is 90° behind the current through C_3, or lagging the RF-tank-circuit voltage by 90°. This lagging current is drawn through the oscillator tank, giving the same effect as though an inductor were connected across the tank. The frequency increases in proportion to the amplitude of the lagging modulator

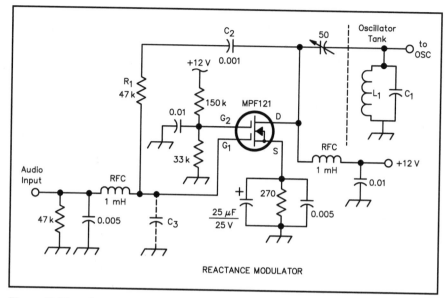

Figure 7-48 — A reactance modulator using a high-transconductance MOSFET.

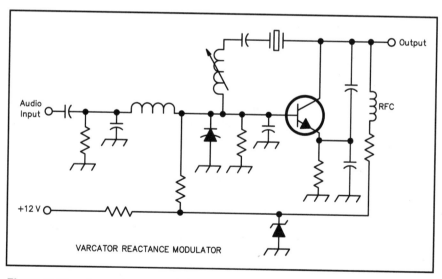

Figure 7-49 — A reactance modulator using a varactor diode.

current. The audio voltage, introduced through a radio-frequency choke, varies the transconductance of the transistor and thereby varies the RF drain current.

The modulator sensitivity (frequency change per unit change in modulating voltage) depends on the transconductance of the modulator transistor. Sensitivity increases when R_1 is made smaller in comparison with the reactance of C_3. It also increases with an increased L/C ratio in the oscillator tank circuit. For highest carrier stability, however, it is desirable to use the largest tank capacitance that will permit you to obtain the required deviation, while keeping within the limits of linear operation.

A change in any of the modulator-transistor voltages will cause a change in RF drain current, and consequently, a frequency change. Therefore, you should use a regulated power supply for both modulator and oscillator.

A reactance modulator can be connected to a crystal oscillator, as shown in Figure 7-49. The resulting signal will be more phase modulated than frequency modulated, however, since varying the frequency of a crystal oscillator will only produce a small amount of frequency deviation. Notice that this particular circuit uses a varactor diode to change the circuit capacitance by means of a bias voltage and the modulating signal. So, you should recognize that a reactance modulator acts as either a variable inductor or a variable capacitor in the tank-circuit oscillator.

Figure 7-50 is a block diagram of a system that uses a reactance modulator to shift an oscillator frequency and generate an FM signal directly. Successive multiplier stages provide output on the desired frequency, which is then amplified by a power amplifier stage.

With any reactance modulator, the modulated oscillator is usually operated on a relatively low frequency, so that a high order of carrier stability can be secured. Frequency multipliers are used to provide the final desired output frequency. It is important to note that when the frequency is multiplied, so is the frequency deviation. So the amount of deviation produced by the modulator must be adjusted carefully to give the proper deviation at the final output frequency.

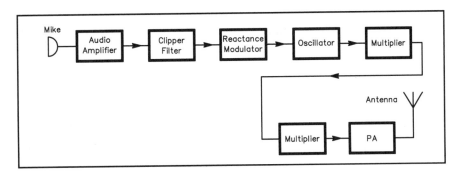

Figure 7-50 — Block diagram of a direct-FM transmitter.

Indirect FM

The same type of reactance-modulator circuit that is used to vary the oscillator-tank tuning for an FM system can be used to vary the amplifier-tank tuning, and thus vary the *phase* of the tank current, to produce *phase modulation (PM)*. Hence, the modulator circuit of Figure 7-49 or Figure 7-50 can be used for PM (emission G3E) if the reactance transistor or tube works on an amplifier tank instead of directly on a self-controlled oscillator. A **phase modulator** varies the tuning of an amplifier tank circuit to produce a PM signal. From a practical view, FM and PM are the same.

The phase shift that occurs when a circuit is detuned from resonance depends on the amount of detuning and the circuit Q. The higher the Q, the smaller the amount of detuning needed to secure a given amount of phase shift. If the Q is at least 10, the relationship between phase shift and detuning (in kilohertz either side of the resonant frequency) will be substantially linear over a phase-shift range of about 25°. From the standpoint of modulator sensitivity, the Q of the tuned circuit on which the modulator operates should be as high as possible. On the other hand, the effective Q of the circuit will not be very high if the amplifier is delivering power to a load, since the load resistance reduces the Q. There must, therefore, be a compromise between modulator sensitivity and RF power output from the modulated amplifier. An optimum Q figure appears to be about 20; this allows reasonable loading of the modulated amplifier, and the necessary tuning variation can be secured from a reactance modulator without difficulty. It is advisable to modulate at a low power level.

Reactance modulation of an amplifier stage usually results in simultaneous amplitude modulation because the modulated stage is detuned from resonance as the phase is shifted. This must be eliminated by feeding the modulated signal through an amplitude limiter or one or more "saturating" stages; that is, amplifiers that are

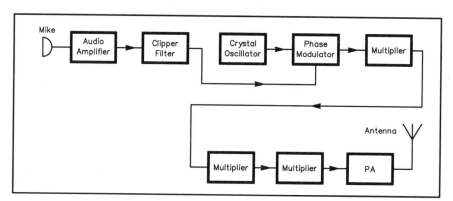

Figure 7-51 — Block diagram of an indirect-FM transmitter.

operated Class C and driven hard enough so that variations in the amplitude of the input excitation produce no appreciable variations in the output amplitude.

The actual frequency deviation increases with the modulating audio frequency in PM. (Higher audio frequencies produce greater frequency deviation.) Therefore, it is necessary to cut off the frequencies above about 3000 Hz before modulation takes place. If this is not done, unnecessary sidebands will be generated at frequencies considerably removed from the carrier frequency.

In an FM system, the frequency deviation does not increase with modulating audio frequency. In this case, an audio shaping network, called a *preemphasis network*, is added to an FM transmitter to attenuate the lower audio frequencies. This spreads the audio signal energy evenly across the audio band. Preemphasis applied to an FM transmitter gives the deviation characteristic of PM. The reverse process, called *deemphasis*, is used at the receiver to restore the audio to its original relative proportions. A PM transmitter does not need this preemphasis network.

The indirect method of generating FM shown in Figure 7-51 is currently popular. Shaped audio is applied to a phase modulator to generate PM. Since the amount of deviation produced is very small, a large number of multiplier stages is necessary to achieve wideband deviation at the operating frequency.

[That completes your study of practical circuits for your Advanced class license exam. Before you go on to the next chapter, though, study questions A7G01 through A7G11 in Chapter 10. Review this section as needed.]

Amplitude modulation — A method of superimposing an information signal on an RF carrier wave in which the amplitude of the RF envelope (carrier and sidebands) is varied in relation to the information signal strength.

Circular polarization — Describes an electromagnetic wave in which the electric and magnetic fields are rotating. If the electric field vector is rotating in a clockwise sense, then it is called right-hand polarization and if the electric field vector is rotating in a counterclockwise sense, it is called left-hand polarization.

Deviation — The peak difference between an instantaneous frequency of the modulated wave and the unmodulated-carrier frequency in an FM system.

Deviation ratio — The ratio of the maximum frequency deviation to the maximum modulating frequency in an FM system.

Dielectric constant — A property of insulating materials that serves as a measure of how much electric charge can be stored in the material with a given voltage.

Dielectric materials — Materials in which it is possible to maintain an electric field with little or no additional energy being supplied. Insulating materials or nonconductors.

Electric field — A region through which an electric force will act on an electrically charged object.

Electric force — A push or pull exerted through space by one electrically charged object on another.

Electromagnetic waves — A disturbance moving through space or materials in the form of changing electric and magnetic fields.

Emission designators — A method of identifying the characteristics of a signal from a radio transmitter using a series of three characters following the ITU system.

Emission types — A method of identifying the signals from a radio transmitter using a "plain English" format that simplifies the ITU **emission designators**.

Facsimile — The process of scanning pictures or images and convening the information into signals that can be used to form a likeness of the copy in another location .

Frequency, f — The number of complete cycles of a wave occurring in a unit of time.

Frequency modulation — A method of superimposing an information signal on an RF carrier wave in which the instantaneous frequency of an RF carrier wave is varied in relation to the information signal strength.

Linear polarization — Describes the orientation of the electric-field component of an electromagnetic wave. The electric field can be vertical or horizontal with respect to the Earth's surface, resulting in either a vertically or a horizontally polarized wave. (Also called **plane polarization**.)

Magnetic field — A region through which a magnetic force will act on a magnetic object .

Magnetic force — A push or pull exerted through space by one magnetically charged object on another.

Modulation index — The ratio of the maximum frequency deviation of the modulated wave to the instantaneous frequency of the modulating signal.

Peak envelope power (PEP) — An expression used to indicate the maximum power level in a signal. It is found by squaring the RMS voltage and dividing by the load resistance.

Peak envelope voltage (PEV) — The maximum peak voltage occurring in a complex waveform.

Peak-to-peak (P-P) voltage — A measure of the voltage taken between the negative and positive peaks on a cycle.

Peak voltage — A measure of voltage on an ac waveform taken from the centerline (0 V) and the maximum positive or negative level.

Period, T — The time it takes to complete one cycle of an ac waveform.

Phase modulation — A method of superimposing an information signal on an RF carrier wave in which the phase of an RF carrier wave is varied in relation to the information signal strength.

Plane polarization — Describes the orientation of the electric-field component of an electromagnetic wave. The electric field can be vertical or horizontal with respect to the Earth's surface, resulting in either a vertically or a horizontally polarized wave. (Also called **linear polarization**.)

Polarization — A property of an electromagnetic wave that describes the orientation of the electric field of the wave.

Root-mean-square (RMS) voltage — A measure of the effective value of an ac voltage.

Sawtooth wave — A waveform consisting of a linear ramp and then a return to the original value. It is made up of sine waves at a fundamental frequency and all harmonics.

Sine wave — A single-frequency waveform that can be expressed in terms of the mathematical sine function.

Single-sideband, suppressed-carrier signal — A radio signal in which only one of the two sidebands generated by amplitude modulation is transmitted. The other sideband and the RF carrier wave are removed before the signal is transmitted.

Slow-scan TV — A TV system used by amateurs to transmit pictures within a signal bandwidth allowed on the HF bands by the FCC.

Square wave — A periodic waveform that alternates between two values, and spends an equal time at each level. It is made up of sine waves at a fundamental frequency and all odd harmonics.

SIGNALS AND EMISSIONS

There are a number of loosely related sections in this chapter. When you have studied the information in each section, use the examination questions from the Element 4A question pool, listed in Chapter 10, to check your understanding of the material. If you are unable to answer a question correctly, go back and review the appropriate part of this chapter.

There will be six questions from the Signals and Emissions subelement on your Advanced class exam, and those questions will be taken from the six question groups in the A8 section of the question pool.

A8A   FCC emission designators versus emission types.
A8B   Modulation symbols and transmission characteristics.
A8C   Modulation methods; modulation index; deviation ratio.
A8D   Electromagnetic radiation; wave polarization; signal-to-noise (S/N) ratio.
A8E   AC waveforms: sine wave, square wave, sawtooth wave.
A8F   AC measurements: peak, peak-to-peak and root-mean-square (RMS) value, peak-envelope-power (PEP) relative to average.

FCC EMISSION DESIGNATORS

The International Telecommunication Union (ITU) has developed a special system of identifiers to specify the types of signals (emissions) permitted to amateurs and other users of the radio spectrum. This system designates emissions according to their necessary bandwidth and their classification. While a complete emission designator might include up to five characters, generally only three of them are used.

The designators begin with a letter that tells what type of modulation is being used. The second character is a number that describes the signal used to modulate the carrier. The third character specifies the type of information being transmitted.

Table 8-1 summarizes the most common characters for each of the three symbols

---

Table 8-1

Partial List of WARC-79 Emissions Designators

(1) First Symbol — Modulation Type

Unmodulated carrier	N
Double sideband full carrier	A
Single sideband reduced carrier	R
Single sideband suppressed carrier	J
Vestigial sidebands	C
Frequency modulation	F
Phase modulation	G
Various forms of pulse modulation	P, K,
L, M, Q, V, W, X	

(2) Second Symbol — Nature of Modulating Signals

No modulating signal	0
A single channel containing quantized or digital information without the use of a modulating subcarrier	1
A single channel containing quantized or digital information with the use of a modulating subcarrier	2
A single channel containing analog information	3
Two or more channels containing quantized or digital information	7
Two or more channels containing analog information	8

(3) Third Symbol — Type of Transmitted Information

No information transmitted	N
Telegraphy — for aural reception	A
Telegraphy — for automatic reception	B
Facsimile	C
Data transmission, telemetry, telecommand	D
Telephony	E
Television	F

that make up an emission designator. Some of the more common combinations are:

- N0N — Unmodulated carrier
- A1A — Morse code telegraphy using amplitude modulation
- A3E — Double-sideband, full-carrier, amplitude-modulated telephony
- J3E — Amplitude-modulated, single-sideband, suppressed-carrier telephony
- J3F — Amplitude-modulated, single-sideband, suppressed-carrier television
- F3E — Frequency-modulated telephony
- G3E — Phase-modulated telephony
- F1B — Telegraphy using frequency-shift keying without a modulating audio tone (FSK RTTY). F1B is designed for automatic reception.
- F2B — Telegraphy produced by modulating an FM transmitter with audio tones (AFSK RTTY). F2B is also designed for automatic reception.
- F1D — FM data transmission, such as packet radio.

You can assemble an emission designator by selecting one character from each of the three sets, based on your knowledge of the transmission system. For example, suppose you know that a certain signal is produced by an AM (double-sideband, full carrier) transmitter. That is represented by the letter A for the first character. If the transmitter is modulated with a single-channel signal containing quantized or digital information without the use of a modulating subcarrier, you would select the number 1 as the second character. Finally, suppose the resulting signal is a telegraphy signal primarily intended for aural (by ear) reception rather than for machine or computer (automatic) copy. Then the third character of our emission symbol will be A. So we have just completely described an A1A signal, which is Morse code telegraphy!

[Review your understanding of ITU emission designators by turning to Chapter 10 and studying question A8A11 and questions A8B01 through A8B11. Review this section as needed.]

Emission Types

Part 97, the FCC Rules governing Amateur Radio, refers to **emission types** rather than **emission designators**. The emission types are CW, phone, RTTY, data, image, MCW (modulated continuous wave), SS (spread spectrum), pulse and test. Any signal may be described by both an emission designator or an emission type.

While emission types are fewer in number and easier to remember, they are a somewhat less descriptive means of identifying a signal. There is still a need for emission designators in amateur work, and Part 97 does reference the designators. In fact, the official FCC definitions for the emission types include references to the symbols that make up acceptable emission designators for each emission type.

The US Code of Federal Regulations, Title 47, consists of telecommunications rules numbered Parts 0 through 200. These Parts contain specific rules for many telecommunications services the FCC administers. Part 2, Section 2.201, *Emission, modulation and transmission characteristics*, spells out the details of the ITU emission designators, as the FCC applies them in the US.

Facsimile and Television Emission Designators

Facsimile is the transmission of fixed images or pictures by electronic means, with the intent to reproduce the images in a permanent (printed) form. By contrast, television is the transmission of transient images of fixed or moving objects. For further discussion of facsimile and **slow-scan television** see Chapter 2.

There are several emission designators that may be used for facsimile and television signals, depending on how the signals are produced:

- A3C for full carrier, amplitude-modulated facsimile signals with a single information channel.
- F3C for frequency-modulated facsimile signals with a single information channel.
- A3F for full-carrier, amplitude-modulated television signals with a single information channel.
- C3F for vestigial-sideband TV signals.
- F3F for frequency-modulated television signals with a single information channel.
- J3F for single-sideband, suppressed-carrier TV signals.

An examination of §97.305 of the FCC Rules reveals that these emission types go together. (See Chapter 1.) Where one is permitted, they are all permitted. Bandwidth restrictions for these modes are covered in §97.307 of the FCC Rules.

When it comes to emissions designators, things are not always what they seem. **Slow-scan TV** signals consist of a series of audio tones that correspond to sync and picture-brightness levels. When SSTV is sent using an SSB transmitter, the emission is actually frequency modulation, although with the designators, J3F is the appropriate symbol. See Chapter 2 for details. Fast-scan TV, by contrast, is normally amplitude modulation or vestigial sideband (like broadcast TV). At microwave frequencies, FM is also used for fast-scan TV in the amateur bands — as it is in the TV satellites.

[Study questions A8A01 through A8A10 in Chapter 10. Review this section as needed.]

MODULATION METHODS

To pass your Advanced class exam, you will need to know what type of modulator circuit is used to produce the various types of radio signals. In Chapter 7, Practical Circuits, we described the circuit and operation of several modulator types. If you find that you don't remember the circuit details for these modulator types, you should go back to the appropriate section in that chapter. Block diagrams of various transmitter systems are included in this section to help you understand how the pieces fit together.

Amplitude Modulation

Double-sideband, full-carrier, **amplitude modulation** (emission A3E) can be realized by simply modulating the supply voltage to an amplifier stage. See Figure 8-1. The two signals will be mixed, so the output from the modulated stage will include the input radio frequency, the modulating frequency, the sum of the two signals (upper sideband), the difference between them (lower sideband). One way to do this is to apply the modulating signal to the plate or collector supply voltage of a Class-C RF amplifier. Other methods require the modulating signal to be applied to the grid or base circuit of a Class-A or -AB amplifier. The main thing to remember is that the amplitude or strength of the output signal is changing in step with the amplitude of a modulating signal. If the frequency of that signal is also changing (as it would be for your voice), then the sideband frequencies are changing.

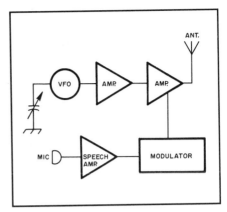

Figure 8-1 — Block diagram of an amplitude-modulated transmitter.

Single Sideband

A **single-sideband, suppressed-carrier signal** (emission J3E) is much like an AM signal, with one important exception. By transmitting only one of the sidebands, and eliminating the carrier, SSB occupies a much smaller bandwidth. It is the most-used method for transmitting voice signals on the HF amateur bands.

SSB signals can be generated in a two-step process. In the first step, a double-sideband, suppressed-carrier signal is generated in a balanced modulator. Remember that in a balanced modulator the input AF and RF signals do not appear at the output — only the sum (upper sideband) and difference (lower sideband) frequencies appear. In the second step, the unwanted sideband is filtered out, leaving only the desired one. Figure 8-2A illustrates the essentials of such a sideband transmitter. Some transmitters use one crystal with the oscillator, but switch in a different filter to eliminate the other sideband.

There is another way to generate an SSB signal, called the phasing method. Instead of using filters to remove the unwanted sideband, the signal phase has to be adjusted carefully so the carrier, audio signal and unwanted sideband are canceled when out-of-phase components are added. This technique was popular in the early days of SSB operation, but designing a phase-shift network that produces the desired 90° phase shift for signals of all frequencies was difficult. Modern filter design has made it possible to build much better filters today than were possible in the early days of SSB experimentation. Therefore, the phasing technique lost popularity and is seldom used.

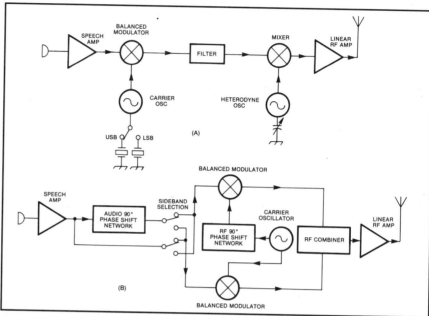

Figure 8-2 — Part A shows the filter method of generating an SSB signal, and part B illustrates one form of the phasing method. The phasing method is not used in modern SSB equipment.

Computer technology and digital signal processing (DSP) techniques may spark some renewed interest in the phasing method of SSB generation, however. In Chapter 7 you learned about the Hilbert-transform filter, which is a DSP filter that introduces a 90° phase shift to all frequency components of a signal. Figure 8-2B shows the basic components of a phasing generator for single-sideband signals.

Frequency or Phase Modulation

Frequency modulation (emission F3E) operates on an entirely different principle than amplitude modulation. With FM, the signal frequency is varied above and below the carrier frequency at a rate equal to the modulating-signal frequency. For example, if a 1000-Hz tone is used to modulate a transmitter, the carrier frequency will vary above and below the center frequency 1000 times per second. The amount of frequency change, however, depends on the instantaneous amplitude of the modulating signal. This frequency change is called **deviation**. A certain signal might produce a 5-kHz deviation. If another signal, with only half the amplitude of the first, were used to modulate the transmitter, it would produce a 2.5-kHz deviation. From this example, you can see that the deviation is proportional to the modulating signal amplitude.

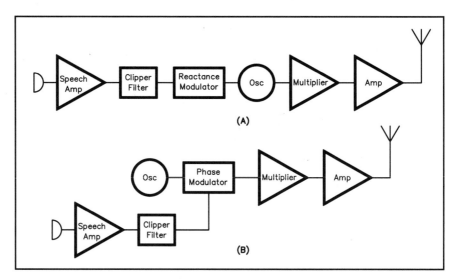

Figure 8-3 — Direct frequency modulation (FM) is shown at A; phase modulation (PM) at B.

Direct FM can be produced by a reactance modulator. The reactance modulator is connected to an oscillator in such a way as to act as a variable inductance or capacitance. When a modulating signal is applied, the oscillator frequency varies. Figure 8-3A shows how such a system is arranged.

Phase modulation (emission G3E, sometimes called indirect FM) can be realized by using a phase modulator. A phase modulator is similar to a reactance modulator in that it appears to be a variable inductance or capacitance when modulation is applied. The difference between the two is where they are found in the transmitter. Whereas the reactance modulator controls an oscillator, the phase modulator acts on a buffer or amplifier stage. This arrangement is shown at Figure 8-3B.

FM Terminology

You will need to understand two terms that refer to FM systems and operation: **deviation ratio** and **modulation index**. They may seem to be almost the same — indeed, they are closely related. Pay special attention to the equations given to calculate these quantities, and you should be able to distinguish between them.

Deviation Ratio

In an FM system, the ratio of the maximum carrier-frequency deviation to the highest modulating frequency is called the deviation ratio. It is a constant value for a given system, calculated by:

$$\text{deviation ratio} = \frac{D_{max}}{M} \qquad \text{(Equation 8-1)}$$

where:

D_{max} = peak deviation in hertz (half the difference between the maximum and minimum carrier-frequency values at 100% modulation).

M = maximum modulating frequency in hertz.

In the case of narrow-band FM, peak deviation at 100% modulation is 5 kHz. The maximum modulating frequency is 3 kHz. Therefore:

$$\text{deviation ratio} = \frac{D_{max}}{M} = \frac{5 \text{ kHz}}{3 \text{ kHz}} = 1.67$$

Notice that since both frequencies were given in kilohertz we did not have to change them to hertz before doing the calculation. The important thing is that they both be in the same units.

Modulation Index

The ratio of the maximum carrier-frequency deviation to the (instantaneous) modulating frequency is called the modulation index. That is:

$$\text{modulation index} = \frac{D_{max}}{m} \qquad \text{(Equation 8-2)}$$

where:

D_{max} = peak deviation in hertz.

m = modulating frequency in hertz at any given instant.

For example, suppose the peak frequency deviation of an FM transmitter is 3000 Hz either side of the carrier frequency. The modulation index when the carrier is modulated by a 1000-Hz sine wave is:

$$\text{modulation index} = \frac{D_{max}}{m} = \frac{3000 \text{ Hz}}{1000 \text{ Hz}} = 3$$

When modulated with a 3000-Hz sine wave with the same peak deviation (3000 Hz), the index would be 1; with a 100-Hz modulating wave and the same 3000-Hz peak deviation, the index would be 30, and so on.

In a phase modulator, the modulation index is constant regardless of the modulating frequency, as long as the amplitude is held constant. In other words, a 2-kHz tone will produce twice as much deviation as a 1-kHz tone if the amplitudes of the tones are equal. This may seem confusing at first; just think of the peak deviation as the variable in Equation 8-2.

By contrast, the modulation index varies inversely with the modulating frequency in a frequency modulator, as Equation 8-2 shows. A higher modulating frequency results in a lower modulation index, if the peak deviation remains the same. The actual deviation depends only on the amplitude of the modulating signal and is independent of frequency. Thus, a 2-kHz tone will produce the same deviation as a 1-kHz tone if the amplitudes of the tones are equal. The modulation index in the case of the 1-kHz tone is double that for the case of the 2-kHz tone.

Notice that with either an FM or a PM system, the deviation ratio and modulation index are independent of the frequency of the modulated RF carrier. It doesn't matter if the transmitter is a 10-meter FM rig or a 2-meter rig.

[At this point you should turn to Chapter 10 and study examination questions A8C01 through A8C11. Review this section as needed.]

ELECTROMAGNETIC WAVES

All **electromagnetic waves** are moving fields of **electric** and **magnetic force**. Their lines of force are at right angles to each other, and are also both perpendicular to the direction of travel. See Figure 8-4. They can have any position with respect to the Earth. The plane containing the continuous lines of electric and magnetic force is called the wave front. Another way of visualizing this concept is to think of the wave front as being a fixed point on the moving wave.

Electromagnetic Radiation

Electricity requires a conductor to carry an electron current through a circuit. Electromagnetic waves move easily through the vacuum of free space. The **electric** and **magnetic fields** that constitute the wave do not require a conductor to carry them.

The medium through which electromagnetic waves travel has a marked influence on their speed of movement. In empty space, electromagnetic waves travel at the same speed as light, approximately 300,000,000 meters per second or about 186,000 miles per second. It is

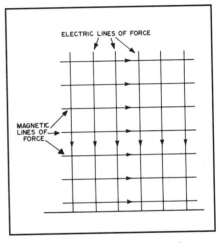

Figure 8-4 — Representation of electric and magnetic lines of force in a radio wave. Arrows indicate instantaneous directions of the fields for a wave traveling toward you, out of the page. Reversing the direction of one set of lines would reverse the direction of travel, but if you reversed the direction of both sets the wave would still be coming out of the page.

slightly less in air, and it varies somewhat with temperature and humidity, depending on the frequency. It is much less in other **dielectric materials** (insulators), where the speed is inversely proportional to the square root of the **dielectric constant** of the material.

Radio waves travel through dielectric materials with ease. Waves cannot penetrate a good conductor, however. Instead of penetrating the conductor as they encounter it, the magnetic field generates current in the conductor surface. These induced currents are called eddy currents.

Wave Polarization

Polarization refers to the direction of the electric lines of force of a radio wave. See Figure 8-5. If the electric lines of force are parallel to the Earth, we call this a horizontally polarized radio wave. In a horizontally polarized wave, the electric lines of force are horizontal and the magnetic lines are vertical. A radio wave is vertically polarized if its electric lines of force are perpendicular to the Earth (vertical). In this case the magnetic lines are horizontal.

For the most part, polarization is determined by the type of transmitting antenna used, and its orientation. On one hand, for example, a Yagi antenna with its elements parallel to the Earth's surface transmits a horizontally polarized radio wave. On the other hand, an amateur mobile whip antenna, mounted vertically on an automobile, radiates a vertically polarized wave.

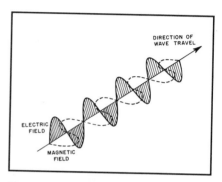

Figure 8-5 — Representation of the magnetic and electric fields of a vertically polarized radio wave. In this diagram, the electric field is in a vertical plane and the magnetic field is in a horizontal plane.

It is possible to generate waves with rotating field lines. This condition, where the electric field lines are continuously rotating through horizontal and vertical orientations, is called **circular polarization**. It is particularly helpful to use circular polarization in satellite communication, where polarization tends to shift.

Polarization that does not rotate is called **linear polarization** or **plane polarization**. Horizontal and vertical polarization are examples of linear polarization. (In space, of course, horizontal and vertical have no convenient reference.) Circular polarization is usable with linearly polarized antennas at the other end of the circuit. There will be some small loss in this case, however. If you use a vertically polarized antenna to receive a horizontally polarized radio wave (or vice versa) over a line-of-sight ground-wave path, you can expect the received signal strength to be reduced by more than 20 dB as compared to using an antenna with the same polarization as the wave. With propagation paths that use sky waves, this effect may disappear completely.

[Now study question numbers A8D01 through A8D10 in Chapter 10. Review this section as needed.]

SIGNAL-TO-NOISE RATIO

Receiver noise performance is established primarily in the RF amplifier and/or mixer stages. Low-noise, active devices should be used in the receiver front end to obtain good performance. The unwanted noise, in effect, masks the weaker signals and makes them difficult or impossible to copy. Noise generated in the receiver front end is amplified in the succeeding stages along with the signal energy. Therefore, in the interest of sensitivity, internal noise should be kept as low as possible.

Don't confuse external noise (man-made and atmospheric noise, which comes in on the antenna lead) with receiver noise during discussions of noise performance. The ratio of external noise to the incoming signal level does have a lot to do with reception. It is because external noise levels are quite high on the 160 through 20-meter bands that emphasis is seldom placed on low-internal-noise receivers for those bands. The primary source of noise heard on an HF receiver with an antenna connected is atmospheric noise. As the operating frequency is increased from 15 meters up through the microwave spectrum, however, the matter of receiver noise becomes a primary consideration. At these higher frequencies the receiver noise almost always exceeds that from external sources, especially at 2 meters and above.

Receiver noise is produced by the movement of electrons in any substance (such as wires, resistors and transistors) in the receiver circuitry. Electrons move in a random fashion colliding with relatively immobile ions that make up the bulk of the material. The final result of this effect is that in most substances there is no net current in any particular direction on a long-term average, but rather a series of random pulses. These pulses produce what is called thermal-agitation noise, or simply *thermal noise.*

Thermal-noise power is directly proportional to bandwidth and absolute temperature (in kelvins). For that reason, narrow-band systems exhibit better noise performance than do wide-band systems. For extremely low-noise operation some amplifiers are cooled with liquid air or nitrogen. That practice is hazardous, however, and is not used by

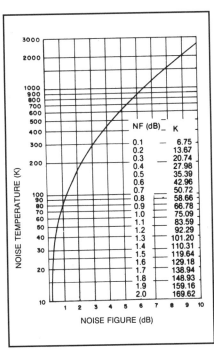

Figure 8-6 — Relationship between noise figure and noise temperature.

amateurs. Noise temperature, noise factor and noise figure are all measures of this thermal noise. Figure 8-6 shows the relationship between noise temperature in kelvins and noise figure in decibels.

[Now study examination questions A8D11 and A8D12 in Chapter 10. Review this section as needed.]

AC WAVEFORMS

Sine Waves

The basic ac waveform is the **sine wave**. A sine wave represents a single **frequency**. To visualize a sine wave, let's imagine a wheel, like a bicycle wheel. We will paint a dot on the edge of the wheel at one point, so we can watch that spot as the wheel spins. If you look at the wheel edge, the spot will just seem to move up and down, as illustrated in Figure 8-7A. B pictures the wheel from one side, with the spot shown stopped at several points around the circle. If you can imagine the pattern at A as the wheel moves sideways to the right, the spot will trace a sine wave, as shown at C. This sine wave also represents the output from an ac generator, or alternator as it is called. One full rotation corresponds to a complete cycle (360 degrees).

Figure 8-7 is not only a mechanical analogy of alternator operation; it is a mathematical model as well. In the mathematical model, a line drawn from the axle to our paint spot is a rotating vector. The graph describes the vector ordinate (value along the Y or vertical axis) as it varies with time. Twice during each cycle a sine wave passes through zero; once while going positive, and once while going negative.

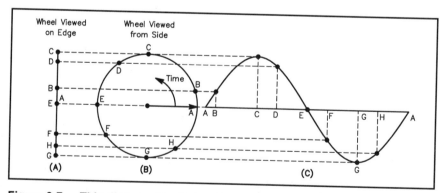

Figure 8-7 — This diagram illustrates the relationship between a sine wave and an object rotating in a circle. You can see how various points on the circle relate to sine-wave values.

The time required to complete one cycle is called the **period, T**. The frequency of the sine wave is the reciprocal of the period:

$$f = \frac{1}{T} \qquad \text{(Equation 8-3)}$$

Sawtooth Waves

A **sawtooth wave**, as shown in Figure 8-8A, is so named because it closely resembles the teeth on a saw blade. It is characterized by a rise time significantly faster than the fall time (or vice versa). A sawtooth wave is made up of a sine wave at the fundamental frequency and sine waves at all the harmonic frequencies as well. When a sawtooth voltage is applied to the horizontal deflection plates of an oscilloscope, the electron beam sweeps slowly across the screen during the slowly changing portion of the waveform and then flies quickly back during the rapidly changing portion of the signal. This type of waveform is desired to obtain a linear sweep in an oscilloscope.

Square Waves

A **square wave** is one that abruptly changes back and forth between two voltage levels and remains an equal time at each level. See Figure 8-8B. (If the wave spends an unequal time at each level, it is known as a rectangular wave.) A square wave is made up of sine waves at the fundamental and all the odd harmonic frequencies.

[Study examination questions A8E01 through A8E11 in Chapter 10. Review this section as needed.]

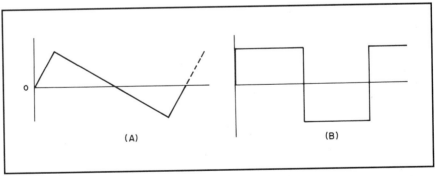

Figure 8-8 — Any waveform that is not a pure sine wave contains harmonics. Sawtooth waves (A) consist of both odd and even harmonics as well as the fundamental. Square waves (B) consist of only fundamental and odd-harmonic frequencies.

AC MEASUREMENTS

The time dependence of alternating-current waveforms raises questions about defining and measuring values of voltage, current and power. Because these parameters change from one instant to the next, one might wonder, for example, which point on the cycle characterizes the voltage or current for the entire cycle. Since the wave is positive for exactly the same time it is negative in value, you might even wonder if the value shouldn't be zero. Actually, the average dc voltage and current are zero! A dc meter connected to an ac voltage would read zero, although you may be able to notice some slight flutter on a sensitive meter. To get some idea of how useful the ac voltage might be, we will have to connect a diode in series with the meter lead and use a specially calibrated scale.

AC Voltage and Current

When viewing a sine wave on an oscilloscope, the easiest dimension to measure is the total vertical displacement, or **peak-to-peak (P-P) voltage**. The maximum positive or negative potential is called the **peak voltage**. In a symmetrical waveform it has half the value of the peak-to-peak amplitude.

When an ac voltage is applied to a resistor, the resistor will dissipate energy in the form of heat, just as if the voltage were dc. The dc voltage that would cause identical heating in the ac-excited resistor is called the **root-mean-square (RMS)** or effective value of the ac voltage. The phrase root-mean-square describes the mathematical process of actually calculating the effective value. The method involves squaring the peak values for a large number of points along the waveform (a calculus procedure), then finding the average of the squared values, and taking the square root of that number. The RMS voltage of any waveform can also be determined by measuring the heating effect in a resistor. For sine waves, the following relationships hold:

$$V_{peak} = V_{RMS} \times \sqrt{2} = V_{RMS} \times 1.414$$

(Equation 8-4)

and

$$V_{RMS} = \frac{V_{peak}}{\sqrt{2}} = V_{peak} \times 0.707$$

(Equation 8-5)

If we consider only the positive or negative half of a cycle, it is possible to calculate an average value for a sine-wave voltage. Meter movements respond to this average value rather than either the peak or RMS values of a voltage because of the inertia inherent in the needle and magnet. Often, the meter has a scale that is calibrated to read RMS values, even though the needle is actually responding to the average value. That is okay, as long as the waveform you are measuring is a pure sine wave. Other, more complex, waveforms will not give a true reading, however. The mathematical relationships between average, peak and RMS values for a sine wave are given by:

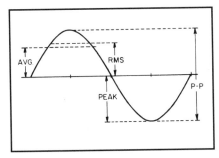

Figure 8-9 — An illustration showing ac voltage and current measurement terms for a pure sine wave.

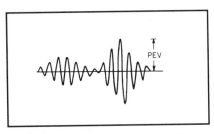

Figure 8-10 — A complex waveform, made up of several individual sine-wave signals. Peak envelope voltage (PEV) is an important parameter for a composite waveform.

$$V_{avg} = V_{peak} \times 0.646 \qquad \text{(Equation 8-6)}$$

and

$$V_{avg} = V_{RMS} \times 0.899 \qquad \text{(Equation 8-7)}$$

Unless otherwise specified or obvious from the context, ac voltage is rendered as an RMS value. For example, the household 120-V ac outlet provides 120-V RMS, 169.7-V peak and 339.4-V P-P. The voltage at your household outlets varies with the amount of load that the power company must supply. It will vary around the nominal 120-V value, and is sometimes even specified as 110. Of course this means that the peak and P-P values also vary. Figure 8-9 illustrates the voltage parameters of a sine wave.

The significant dimension of a multitone signal (a complex waveform) is the **peak envelope voltage (PEV)**, shown in Figure 8-10. PEV is important in calculating the power in a modulated signal, such as that from an amateur SSB transmitter.

All that has been said about voltage measurements applies also to current (provided the load is resistive) because the waveshapes are identical.

AC Power

The terms RMS, average and peak have different meanings when they refer to ac power. The reason is that while voltage and current are sinusoidal functions of time, power is the product of voltage and current, and this product is a sine squared function. The calculus operations that define RMS, average and peak values will naturally yield different results when applied to this new function. The relationships between ac voltage, current and power are as follows:

$$V_{RMS} \times I_{RMS} = P_{avg} \qquad \text{(Equation 8-8)}$$

Note that this calculation does not give a value for RMS power. The average power used to heat a resistor is equal to the dc power required to produce the same heat. RMS power has no physical significance!

For continuous sine wave signals:

$$V_{peak} \times I_{peak} = P_{peak} = 2 \times P_{avg}$$ (Equation 8-9)

Unfortunately, the situation is more complicated in radio work. We seldom have a steady sine-wave signal being produced by a transmitter. The waveform varies with time, in order to carry some useful information for us. The peak power output of a radio transmitter, then, is the power averaged over the RF cycle having the greatest amplitude. Modulated signals are not purely sinusoidal because they are composites of two or more audio tones. The cycle-to-cycle variation is small enough, however, that sine-wave measurement techniques produce accurate results. In the context of radio signals, then, peak power means maximum average power. **Peak envelope power (PEP)** is the parameter most often used to express the maximum signal level. To compute the PEP of a waveform such as that sketched in Figure 8-10, multiply the PEV by 0.707 to obtain the RMS value, square the result and divide by the load resistance.

SSB Power

Envelope peaks occur only sporadically during voice transmission and have no relationship with meter readings. The meters respond to the amplitude (current or voltage) of the signal averaged over several cycles of the modulation envelope.

The ratio of peak-to-average amplitude varies widely with voices of different characteristics. In the case shown in Figure 8-11, the average amplitude (found

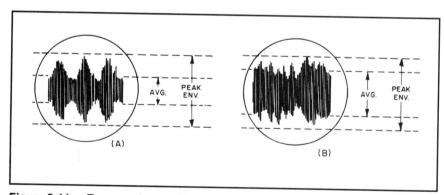

Figure 8-11 — Two envelope patterns that show the difference between average and peak levels. In each case, the RF amplitude (current or voltage) is plotted as a function of time. In B, the average level has been increased. That will raise the average output power compared to the peak value.

graphically) is such that the peak-to-average ratio of amplitudes is almost 3:1. Typical ratio values range from 2:1 to more than 10:1. So the PEP of an SSB signal may be about 2 or 3 times greater than the average power output. It may even be more than that, depending on the voice characteristics of the person speaking into the microphone.

[Turn to Chapter 10 and study questions A8F01 through A8F08. Review this section as needed.]

**CHAPTER 9
KEYWORDS
KEYWORDS
KEYWORDS**

Antenna — An electric circuit designed specifically to radiate the energy applied to it in the form of electromagnetic waves. An antenna is reciprocal; a wave moving past it will induce a current in the circuit also. Antennas are used to transmit and receive radio waves.

Antenna bandwidth — A range of frequencies over which the antenna SWR will be below some specified value.

Antenna efficiency — The ratio of the radiation resistance to the total resistance of an antenna system, including losses.

Base loading — The technique of inserting a coil at the bottom of an electrically short vertical antenna in order to cancel the capacitive reactance of the antenna, producing a resonant antenna system.

Beamwidth — As related to directive antennas, the width (measured in degrees) of a major lobe between the two directions at which the relative power is one half (– 3 dB) its value at the peak of the lobe.

Center loading — A technique for adding a series inductor at or near the center of an antenna element in order to cancel the capacitive reactance of the antenna. This technique is usually used with elements that are less than $1/4$ wavelength.

Dielectric — An insulating material. A dielectric is a medium in which it is possible to maintain an electric field with little or no additional direct-current energy supplied after the field has been established.

Dielectric constant (ε) — Relative figure of merit for an insulating material. This is the property that determines how much electric energy can be stored in a unit volume of the material per volt of applied potential.

Dipole — An antenna with two elements in a straight line that are fed in the center; literally, two poles. For amateur work, dipoles are usually operated near half-wave resonance.

Folded dipole — An antenna consisting of two (or more) parallel, closely spaced halfwave wires connected at their ends. One of the wires is fed at its center.

Frequency — A property of an electromagnetic wave that refers to the number of complete alternations (or oscillations) made in one second.

Gain — An increase in the effective power radiated by an antenna in a certain desired direction. This is at the expense of power radiated in other directions.

Loading coil — An inductor that is inserted in an antenna element or transmission line for the purpose of producing a resonant system at a specific frequency.

Major lobe of radiation — A three-dimensional area that contains the maximum radiation peak in the space around an antenna. The field strength decreases from the peak level, until a point is reached where it starts to increase again. The area described by the radiation maximum is known as the major lobe.

Minor lobe of radiation — Those areas of an antenna pattern where there is some increase in radiation, but not as much as in the major lobe. Minor lobes normally appear at the back and sides of the antenna.

Radiation resistance — The equivalent resistance that would dissipate the same amount of power as is radiated from an antenna. It is calculated by dividing the radiated power by the square of the RMS antenna current.

Reflection coefficient (ρ) — The ratio of the reflected voltage at a given point on a transmission line to the incident voltage at the same point. The reflection coefficient is also equal to the ratio of reflected and incident currents.

Top loading — The addition of inductive reactance (a coil) or capacitive reactance (a capacitance hat) at the end of a driven element opposite the feed point. It is intended to increase the electrical length of the radiator.

Traps — Parallel LC networks inserted in an antenna element to provide multiband operation.

Velocity factor — An expression of how fast a radio wave will travel through a material. It is usually stated as a fraction of the speed the wave would have in free space (where the wave would have its maximum velocity). Velocity factor is also sometimes specified as a percentage of the speed of a radio wave in free space.

Wavelength — The distance between two points that describe one complete cycle of a wave.

ANTENNAS AND FEED LINES

This chapter covers various antenna and feed-line topics. When you have studied the information in each section, use the examination questions to check your understanding of the material. If you are unable to answer a question correctly, go back and review the appropriate part of this chapter. Your Advanced class license (Element 4A) exam will include five questions from the Antennas and Feed Lines subelement. Those five questions will be taken from the 5 groups in this subelement.

A9A   Basic antenna parameters: radiation resistance and reactance (including wire dipole, folded dipole), gain, beamwidth, efficiency.

A9B   Free-space antenna patterns: E and H plane patterns (ie, azimuth and elevation in free-space); gain as a function of pattern; antenna design (computer modeling of antennas).

A9C   Antenna patterns: elevation above real ground, ground effects as related to polarization, take-off angles as a function of height above ground.

A9D   Losses in real antennas and matching: resistivity losses, losses in resonating elements (loading coils, matching networks, etc. {ie, mobile, trap}); SWR bandwidth; efficiency.

A9E   Feed lines: coax versus open-wire; velocity factor; electrical length; transformation characteristics of line terminated in impedance not equal to characteristic impedance.

RADIATION RESISTANCE

The energy supplied to an **antenna** is dissipated in the form of radio waves and in heat losses in the wire and nearby dielectrics. The radiated energy is the useful part, and as far as the antenna is concerned it represents a loss just as much as the energy used in heating the wire is a loss. In either case the dissipated power is equal to I^2R. In the case of heat losses, R is a real resistance, but in the case of radiation, R is an assumed resistance, which, if present, would dissipate the power actually radiated from the antenna. This assumed resistance is called the **radiation resistance**. The total power loss in the antenna is therefore equal to $I^2 (R_R + R)$, where R_R is the radiation resistance and R is the real, or ohmic, resistance.

In the ordinary half-wave **dipole** antenna operated at amateur frequencies, the power lost as heat in the conductor does not exceed a few percent of the total power supplied to the antenna. This is because the RF resistance of copper wire even as small as number 14 is very low compared with the radiation resistance of an antenna that is reasonably clear of surrounding objects and is not too close to the ground. Therefore, we can assume that the ohmic loss in a reasonably well located antenna is negligible, and that all of the resistance shown by the antenna is radiation resistance. As a radiator of electromagnetic waves, such an antenna is a highly efficient device.

The value of radiation resistance, as measured at the center of a half-wave antenna, depends on a number of factors. One is the location of the antenna with respect to other objects, particularly the earth. Another is the length/diameter ratio of the conductor used. In free space — with the antenna remote from everything else — the radiation resistance of a resonant $^1/_2$-wavelength dipole antenna made of an infinitely thin conductor is approximately 73 Ω. The concept of a free-space antenna forms a convenient basis for calculation because the modifying effect of the ground can be taken into account separately. If the antenna is at least several wavelengths away from ground and other objects, it can be considered to be in free space insofar as its own electrical properties are concerned. This condition can be met easily with antennas in the VHF and UHF range. At these frequencies, antennas are small and a wavelength may be only a few feet (or less) so it is easy to mount the antenna several wavelengths above ground.

As the antenna is made thicker, the radiation resistance decreases. For most wire antennas it is close to 65 Ω. The radiation resistance will usually lie between 55 and 60 Ω for antennas constructed of rod or tubing.

The actual value of the radiation resistance has no appreciable effect on the radiation efficiency of a practical antenna. This is because the ohmic resistance is only on the order of 1 Ω with the conductors used for thick antennas. The ohmic resistance does not become important until the radiation resistance drops to very low values — say less than 10 Ω — as may be the case when several antenna elements are coupled to form an array.

The radiation resistance of a resonant antenna is the "load" for the transmitter or for the RF transmission line connecting the transmitter and antenna. Its value is important, therefore, in determining the way in which the antenna and transmitter or line are coupled. Most modern transmitters require a 50-Ω load. To transfer the maximum amount of power possible, the transmitter output impedance, the transmission-line characteristic impedance and the radiation resistance should all be equal, or matched by means of an appropriate impedance-matching network.

ANTENNA EFFICIENCY

As you try to optimize your Amateur Radio station, you may want to begin with the antenna system. Improvements to the antenna system will probably have a greater impact on your station than any other changes you can make, considering cost and time invested. **Antenna efficiency** may be one of your first considerations. The efficiency of an antenna is given by:

$$\text{Efficiency} = \frac{R_R}{R_T} \times 100\% \qquad\qquad \text{(Equation 9-1)}$$

where:

R_R = radiation resistance.
R_T = total resistance.

The total resistance includes radiation resistance, resistance in conductors and dielectrics (including the resistance of loading coils, if used), and the resistance of the grounding system, usually referred to as "ground resistance."

A half-wave antenna operates at very high efficiency because the conductor resistance is negligible compared with the radiation resistance. In the case of a $^1/_4$-wavelength vertical antenna, with one side of the feed line connected to ground, the ground resistance usually is not negligible. If the antenna is short (compared with a quarter wavelength) the resistance of the necessary loading coil may become appreciable. To attain an efficiency comparable with that of a half-wave antenna in a grounded one having a height of $^1/_4$ wavelength or less, great care must be used to reduce both ground resistance and the resistance of any required loading inductors. Without a fairly elaborate grounding system, the efficiency is not likely to exceed 50% and may be much less, particularly at heights below $^1/_4$ wavelength. A $^1/_4$-wavelength ground-mounted vertical antenna normally includes many radial wires, laid out as spokes on a wheel, with the antenna element in the center.

If a half-wave dipole antenna has a radiation resistance of 70 Ω and a total resistance of 75 Ω, Equation 9-1 tells us the efficiency of the antenna:

$$\text{Efficiency} = \frac{R_R}{R_T} \times 100\% = \frac{70\ \Omega}{75\ \Omega} \times 100\% = 93\%$$

As another example, let's calculate the efficiency of a ground-mounted vertical antenna that has a radiation resistance of 25 Ω and a total resistance of 70 Ω. Again using Equation 9-1, we find that the efficiency of this antenna is 36%.

[Before proceeding, turn to Chapter 10 and study questions A9A01 through A9A06 and questions A9A12 and A9A13. Review this section as needed.]

FOLDED DIPOLE ANTENNAS

A **folded dipole** antenna is a wire antenna consisting of two (or more) parallel wires that are closely spaced and connected at their ends. One of the wires is fed at its center.

In Figure 9-1, suppose for the moment that the upper conductor between points B and C is disconnected and removed. The system is then a simple center-fed dipole, and the current direction along the antenna and line at a given instant is as shown by the arrows. Next, restore the upper conductor between B and C. The current in the top section will flow away from B and toward C at the instant shown for the antenna current.

This may seem confusing and be opposite to the direction you would expect the current to flow on that portion of the line. Just remember that for a sine wave, the current direction is reversed in alternate half-wave sections along a wire. Because of the way the second wire is "folded," however, the currents in the two conductors of the antenna are actually flowing in the same direction. Although the antenna physically resembles a transmission line, it is not actually a line. The antenna element merely consists of two parallel conductors carrying current in the same direction. If it were acting like a transmission line, the currents would be flowing in opposite directions. The connections at the ends of the two conductors are assumed to be of negligible length.

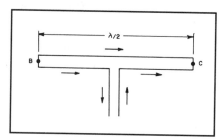

Figure 9-1 — This diagram illustrates the direction of current in a folded dipole.

A half-wave dipole formed in this way will have the same directional properties and total radiation resistance as an ordinary dipole. The transmission line is connected to only one of the conductors, however. You should expect that the antenna will "look" different, with respect to its input impedance, as viewed by the line.

The effect on the impedance at the antenna input terminals can be visualized quite readily. The center impedance of the dipole as a whole is the same as the impedance of a single-conductor dipole — that is, approximately 73 Ω. A given amount of power will therefore cause a definite value of current, I. In the ordinary half-wave dipole this current flows at the junction of the line and antenna. In the folded dipole the same total current also flows, but is equally divided between two conductors in parallel. The current in each conductor is therefore I/2. Consequently, the line "sees" a higher impedance because it is delivering the same power at only half the current.

Ohm's Law reveals that the new value of impedance is equal to four times the impedance of a simple dipole. The input-terminal impedance at the center of a two-conductor folded dipole is about 300 Ω. If more wires are added in parallel the current continues to divide between them and the terminal impedance is raised still more. This explanation is a simplified one based on the assumption that the conductors are close together and have the same diameter.

Another advantage of the folded dipole is that it has a low SWR over a wider frequency range than a normal dipole. The term **antenna bandwidth** refers generally to the range of frequencies over which an antenna can be used to obtain good performance. The bandwidth is often referenced to some SWR value, such as, "The 2:1 SWR bandwidth is 3.5 to 3.8 MHz." This means that the SWR between 3.5 and 3.8 MHz will be 2:1 or lower.

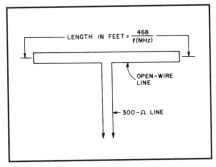

Figure 9-2 — This diagram shows the construction information for a half-wave folded dipole.

The two-wire system shown in Figure 9-2 is an especially useful one because the input impedance is so close to 300 Ω that it can be fed directly with 300-Ω twin-lead or open-wire line without any other matching arrangement.

[Study questions A9A07, A9A08, A9A10 and A9D01 in Chapter 10. Review this section as needed.]

TRAP ANTENNAS

By using tuned circuits of appropriate design strategically placed in a dipole, the antenna can be made to show what is essentially fundamental resonance at a number of different frequencies. The general principle is illustrated by Figure 9-3. The two inner lengths of wire, X, together form a simple dipole resonant at the highest band desired, say 14 MHz. The tuned circuit L_1-C_1 (called a **trap**) is also resonant at this frequency, and when connected as shown offers a very high impedance to RF current of that frequency. Effectively, therefore, these two tuned circuits act as insulators for the inner dipole, and the outer sections beyond L_1-C_1 are inactive.

On the next lower frequency band of interest, say 7 MHz, L_1-C_1 shows an inductive reactance and is the electrical equivalent of a coil. The two sections marked Y are now added electrically to X-X so that, together with the loading coils represented by the inductive reactance of L_1-C_1, the system is resonant at 7 MHz out to the ends of the Y sections. This part of the antenna is equivalent to a loaded dipole on 7 MHz and will exhibit about the same impedance at the feed point as a simple dipole for that band. The tuned circuit L_2-C_2 is resonant at 7 MHz and acts as a high impedance for this frequency, so the 7-MHz dipole is in turn insulated, for all

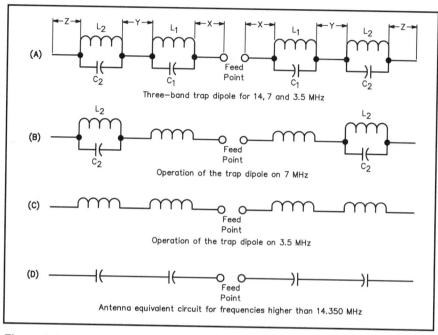

Figure 9-3 — Part A shows the basic construction of an antenna that uses two sets of traps, for operation on three frequency bands. Parts B and C show the inductive loading of the antenna on successively lower-frequency bands. D shows the capacitive loading that would result if the antenna were operated on a higher frequency, although the antenna will not normally be used on frequencies higher than the inner-dipole-section resonance.

practical purposes, from the remaining outer parts of the antenna.

Carrying the same reasoning one step further, L_2-C_2 shows inductive reactance on the next lower frequency band, 3.5 MHz, and is equivalent to a coil on that band. The length of the added sections, Z-Z, together with the two sets of equivalent loading coils indicated in part C, makes the whole system resonant as a loaded dipole on 3.5 MHz. A single transmission line having a characteristic impedance of the same order as the feed-point impedance of a simple dipole can be connected at the center of the antenna. This line will be satisfactorily matched on all three bands, and so will operate at a low SWR on all three. A line of 50-Ω impedance will work just fine.

Since the tuned circuits have some inherent losses, the efficiency of this system depends on the Q of the tuned circuits. Low-loss (high-Q) coils should be used, and the capacitor losses likewise should be kept as low as possible. With tuned circuits that are good in this respect — comparable with the low-loss components used in transmitter tank circuits, for example — the reduction in efficiency as compared with the efficiency of a simple dipole is small, but tuned circuits of low Q can lose an appreciable portion of the power supplied to the antenna.

The lengths of the added antenna sections Y and Z must, in general, be determined experimentally. The length required for resonance in a given band depends on the length/diameter ratio of the antenna conductor and on the L/C ratio of the trap acting as a loading coil. The effective reactance of an LC circuit at half the frequency to which it is resonant is equal to $^2/_3$ the reactance of the inductance at the resonant frequency. For example, if L_1-C_1 resonates at 14 MHz and L_1 has an inductive reactance of 300 Ω at that frequency, the inductive reactance of the circuit at 7 MHz will be equal to $^2/_3 \times 300 = 200$ Ω. The added antenna section, Y, would have to be cut to the proper length to resonate at 7 MHz with this amount of loading. Since any reasonable L/C ratio can be used in the trap without affecting its performance materially at the resonant frequency, the L/C ratio can be varied to control the added antenna length required. The added section will be shorter with high-inductance trap circuits and longer with high-capacitance traps.

Trap dipoles have two major disadvantages. Because the trap dipole is a multi-band antenna, it can do a good job of radiating harmonics. Further, during operation on the lower frequency bands, the series inductance (loading) from the traps raises the antenna Q. That means less bandwidth for a given SWR limit.

Of course quarter-wavelength vertical antennas can also use trap construction. Only one trap is needed for each band in that case, since only the vertical portion of the antenna uses traps. The ground system under the antenna does not use traps.

MOBILE ANTENNAS FOR HF

Mobile antennas are usually vertically mounted whip antennas 8 feet or less in length. As the operating frequency is lowered, the feed-point impedance of a fixed-length antenna appears to be a decreasing resistance in series with an increasing capacitive reactance. This capacitive reactance must be tuned out, which indicates the use of a series inductive reactance, or **loading coil**. The amount of inductance required is determined by the desired operating frequency and where the coil is placed in the antenna.

Base loading requires the lowest value of inductance for a given antenna length, and as the coil is moved farther up the whip, the necessary value increases. This is because the capacitance between the portion of the whip above the coil and the car body decreases (higher capacitive reactance), requiring more inductance to tune the antenna to resonance. One advantage of placing the coil at least part way up the whip, however, is that the current distribution is improved, and that increases the radiation resistance. The major disadvantage is that the requirement for a larger loading coil means that the coil losses will be greater, although this is offset somewhat by lower current through the larger coil. **Center loading** has been generally accepted as a good compromise in mobile-antenna design.

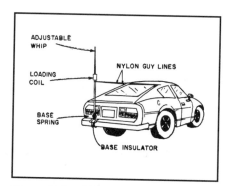

Figure 9-4 — This drawing shows a typical bumper-mounted HF-mobile antenna. Note the nylon guy lines.

Figure 9-4 shows a typical bumper-mounted, center-loaded whip antenna suitable for operation in the HF range. The antenna could also be mounted on the car body proper (such as a fender). The base spring acts as a shock absorber for the base of the whip, since the continual flexing while the car is in motion would otherwise weaken the antenna. A short, heavy, mast section is mounted between the base spring and loading coil. Some models have a mechanism that allows the antenna to be tipped over for adjustment or for fastening to the roof of the car when not in use.

It is also advisable to extend a couple of guy lines from the base of the loading coil to clips or hooks fastened to the rain trough on the roof of the car. Nylon fishing line of about 40-pound test strength is suitable for this purpose. The guy lines act as safety cords and also reduce the swaying motion of the antenna considerably. The feed line to the transmitter is connected to the bumper and base of the antenna. Good low-resistance connections are important here.

Tune-up of the antenna is usually accomplished by changing the height of the adjustable whip section above the precut loading coil. A noise bridge or "SWR analyzer" can aid the process of adjusting the antenna for operation on a particular frequency. Without one of those devices, you can tune the receiver and try to determine where the signals seem to peak up. Once you find this frequency, check the SWR with the transmitter on, and find the frequency of lowest SWR. Shorten the adjustable section to increase the resonant frequency or make it longer to lower the frequency. It is important that the antenna be 10 feet or more away from surrounding objects such as overhead wires, since considerable detuning can occur. Once you find the setting where the SWR is lowest at the center of the desired operating frequency range, record the length of the adjustable section.

Loading Coils

The difficulty in constructing suitable loading coils increases as the operating frequency is lowered for typical antenna lengths used in mobile work. Since the required resonating inductance gets larger and the radiation resistance decreases at lower frequencies, most of the power may be dissipated in the coil resistance and in other ohmic losses. This is one reason why it is advisable to buy a commercially made loading coil with the highest power rating possible, even though you may only be considering low-power operation. Percentage-wise, the coil losses in the higher power loading coils are usually less, with subsequent improvement in radiating efficiency, regardless of the power level used. Of course, this same philosophy also applies to homemade loading coils.

The primary goal here is to provide a coil with the highest possible Q. This means the coil should have a high ratio of reactance to resistance, so that heating losses will be minimized. High-Q coils require a large conductor, "air-wound" construction, large spacing between turns, the best insulating material available, a diameter not less than half the length of the coil (this is not always mechanically feasible) and a minimum of metal in the field.

Once the antenna is tuned to resonance, the input impedance at the antenna terminals will look like a pure resistance. Neglecting losses, this value drops from nearly 15 Ω on 15 meters to 0.1 Ω on 160 meters for an 8-foot whip. When coil and other losses are included, the input resistance increases to approximately 20 Ω on 160 meters and 16 Ω on 15 meters. These values are for relatively high-efficiency systems. From this, you can see that the radiating efficiency is much poorer on 160 meters than on 15 meters under typical conditions.

Since most modern gear is designed to operate into a 50-Ω impedance, a matching network may be necessary with some mobile antennas. This can take the form of either a broad-band transformer, a tapped coil or an LC matching network. With homemade or modified designs, the tapped-coil arrangement is perhaps the easiest to build, while the broad-band transformer requires no adjustment. As the losses go up, so does the input resistance, and in less efficient systems the matching network may be eliminated.

The Equivalent Circuit of a Typical Mobile Antenna

Antenna resonance is defined as the frequency at which the input impedance at the antenna terminals is a pure resistance. The shortest length at which this occurs for a vertical antenna over a ground plane is a quarter wavelength at the operating frequency; the impedance value for this length (neglecting losses) is about 36 Ω. The idea of resonance can be extended to antennas shorter (or longer) than a quarter wavelength, and means only that the input impedance is purely resistive. When the frequency is reduced, the antenna looks like a series RC circuit, as shown in Figure 9-5.

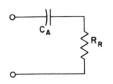

Figure 9-5 — At frequencies below the resonant frequency, the whip antenna will show capacitive reactance as well as resistance. R_R is the radiation resistance, and C_A represents the capacitive reactance.

The capacitive reactance can be canceled out by connecting an equivalent inductive reactance, L_L, in series as shown in Figure 9-6. This arrangement tunes the system to resonance at a particular frequency. The price you pay for the shortened antenna is decreased bandwidth.

Mobile Antenna Efficiency

Antenna efficiency was discussed earlier in this chapter. In that section the importance of minimizing ohmic losses was discussed. If you have trouble understanding the material here, review that section.

For lowest loss, an amateur would use a self-resonant antenna, such as a dipole or a quarter-wavelength vertical. That is not possible, of course, for HF mobile operation, except at the upper end of the range.

Mobile antenna losses at HF, for the most part, are caused by two factors: ground return resistance and loading coil losses. To minimize ground losses, the transmission line should be connected to the metal automobile body through a low resistance and low reactance connection. A good way to do this is with a short length of ground strap or coaxial-cable braid.

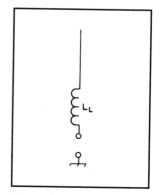

Figure 9-6 — At frequencies lower than the resonant frequency, the capacitive reactance of a whip antenna can be canceled by adding an equivalent inductive reactance, in the form of a loading coil, in series with the antenna.

Another method that can be used successfully to reduce loading-coil losses is a technique called **top loading**. This method calls for a "capacitive hat" to be added above the loading coil, either just above the coil or near the top of the antenna whip.

The added capacitance at the top of the whip allows a smaller value of load inductance. This reduces the amount of loss in the system, and improves the antenna radiation efficiency.

[Now turn to Chapter 10 and study questions A9D02 through A9D07, A9D11 and A9D12. Review this section as needed.]

ANTENNA GAIN, BEAMWIDTH AND RADIATION PATTERNS

In a perfect directional antenna, the radio energy would be concentrated in the *forward* direction only. This is known as the **major lobe of radiation**. (Most beams also have **minor lobes** in the back and side directions.)

Figure 9-7 is an example of a radiation pattern for a beam antenna, illustrating major and minor pattern lobes. The greater the number of elements and the longer the distance between elements (up to an optimum spacing), the narrower the radiated beam. By reducing radiation in the side and back directions and concentrating it instead into a narrow beam in the forward direction, a beam antenna can have more effective radiated power than a dipole. The ratio (expressed in decibels) between the signal radiated from a beam in the direction of the main lobe and the signal radiated from a reference antenna (usually a dipole) in the same direction and at the same transmitting location is called the **gain** of the beam. A typical beam might have 6 dB of gain compared to a dipole, which means that it makes your signal sound four times (6 dB) louder than if you were using a dipole with the same transmitter .

The gain of directional antennas is the result of concentrating the radio wave in one direction at the expense of radiation in other directions. Since practical antennas are not perfect, there is always some radiation in undesired directions as well. A plot of relative field strength in all horizontal directions is called the horizontal-radiation pattern. The vertical-radiation pattern is a similar plot of the field strength in the vertical plane.

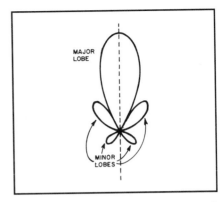

Figure 9-7 — This diagram represents the radiation pattern for a hypothetical beam antenna, illustrating major and minor radiation lobes.

Measuring an antenna horizontal radiation pattern accurately requires that the antenna be placed well away from any obstructions for a distance of at least several wavelengths. On a typical antenna test range, a low-power transmitter feeds a signal to a standard antenna. The antenna to be tested is connected to a receiver, and the received signal strengths are recorded as the test antenna is rotated through 360°. The received signal strengths at various points are plotted, and a radiation pattern results. Vertical radiation patterns are more difficult to measure.

Modern computers make it possible to analyze antenna designs by mathematically modeling the antenna. Computer analysis allows us to study how performance changes as the height of the antenna changes or what effects different ground conditions will have. There are a number of programs in common use for this type of analysis. Most of them are derived from a program developed at US government laboratories, called *NEC*, short for "Numerical Electromagnetics Code." This complex program, written for mainframe computers, uses a modeling technique called the "method of moments." The antenna wire (or tubing elements) is modeled as a series of segments, and a distinct value of current through each segment is computed. The field resulting from the RF current in each segment is evaluated, along with the effects from other mutually coupled segments.

In the early 1980s, *MININEC* was written in BASIC for use on personal computers. Several simplifying assumptions were necessary because of memory limitations and computing speed. Most of the programs available to amateurs for modeling antenna performance are based on *MININEC*, or on the full *NEC* program as personal computers gain computing speed and memory.

Programs that use the method of moments analysis can predict antenna performance, provided the antenna is modeled properly for the program input. Most of the programs can even produce a set of radiation patterns for the antenna.

If the radiation pattern is calculated for the plane of the antenna elements, it is called an E-plane radiation pattern. (It is called the E-plane because the electric field of the signal radiated from the antenna is oriented in this plane.) If the radiation pattern is calculated for the plane perpendicular to the elements, it is called the H-plane radiation pattern. (The symbol H is used to represent the magnetic field, which is oriented perpendicular — 90° — to the electric field.) The free-space polarization of an antenna is determined by the orientation of the electric field of the signal radiated from an antenna.

When we consider real antennas near the Earth, we usually refer to the antenna polarization relative to the ground. For example, if the electric field of the radiation from an antenna is oriented parallel to the surface of the Earth, we say the antenna is *horizontally polarized*. If the electric field is oriented perpendicular to the surface of the Earth, the antenna is *vertically polarized*. The simplest way to determine the polarization of an antenna is by looking at the orientation of the elements. The electric field will be oriented in the same direction as the antenna elements. So if a Yagi antenna has elements oriented parallel to the ground, it will produce horizontally polarized radiation, and if the antenna elements are oriented perpendicular to the ground, it will produce vertically polarized radiation.

Analyzing Radiation Patterns

An antenna radiation pattern contains a wealth of information about the antenna and its expected performance. By learning to recognize certain types of patterns and what they represent, you will be able to identify many important antenna characteristics, and to compare design variations.

Always look at the signal-strength scale on a pattern. The scales are usually expressed in decibels, but the actual values used can vary quite a bit. The scale is usually selected to show desired pattern details. Different scales change the shape of the pattern, however, so it is difficult or misleading to compare radiation patterns of several antennas unless the scales are the same.

The radiation pattern is usually drawn so it just touches the outer circle at its maximum strength. The scale can then be used to measure the relative strength of signals radiated in any direction. The signal-strength scale represents a relative comparison then. To compare one antenna with another, you should also look for a note indicating the basis for the outer ring, or 0 dB circle on the pattern. A dipole pattern plotted on a scale that uses 0 dB = +2.15 dBi can't be compared directly with a Yagi pattern plotted on a scale that uses 0 dB = +6 dBi. (The units dBi tell us the signal powers are compared to an *isotropic radiator*, which is a theoretical ideal point-source antenna located in free space. An isotropic radiator transmits signals equally in all directions, so the radiation pattern for this antenna is a perfect sphere.)

The **beamwidth** is the angular distance between the points on either side of the main lobe, at which the gain is 3 dB below the maximum. See Figure 9-8. A three-element beam, for example, might be found to have a beamwidth of 50°. This means if you turn your beam plus or minus 25° from the optimum heading, the signal you receive (and the signal received from your transmitter) will drop by 3 dB.

Take a look at Figure 9-9. This is a free-space pattern, meaning it is taken as being completely isolated from any Earth effects. If the same pattern were considered relative to Earth, however, we would call it an *azimuthal pattern*, or a *horizontal pattern*. (Azimuth refers to compass directions.) In that case we would probably label the angles from 0° to 360° instead of 0° to + and − 180°. Keep in mind that the pattern doesn't tell us the polarization of the radiation from the antenna. Calling it an azimuthal pattern helps avoid any confusion about the antenna polarization.

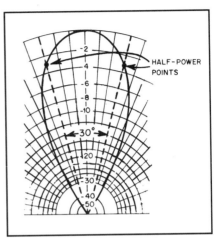

Figure 9-8 — The width of a beam is the angular distance between the directions at which the received or transmitted power is one half the maximum power (−3 dB).

The major lobe of radiation from the antenna of Figure 9-9 points to the right, and is centered along the 0° axis. You can make a pretty good estimate of the beamwidth of this antenna by carefully reading the graph. Notice that angles are marked off every 15°, and the −3 dB circle is the first one inside the outer circle. It looks like the pattern crosses the −3 dB circle at points about 25° either side of 0°. So we can estimate the beamwidth of this antenna as 50°.

We are usually interested in the *front-to-back ratio* of our beam antennas. This is a comparison of the signal strength in the forward direction compared to the strength in a direction 180° from that. This number gives us a good idea of how well an antenna will reject an interfering signal coming in from a direction opposite the desired direction. In the case of the example shown in Figure 9-9, we can read the front-to-back ratio by finding the maximum value of the minor lobe at 180°. This maximum appears to be about half way between the −12 dB and −24 dB circles, so we would estimate it to be about 18 dB below the main lobe.

Front-to-side ratio is another antenna parameter that is often of interest. You can probably guess that this refers to the strength of signals coming in from a direction 90° away from the main lobe. We can estimate the front-to-side ratio of the pattern shown in Figure 9-9 by reading the value off the graph. You can see there is a minor radiation lobe off each side of the antenna, and the maximum strength of

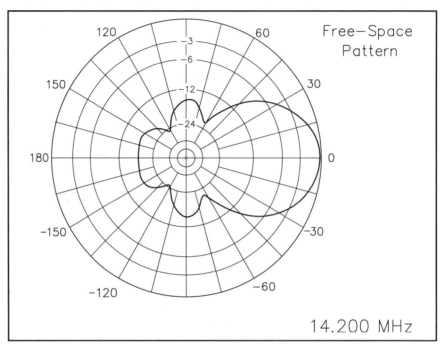

Figure 9-9 — This free-space radiation pattern is the H-plane pattern of a beam antenna for operation on the 20-meter band. Viewed in relation to the Earth, this pattern represents an azimuthal, or horizontal radiation pattern. The text describes how to read front-to-back and front-to-side ratios from the graph.

each lobe is a bit more than 12 dB below the main-lobe maximum. A front-to-side ratio of 14 dB looks like a pretty good estimate for this pattern.

When we look at the radiation pattern in a single plane we are not getting the whole picture. Antennas radiate signals in all directions, so we would have to look at the radiation pattern in three dimensions to get the complete picture. Showing such patterns on a two-dimensional page is rather complicated, however. Because there is usually some symmetry to the pattern, we usually take a second view of the radiation in a plane perpendicular to the first plane. In the case of an antenna in free space, we might call this second view an H-plane pattern, because it is taken in the plane of the radio wave magnetic field. If we consider the antenna in relation to Earth, the second view is called a *vertical radiation pattern*, or an *elevation pattern*.

An elevation pattern over real ground is usually shown for half the circle. Any radiation that would have gone down into the ground is reflected back into space above the Earth, and that energy is added to the pattern above the antenna. Figure 9-10 shows an elevation pattern over real ground.

One of the first things you should notice about the pattern shown in Figure 9-10 is that there are four lobes in the forward direction, at various angles above the horizon. The main radiation lobe is at an angle of about 7.5° above the horizon. The other forward-direction lobes appear at about 25°, 45° and 75°.

There are three minor radiation lobes off the back of the antenna shown in Figure 9-10. The graph may be a bit difficult to read accurately here, but the largest of these lobes appears to cross the –30 dB circle of the pattern, so we can estimate the front-to-back ratio of the antenna at about 28 dB for this antenna.

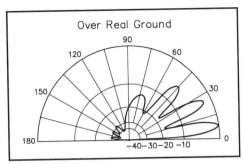

Figure 9-10 — This radiation pattern for a beam antenna over real ground represents an elevation pattern. Notice that there are four radiation lobes in the forward direction and three in the rearward direction. The main lobe in the forward direction is about 7.5° above the horizon.

[We've covered quite a bit of material since a review in the question pool. This is a good time to go to Chapter 10 and study questions A9A09, A9A11, A9B01 through A9B04, A9B09 through A9B11, A9C01, A9C02 and A9C09 through A9C11 before proceeding. If you are uncertain about any of the answers to these questions, be sure to review this section.]

Antenna Optimization

Any antenna design represents some compromises. You may be able to modify the design of a particular antenna to improve some desired characteristic, if you are aware of the trade-offs. As we mentioned earlier, the method of moments computer modeling techniques that have become popular can be a great help in deciding which design modifications will produce the "best" antenna for your situation.

When you evaluate the gain of an antenna, you (or the computer modeling program) will have to take into account a number of parameters. You will have to include the antenna radiation resistance, any loss resistance in the elements and impedance-matching components as well as the E-field and H-field radiation patterns.

You should also evaluate the antenna across the entire frequency band for which it is designed. You may discover that while the gain seems high, it may decrease rapidly as you move away from the single design frequency, and the performance may be poor over the remainder of the band. (You may be willing to make that trade-off if all your operating on that band is within a narrow frequency range.) You may also discover that the feed-point impedance varies widely as you change frequency across the desired band. This will make it difficult to design a single impedance-matching system for the antenna. You may also discover that the front-to-back ratio varies excessively across the band, resulting in too much variation in the rearward pattern lobes.

The forward gain of a Yagi antenna can be increased by using a longer boom, spreading the elements farther apart or adding more elements. Of course there are practical limitations on how long you can make the boom for any antenna! The element lengths will have to be adjusted to retune them when the boom length changes. You can design a Yagi antenna for maximum forward gain, but in that case the feed-point impedance usually becomes too low, and the SWR bandwidth will decrease.

Effects of Ground on Radiation Patterns

The radiation pattern of an antenna over real ground is always affected by the electrical conductivity and dielectric constant of the soil, and most importantly by the height of a horizontally polarized antenna over ground. Signals reflected from the ground combine with the signals radiated directly from the antenna. If the signals are in phase when they combine, the signal strength will be increased, but if they are out of phase, the strength will be decreased. These ground reflections affect the radiation pattern for many wavelengths out from the antenna.

This is especially true of the far-field pattern of a vertically polarized antenna, which in theory will produce a low takeoff angle for the radiation. (The *far-field pattern* refers to signal strengths measured several wavelengths away from the antenna. Signals reflected from the ground as far as 100 wavelengths from the antenna will effect the far-field pattern.) The low-angle radiation from a vertically polarized antenna mounted over seawater will be much stronger than for a similar antenna mounted over rocky soil, for example. (The far-field, low-angle radiation pattern for a horizontally polarized antenna will not be significantly affected, however.)

There are several things you can do to reduce the near-field ground losses of a vertically polarized antenna system. Adding more radials is one common technique. If you can only manage a modest on-ground radial system under an eighth-wavelength, inductively loaded vertical antenna, a wire-mesh screen about an eighth-wavelength square is a good compromise.

Many amateurs prefer to raise the vertical antenna above ground and use an elevated-radial *counterpoise* system. The counterpoise takes the place of a direct connection to the ground. An elevated counterpoise under a vertically polarized antenna can reduce the near-field ground losses, as compared to on-ground radial systems that are much more extensive.

There is little you can do to improve the far-field, low-angle radiation pattern of a vertically polarized antenna if the ground under the antenna has poor conductivity. Most measures that people might try will only be applied close to the antenna, and will only affect the near-field pattern. For example, you can water the ground under the antenna, but unless you watered the ground for 100 or more wavelengths, it won't improve the ground conductivity for these distant ground reflections. (Even then, it would only be a temporary fix, or would have to be repeated often enough to keep the ground wet.) Adding more radials or extending the radials more than a quarter wavelength won't help either, unless you are able to build an extensive ground screen under the antenna out to 10 or more wavelengths!

[Turn to Chapter 10 now and study questions A9B05 through A9B08 and questions A9C03 through A9C08. Review this section if you have difficulty with any of these questions.]

DESIGNING IMPEDANCE-MATCHING SYSTEMS

Many amateur antenna systems use some type of impedance matching network in the station, between the transmitter and the antenna feed line. This method

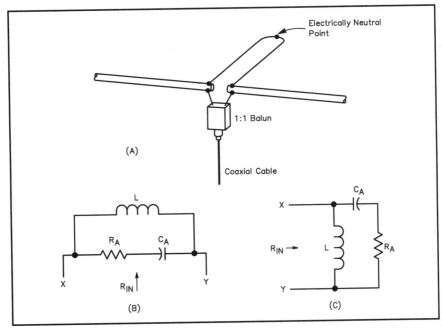

Figure 9-11 — The driven element of a Yagi antenna can be fed with a hairpin matching system, as shown in Part A. B shows the lumped-constant equivalent circuit, where R_A and C_A represent the antenna feed-point impedance, and L represents the parallel inductance of the hairpin. Points X and Y represent the feed-line connection. When the equivalent circuit is redrawn as shown in Part C, we can see that L and C_A form a simple L network to match the feed-line characteristic impedance to the antenna resistance, R_A.

has the advantage of putting the operating controls in the shack, close to the operator, and so is quite convenient. For best efficiency, however, it is desirable to put any impedance-matching network at the antenna, between the antenna and feed line.

It is often possible to design a system that will provide a good impedance match over all or most of a band without further adjustments once it has been set. Of course it is also possible to use some type of remote tuning, such as a motor-driven variable capacitor to adjust the match as you change operating frequency.

You will have to know several pieces of information to design an impedance-matching system. The feed-point radiation resistance and any loss resistance must be known, as well as the reactance at the antenna feed point. You must also take the characteristic impedance of the feed line into account.

There are several popular matching systems used with Yagi antennas. The *hairpin* matching system is one popular method. Figure 9-11A illustrates this technique. To use a hairpin match, the driven element must be tuned so it has a capacitive reactance at the desired operating frequency. (This means the element is a little too short for resonance.) The hairpin adds some inductive reactance to match the feed-point impedance.

Figure 9-11B shows the equivalent lumped-constant network for a typical hairpin matching system on a 3-element Yagi. R_A and C_A represent the antenna feed-point impedance. L is the parallel inductance of the hairpin. When the network is redrawn, as shown in Figure 9-11C, you can see that the circuit is the equivalent of an L network.

[Study questions A9D08 through A9D10 in Chapter 10 before you go on to the next section. Review this section as needed.]

VELOCITY FACTOR AND ELECTRICAL LENGTH OF A TRANSMISSION LINE

Radio waves travel through space at a speed of about 300,000,000 meters per second (approximately 186,000 miles per second). If a wave travels through anything but a vacuum, its speed is always less than that.

Waves used in radio communication may have frequencies from about 10,000 to several billion hertz (Hz). Suppose the **frequency** of the wave shown in Figure 9-12 is 30,000,000 Hz, or 30 megahertz (MHz). One cycle is completed in 1/30,000,000 second (33 nanoseconds). This time is called the *period* of the wave. The wave is traveling at 300,000,000 meters per second, so it will move only 10 meters during the time the current is going through one complete cycle. The electromagnetic field 10 meters away from the source is caused by the current that was flowing one period earlier in time. The field 20 meters away is caused by the current that was flowing two periods earlier, and so on.

Wavelength is the distance between two points of the same phase (for example, peaks) in consecutive cycles. This distance must be measured along the

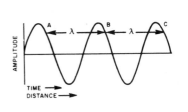

Figure 9-12 — The instantaneous amplitude of both fields (electric and magnetic) varies sinusoidally with time, as shown in this graph. Since the fields travel at constant velocity, the graph also represents the instantaneous distribution of field intensity along the wave path. The distance between two points of equal phase, such as A-B and B-C, is the length of the wave.

direction of wave travel. In the example found in the previous paragraph, the wavelength is 10 meters. The formula for wavelength is:

$$\lambda = \frac{v}{f}$$ (Equation 9-2)

where:
  λ = wavelength.
  v = velocity of wave.
  f = frequency of wave.

Equations normally specify basic units, such as velocity in m/s or ft/s and frequency in Hz. When we are working with very large or very small numbers it is sometimes more convenient to learn a formula that includes a smaller number and specifies units in more common multiples, such as frequency in MHz. For waves traveling in free space, the formula to calculate wavelength is:

$$\lambda(\text{meters}) = \frac{300}{f(\text{MHz})}$$ (Equation 9-3)

or

$$\lambda(\text{feet}) = \frac{984}{f(\text{MHz})}$$ (Equation 9-4)

Wavelength in a Wire

An alternating voltage applied to a feed line would give rise to the sort of current shown in Figure 9-13. (Feed lines are also known as *transmission lines*, because they transfer the radio energy from the transmitter to the antenna — and from the antenna to the receiver.) If the frequency of the ac voltage is 10 MHz, each cycle will take 0.1 microsecond. Therefore, a complete current cycle will be present along each 30 meters of line (assuming free-space velocity). This distance

is one wavelength. Current observed at
B occurs just one cycle later in time than
the current at A. To put it another way,
the current initiated at A does not appear
at B, one wavelength away, until the applied voltage has had time to go through
a complete cycle.

In Figure 9-13, the series of drawings shows how the instantaneous current might appear if we could take snap-
shots of it at quarter-cycle intervals. The
current travels out from the input end of
the line in waves.

At any selected point on the line, the
current goes through its complete range
of ac values in the time of one cycle just
as it does at the input end.

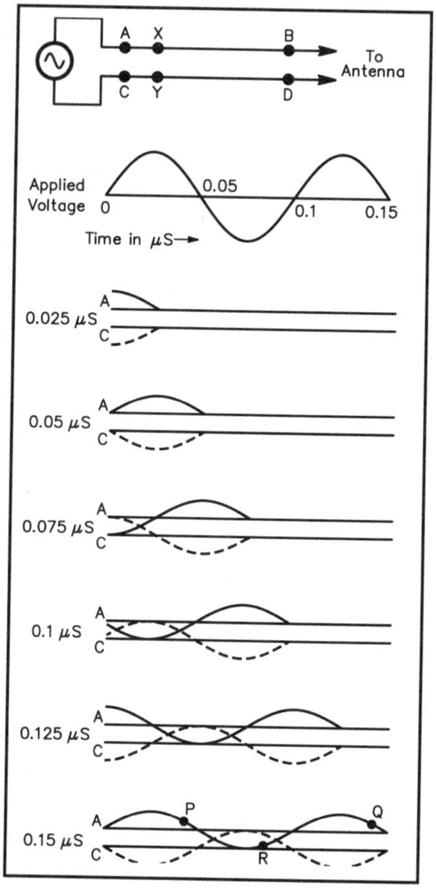

Velocity of Propagation

In the previous example, we assumed that energy traveled along the line
at the velocity of light. The actual velocity is very close to that of light if the
insulation between the conductors of the
line is solely air. The presence of **dielectric** materials other than air reduces the
velocity, since electromagnetic waves
travel more slowly in materials other
than a vacuum. Because of this, the
length of line in one wavelength will
depend on the velocity of the wave as it
moves along the line.

The ratio of the actual velocity at
which a signal travels along a line to the
speed of light in a vacuum is called the
velocity factor.

Figure 9-13 — **Instantaneous current
along a transmission line at
successive time intervals. The period
of the wave (the time for one cycle)
is 0.1 microsecond.**

$$V = \frac{\text{speed of wave (in line)}}{\text{speed of light (in vacuum)}}$$

(Equation 9-5)

where V = velocity factor.

The velocity factor is also related to the **dielectric constant**, ε, by:

$$V = \frac{1}{\sqrt{\varepsilon}}$$

(Equation 9-6)

where:

    V = velocity factor.

    ε = dielectric constant.

    For example, several popular types of coaxial cable have a polyethylene dielectric, which has a dielectric constant of 2.3. For those types of coaxial cable, we can use Equation 9-6 to calculate the velocity factor of the line.

$$V = \frac{1}{\sqrt{\varepsilon}} = \frac{1}{\sqrt{2.3}} = \frac{1}{1.5} = 0.66$$

Electrical Length

    The electrical length of a transmission line (or antenna) is not the same as its physical length. The electrical length is measured in wavelengths at a given frequency. To calculate the physical length of a transmission line that is electrically one wavelength, use the formulas:

$$\text{Length (meters)} = \frac{300}{f(\text{MHz})} \times V \qquad \text{(Equation 9-7)}$$

or

$$\text{Length (feet)} = \frac{984}{f(\text{MHz})} \times V \qquad \text{(Equation 9-8)}$$

where:

    f = operating frequency (in MHz).

    V = velocity factor.

    Suppose you want a section of RG-8 coaxial cable that is one-quarter wavelength at 14.1 MHz. What is its physical length? The answer depends on the dielectric used in the coaxial cable. RG-8 is manufactured with polyethylene or foamed polyethylene dielectric; velocity factors for the two versions are 0.66 and 0.80, respectively. We'll use the polyethylene line with a velocity factor of 0.66 for our example. The length in meters of one wavelength is given by Equation 9-7:

$$\text{Length (meters)} = \frac{300}{f(\text{MHz})} \times V = \frac{300}{14.1} \times 0.66 = 21.3 \times 0.66 = 14.1 \text{ m}$$

To find the physical length for a quarter wavelength section of line, we must divide this value by 4. A quarter wavelength section of this coax is 3.52 meters.

    Table 9-1 lists velocity factors and other characteristics for some other common feed lines. You can calculate the physical length of a section of any type of feed line, including twin lead and ladder line, at some specific frequency as long as you know the velocity factor.

    In review, the lower the velocity factor, the slower a radio-frequency wave moves through the line. The lower the velocity factor, the shorter a line is for the

Table 9-1

Characteristics of Commonly Used Transmission Lines

Type of line	Z_o Ohms	Vel %	pF per foot	OD (inches)	Diel. Material	Max Operating volts (RMS)	Loss in dB per 100 ft at 30 MHz
RG-8	52.0	66	29.5	0.405	PE	4000	1.4
RG-8 Foam	50.0	80	25.4	0.405	Foam PE	1500	0.9
RG-8X	52.0	75	26.0	0.242	Foam PE	—	—
RG-58	53.5	66	28.5	0.195	PE	1900	2.4
RG-58 Foam	53.5	79	28.5	0.195	Foam PE	600	2.1
RG-59	73.0	66	21.0	0.242	PE	2300	2.1
RG-59 Foam	75.0	79	16.9	0.242	Foam PE	800	1.6
RG-141	50.0	70	29.4	0.190	PTFE	1900	—
RG-174	50.0	66	30.8	0.100	PE	1500	6.5
Aluminum Jacket Foam Dielectric							
1/2 inch	50.0	81	25.0	0.500		2500	0.4
3/4 inch	50.0	81	25.0	0.750		4000	0.32
7/8 inch	50.0	81	25.0	0.875		4500	0.26
1/2 inch	75.0	81	16.7	0.500		2500	0.49
3/4 inch	75.0	81	16.7	0.750		3500	0.34
7/8 inch	75.0	81	16.7	0.875		4000	0.33
Parallel Conductor							
Open wire, #12	600	97	—	—		—	0.1
75-Ω transmitting twin lead	75.0	67	19.0	—		—	—
300-Ω twin lead	300.0	80	5.8	—		—	—
300-Ω tubular	300.0	77	4.6	—		—	0.59
Open wire, "Window" Type Ladder Line							
1/2 inch	300.0	95	—	—		—	0.18
1 inch	450.0	95	—	—		—	0.18

Dielectric Designation	Name	Temperature Limits
PE	Polyethylene	−65° to +80°C
Foam PE	Foamed Polyethylene	−65° to +80°C
PTFE	Polyetrafluoroethylene (Teflon)	−65° to +250°C

same electrical length at a given frequency. One wavelength in a practical line is always shorter than a wavelength in free space.

Feed Line Loss

When you select a feed line for a particular antenna, you must consider some conflicting factors and make a few trade-offs. For example, most amateurs want to use a relatively inexpensive feed line. We also want a feed line that does not lose an appreciable amount of signal energy, however. For many applications, coaxial cable seems to be a good choice, but parallel-conductor feed lines generally have lower loss and may provide some other advantages, such as operation with high SWR conditions on the line.

Line loss increases as the operating frequency increases, so on the lower-frequency HF bands, you may decide to use a less-expensive coaxial cable with a higher loss than you would on 10 meters. Open-wire or ladder-line feed lines generally have lower loss than coaxial cables at any particular frequency. Table 9-1 includes approximate loss values for 100 feet of the various feed lines at 30 MHz. Again, these values vary significantly as the frequency changes.

The line loss values given in Table 9-1 are for a feed line with load and input impedances that are matched to the feed line characteristic impedance. (A perfectly matched line condition means the SWR will be 1:1.) If the impedances are not matched, the loss will be somewhat larger than the values listed. (In the case of a mismatched line, the SWR will be greater than 1:1.)

The voltage **reflection coefficient** is the ratio of the reflected voltage at some point on a feed line to the incident voltage at the same point. It is also equal to the ratio of reflected current to incident current at some point on the line. The reflection coefficient is determined by the relationship between the feed line characteristic impedance and the actual load impedance. The reflection coefficient is a complex quantity, having both amplitude and phase. It is generally designated by the lower case Greek letter ρ (rho), although some professional literature uses the capital Greek letter Γ (gamma). The reflection coefficient is a good parameter to describe the interactions at the load end of a mismatched transmission line.

[Before proceeding, study examination questions A9E01 through A9E12 in Chapter 10. Review this section as needed.]

ADVANCED CLASS (ELEMENT 4A) QUESTION POOL — WITH ANSWERS

DON'T START HERE!

This chapter contains the complete question pool for the Element 4A exam. Element 4A is the Advanced class Amateur Radio license exam. To earn an Advanced class license, you must pass the Element 2 (Novice), Element 3A (Technician) and Element 3B (General) written exams, in addition to the Element 1B 13-wpm Morse-code exam. You have credit for passing the Element 2 exam if you have a Novice license. You also have credit for the Element 3A exam if you hold a Technician or Technician Plus class license, and credit for the Element 3B and 1B exams if you have a General class license. If you do not have a Novice, Technician, Technician Plus or General license you will have to pass those written exams and the code exam. *In that case, you should be studying a copy of the ARRL publication*

Now You're Talking! *for the Novice and Technician exams or* The ARRL General Class License Manual *for the General class exam.*

Before you read the questions and answers printed in this chapter, be sure to read the text in the previous chapters. Use these questions as review exercises, when the text tells you to study them. (Paper clips make excellent place markers to help you find your place in the text and the question pool as you study.) Don't try to memorize all the questions and answers.

The material presented in this book has been carefully written and presented to guide you step-by-step through the learning process. By understanding the electronics principles and Amateur Radio concepts as they are presented, your insight into our hobby and your appreciation for the privileges granted by an Amateur Radio license will be greatly enhanced.

This question pool, released by the Volunteer Examiner Coordinators' Question Pool Committee in December 1994, will be used on exams beginning July 1, 1995. The pool is scheduled to be used until June 30, 1999. Changes to FCC Rules and other factors may result in earlier revisions. Such changes will be announced in *QST* and other Amateur Radio publications.

How Many Questions?

The FCC specifies that an Element 4A exam must include 50 questions. Nine subelements make up each exam question pool. The FCC also specifies the number of questions from each subelement that must appear on the exam. For example, there must be one question from the Operating Procedures subelement and six questions from the Circuit Components subelement. Table 10-1 summarizes the number of questions to be selected from each subelement on the Advanced class (Element 4A) exam. These numbers are also included at the beginning of each subelement in the question pool.

Table 10-1

Advanced Class Exam Content

Subelement	Topic	Number of questions
A1	Commission's Rules	6
A2	Operating Procedures	1
A3	Radio-Wave Propagation	2
A4	Amateur Radio Practices	4
A5	Electrical Principles	10
A6	Circuit Components	6
A7	Practical Circuits	10
A8	Signals and Emissions	6
A9	Antennas and Feed Lines	5

The Volunteer Examiner Coordinators' Question Pool Committee has broken the subelements into smaller groups. Each subelement has the same number of groups as there are questions from that subelement on the exam. This means there are six groups for the Commission's Rules subelement and ten groups for the Electrical Principles subelement. The Question Pool Committee suggests one question from each group be used on the exam. These groups are called out with a list of topics given in bold type in the question pool. They are labeled alphabetically within each subelement. The six groups in the Commission's Rules subelement are labeled A1A through A1F. This list of topics for each subelement forms the syllabus, or study guide for the exam element.

The question numbers used in the question pool relate to the syllabus or study guide printed at the end of the Introduction, before Chapter 1 of this book. The syllabus is an outline of topics covered by the exam. Each question number begins with an A to indicate the question is from the Advanced question pool. Next is a number to indicate which of the nine subelements the question is from. A letter following this number indicates which subelement group the question is from. Each question number ends with a two-digit number to specify its position in the set. So question A2A01 is the first question in the A group of the second subelement. Question A9C08 is the eighth question in the C group of the ninth subelement.

This question pool contains more than ten times the number of questions necessary to make up an exam. This ensures that the examiners have sufficient questions to choose from when they make up an exam.

Who Picks the Questions?

The FCC allows Volunteer Examiner teams to select the questions that will be used on amateur exams. If your test is coordinated by the ARRL/VEC, however, your test will be prepared by the VEC, or by using a computer program supplied by the VEC.

Question Pool Format

The rest of this chapter contains the entire Element 4A question pool. We have printed the answer key to these questions along the edge of the page. There is a line to indicate where you should fold the page to hide the answer key while you study. (Fold the edge of the page *under* rather than *over* the page, so you don't cover part of the questions on the page.) After making your best effort to answer the questions, you can look at the answers to check your understanding.

We have also included page references along with the answers. These page numbers indicate where in this book you will find the text discussion related to each question. If you have any problems with a question, refer to the page listed for that question. You may have to study beyond the listed page number to review all the related material. With the "Commission's Rules" questions of Subelement 1, we have also included references to the sections of Part 97 of the FCC Rules. Chapter 1 of this book includes the text of the rules for these questions, along with an explanation to help you answer the questions. Chapter 1 does *not* include a complete copy of Part 97, however. For the complete text of the FCC Rules governing Amateur Radio, we recommend a copy of *The FCC Rule Book*, published by ARRL. That book includes the complete text for Part 97, along with text to explain all the rules. *The FCC Rule Book* is updated regularly to reflect changes to the Rules.

The VEC Question Pool Committee included all the drawings for the question pool on a single page. We placed a copy of this page at the beginning of the question pool, for your information. You may receive a page like this with your exam papers, so you will have the figures available if your exam includes any of the questions that reference them. Some Volunteer Examiner Coordinators (such as the ARRL/VEC) may place the appropriate figures with those exam questions that require them, rather than use the entire graphics sheet with each exam. Either way, you will have the necessary drawings for reference if your exam includes one of those questions. For your convenience, we also placed the individual figures near the questions that refer to them in the question pool.

Good luck with your studies.

Element 4A (Advanced)

Figure A5-1

Figure A6-1

Figure A6-2

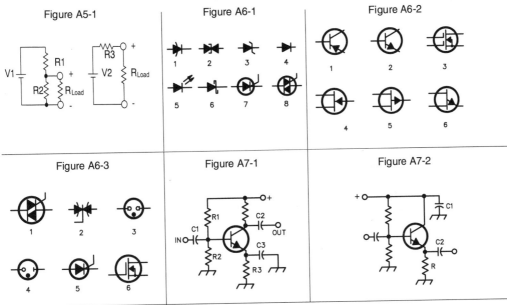

Figure A6-3

Figure A7-1

Figure A7-2

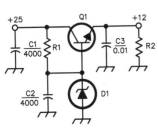

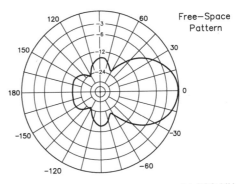

Figure A7-3

Figure A9-1

Free—Space
Pattern

14.200 MHz

Figure A9-2

Over Real Ground

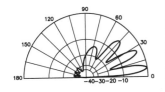

As prepared by Question Pool Committee
for all examinations administered
July 1, 1995 through June 30, 1999

Answer
Key

Page
numbers
tell you
where to
look in this
book for
more
information.

Subelement
A1

Numbers
in [square
brackets]
indicate
sections in
Part 97,
the
Amateur
Radio
Rules.

ELEMENT 4A (ADVANCED CLASS) QUESTION POOL

As released by The Question Pool Committee, National Conference of
Volunteer Examiner Coordinators

December 1, 1994

Subelement A1 — COMMISSION'S RULES

[6 Exam Questions — 6 Groups]

A1A Advanced control operator frequency privileges; station identification; emissions standards

A1A01
(A)
[97.301c]
Page 1-3

A1A01
What are the frequency limits for Advanced class operators in the
75/80-meter band (ITU Region 2)?
A. 3525 - 3750 kHz and 3775 - 4000 kHz
B. 3500 - 3525 kHz and 3800 - 4000 kHz
C. 3500 - 3525 kHz and 3800 - 3890 kHz
D. 3525 - 3775 kHz and 3800 - 4000 kHz

A1A02
(B)
[97.301c]
Page 1-3

A1A02
What are the frequency limits for Advanced class operators in the 40-meter
band (ITU Region 2)?
A. 7000 - 7300 kHz
B. 7025 - 7300 kHz
C. 7025 - 7350 kHz
D. 7000 - 7025 kHz

A1A03
(D)
[97.301c]
Page 1-3

A1A03
What are the frequency limits for Advanced class operators in the 20-meter
band?
A. 14000 - 14150 kHz and 14175 - 14350 kHz
B. 14025 - 14175 kHz and 14200 - 14350 kHz
C. 14000 - 14025 kHz and 14200 - 14350 kHz
D. 14025 - 14150 kHz and 14175 - 14350 kHz

A1A04
What are the frequency limits for Advanced class operators in the 15-meter band?
A. 21000 - 21200 kHz and 21250 - 21450 kHz
B. 21000 - 21200 kHz and 21300 - 21450 kHz
C. 21025 - 21200 kHz and 21225 - 21450 kHz
D. 21025 - 21250 kHz and 21270 - 21450 kHz

A1A05
If you are a Technician Plus licensee with a Certificate of Successful Completion of Examination (CSCE) for Advanced privileges, how do you identify your station when transmitting on 14.185 MHz?
A. Give your call sign followed by the name of the VEC who coordinated the exam session where you obtained the CSCE
B. Give your call sign followed by the slant mark "/" followed by the identifier "AA"
C. You may not use your new frequency privileges until your license arrives from the FCC
D. Give your call sign followed by the word "Advanced"

A1A06
How must an Advanced class operator using Amateur Extra frequencies identify during a contest, assuming the contest control operator holds an Amateur Extra class license?
A. With his or her own call sign
B. With the control operator's call sign
C. With his or her own call sign followed by the identifier "AE"
D. With the control operator's call sign followed by his or her own call sign

A1A07
How must an Advanced class operator using Advanced frequencies identify from a Technician Plus class operator's station?
A. With either his or her own call sign followed by the identifier "KT", or the Technician Plus call sign followed by the identifier "AA"
B. With the Technician Plus call sign
C. The Advanced class operator cannot use Advanced frequencies while operating the Technician Plus station
D. With either his or her own call sign only, or the Technician Plus call sign followed by his or her own call sign

A1A08
What is the maximum mean power permitted for any spurious emission from a transmitter or external RF power amplifier transmitting on a frequency below 30 MHz?
A. 50 mW
B. 100 mW
C. 10 mW
D. 10 W

A1A04
(C)
[97.301c]
Page 1-3

A1A05
(B)
[97.119e3]
Page 1-6

A1A06
(B)
[97.119a]
Page 1-7

A1A07
(D)
[97.119d]
Page 1-7

A1A08
(A)
[97.307d]
Page 1-8

A1A09
(B)
[97.307d]
Page 1-8

A1A09
How much below the mean power of the fundamental emission must any
spurious emissions from a station transmitter or external RF power amplifier
transmitting on a frequency below 30 MHz be attenuated?
A. At least 10 dB
B. At least 40 dB
C. At least 50 dB
D. At least 100 dB

A1A10
(C)
[97.307e]
Page 1-8

A1A10
How much below the mean power of the fundamental emission must any
spurious emissions from a transmitter or external RF power amplifier
transmitting on a frequency between 30 and 225 MHz be attenuated?
A. At least 10 dB
B. At least 40 dB
C. At least 60 dB
D. At least 100 dB

A1A11
(D)
[97.307e]
Page 1-8

A1A11
What is the maximum mean power permitted for any spurious emission
from a transmitter having a mean power of 25 W or less on frequencies
between 30 and 225 MHz?
A. 5 microwatts
B. 10 microwatts
C. 20 microwatts
D. 25 microwatts

**A1B Definition and operation of remote control and automatic control;
control link**

A1B01
(D)
[97.3a36]
Page 1-10

A1B01
What is meant by a remotely controlled station?
A. A station operated away from its regular home location
B. Control of a station from a point located other than at the station
   transmitter
C. A station operating under automatic control
D. A station controlled indirectly through a control link

A1B02
(D)
[97.3a6]
Page 1-11

A1B02
What is the term for the control of a station that is transmitting without the
control operator being present at the control point?
A. Simplex control
B. Manual control
C. Linear control
D. Automatic control

A1B03
Which kind of station operation may not be automatically controlled?
A. Control of a model craft
B. Beacon operation
C. Auxiliary operation
D. Repeater operation

A1B04
Which kind of station operation may be automatically controlled?
A. Stations without a control operator
B. Stations in repeater operation
C. Stations under remote control
D. Stations controlling model craft

A1B05
What is meant by automatic control of a station?
A. The use of devices and procedures for control so that a control operator does not have to be present at a control point
B. A station operating with its output power controlled automatically
C. Remotely controlling a station such that a control operator does not have to be present at the control point at all times
D. The use of a control link between a control point and a locally controlled station

A1B06
How do the control operator responsibilities of a station under automatic control differ from one under local control?
A. Under local control there is no control operator
B. Under automatic control a control operator is not required to be present at a control point
C. Under automatic control there is no control operator
D. Under local control a control operator is not required to be present at a control point

A1B07
What frequencies in the 10-meter band are available for repeater operation?
A. 28.0 - 28.7 MHz
B. 29.0 - 29.7 MHz
C. 29.5 - 29.7 MHz
D. 28.5 - 29.7 MHz

A1B08
What frequencies in the 6-meter band are available for repeater operation (ITU Region 2)?
A. 51.00 - 52.00 MHz
B. 50.25 - 52.00 MHz
C. 52.00 - 53.00 MHz
D. 51.00 - 54.00 MHz

A1B03
(A)
[97.201d,
97.203d,
97.205d]
Page 1-11

A1B04
(B)
[97.205d]
Page 1-11

A1B05
(A)
[97.3a6]
Page 1-11

A1B06
(B)
[97.3a6]
Page 1-11

A1B07
(C)
[97.205b,
97.301b,
c,d]
Page 1-12

A1B08
(D)
[97.205b,
97.301a]
Page 1-12

A1B09
(A)
[97.205b,
97.301a]
Page 1-12

A1B09
What frequencies in the 2-meter band are available for repeater operation (ITU Region 2)?
A. 144.5 - 145.5 and 146 - 148 MHz
B. 144.5 - 148 MHz
C. 144 - 145.5 and 146 - 148 MHz
D. 144 - 148 MHz

A1B10
(B)
[97.205b,
97.301a]
Page 1-12

A1B10
What frequencies in the 1.25-meter band are available for repeater operation (ITU Region 2)?
A. 220.25 - 225.00 MHz
B. 222.15 - 225.00 MHz
C. 221.00 - 225.00 MHz
D. 223.00 - 225.00 MHz

A1B11
(A)
[97.205b,
97.301a]
Page 1-12

A1B11
What frequencies in the 70-cm band are available for repeater operation (ITU Region 2)?
A. 420 - 431, 433 - 435 and 438 - 450 MHz
B. 420 - 440 and 445 - 450 MHz
C. 420 - 435 and 438 - 450 MHz
D. 420 - 431, 435 - 438 and 439 - 450 MHz

A1B12
(C)
[97.301a]
Page 1-12

A1B12
What frequencies in the 23-cm band are available for repeater operation?
A. 1270 - 1300 MHz
B. 1270 - 1295 MHz
C. 1240 - 1300 MHz
D. Repeater operation is not permitted in the band

A1B13
(C)
[97.213b]
Page 1-10

A1B13
If the control link of a station under remote control malfunctions, how long may the station continue to transmit?
A. 5 seconds
B. 10 minutes
C. 3 minutes
D. 5 minutes

A1B14
(C)
[97.3a36,
97.3a37,
97.213a]
Page 1-10

A1B14
What is a control link?
A. A device that automatically controls an unattended station
B. An automatically operated link between two stations
C. The means of control between a control point and a remotely controlled station
D. A device that limits the time of a station's transmission

A1B15
What is the term for apparatus to effect remote control between a control
point and a remotely controlled station?
A. A tone link
B. A wire control
C. A remote control
D. A control link

A1C Type acceptance of external RF power amplifiers and external RF power amplifier kits

A1C01
How many external RF amplifiers of a particular design capable of
operation below 144 MHz may an unlicensed, non-amateur build or modify
in one calendar year without obtaining a grant of FCC type acceptance?
A. 1
B. 5
C. 10
D. None

A1C02
If an RF amplifier manufacturer was granted FCC type acceptance for one
of its amplifier models for amateur use, what would this allow the
manufacturer to market?
A. All current models of their equipment
B. Only that particular amplifier model
C. Any future amplifier models
D. Both the current and any future amplifier models

A1C03
Under what condition may an equipment dealer sell an external RF power
amplifier capable of operation below 144 MHz if it has not been FCC type
accepted?
A. If it was purchased in used condition from an amateur operator and is
sold to another amateur operator for use at that operator's station
B. If it was assembled from a kit by the equipment dealer
C. If it was imported from a manufacturer in a country that does not require
type acceptance of RF power amplifiers
D. If it was imported from a manufacturer in another country, and it was
type accepted by that country's government

A1B15
(D)
[97.3a36,
97.3a37,
97.213a]
Page 1-10

A1C01
(D)
[97.315a]
Page 1-12

A1C02
(B)
[97.315c]
Page 1-12

A1C03
(A)
[97.315b5]
Page 1-15

A1C04
(D)
[97.317a1]
Page 1-12

A1C04
Which of the following is one of the standards that must be met by an external RF power amplifier if it is to qualify for a grant of FCC type acceptance?
A. It must produce full legal output when driven by not more than 5 watts of mean RF input power
B. It must be capable of external RF switching between its input and output networks
C. It must exhibit a gain of 0 dB or less over its full output range
D. It must satisfy the spurious emission standards when operated at its full output power

A1C05
(D)
[97.317a2]
Page 1-12

A1C05
Which of the following is one of the standards that must be met by an external RF power amplifier if it is to qualify for a grant of FCC type acceptance?
A. It must produce full legal output when driven by not more than 5 watts of mean RF input power
B. It must be capable of external RF switching between its input and output networks
C. It must exhibit a gain of 0 dB or less over its full output range
D. It must satisfy the spurious emission standards when placed in the "standby" or "off" position, but is still connected to the transmitter

A1C06
(C)
[97.317b]
Page 1-12

A1C06
Which of the following is one of the standards that must be met by an external RF power amplifier if it is to qualify for a grant of FCC type acceptance?
A. It must produce full legal output when driven by not more than 5 watts of mean RF input power
B. It must exhibit a gain of at least 20 dB for any input signal
C. It must not be capable of operation on any frequency between 24 MHz and 35 MHz
D. Any spurious emissions from the amplifier must be no more than 40 dB stronger than the desired output signal

A1C07
(B)
[97.317a3]
Page 1-12

A1C07
Which of the following is one of the standards that must be met by an external RF power amplifier if it is to qualify for a grant of FCC type acceptance?
A. It must have a time-delay circuit to prevent it from operating continuously for more than ten minutes
B. It must satisfy the spurious emission standards when driven with at least 50 W mean RF power (unless a higher drive level is specified)
C. It must not be capable of modification by an amateur operator without voiding the warranty
D. It must exhibit no more than 6 dB of gain over its entire operating range

A1C08
Which of the following would disqualify an external RF power amplifier from being granted FCC type acceptance?
A. Any accessible wiring which, when altered, would permit operation of the amplifier in a manner contrary to FCC Rules
B. Failure to include a schematic diagram and theory of operation manual that would permit an amateur to modify the amplifier
C. The capability of being switched by the operator to any amateur frequency below 24 MHz
D. Failure to produce 1500 watts of output power when driven by at least 50 watts of mean input power

A1C08
(A)
[97.317c1]
Page 1-15

A1C09
Which of the following would disqualify an external RF power amplifier from being granted FCC type acceptance?
A. Failure to include controls or adjustments that would permit the amplifier to operate on any frequency below 24 MHz
B. Failure to produce 1500 watts of output power when driven by at least 50 watts of mean input power
C. Any features designed to facilitate operation in a telecommunication service other than the Amateur Service
D. The omission of a schematic diagram and theory of operation manual that would permit an amateur to modify the amplifier

A1C09
(C)
[97.317c8]
Page 1-15

A1C10
Which of the following would disqualify an external RF power amplifier from being granted FCC type acceptance?
A. The omission of a safety switch in the high-voltage power supply to turn off the power if the cabinet is opened
B. Failure of the amplifier to exhibit more than 15 dB of gain over its entire operating range
C. The omission of a time-delay circuit to prevent the amplifier from operating continuously for more than ten minutes
D. The inclusion of instructions for operation or modification of the amplifier in a manner contrary to the FCC Rules

A1C10
(D)
[97.317c3]
Page 1-15

A1C11
Which of the following would disqualify an external RF power amplifier from being granted FCC type acceptance?
A. Failure to include a safety switch in the high-voltage power supply to turn off the power if the cabinet is opened
B. The amplifier produces 3 dB of gain for input signals between 26 MHz and 28 MHz
C. The inclusion of a schematic diagram and theory of operation manual that would permit an amateur to modify the amplifier
D. The amplifier produces 1500 watts of output power when driven by at least 50 watts of mean input power

A1C11
(B)
[97.317b2]
Page 1-15

A1D Definition and operation of spread spectrum; auxiliary station operation

A1D01
(C)
[97.3c8]
Page 1-15

A1D01

What is the name for emissions using bandwidth-expansion modulation?
A. RTTY
B. Image
C. Spread spectrum
D. Pulse

A1D02
(C)
[97.311c]
Page 1-15

A1D02

What two spread spectrum techniques are permitted on the amateur bands?
A. Hybrid switching and direct frequency
B. Frequency switching and linear frequency
C. Frequency hopping and direct sequence
D. Logarithmic feedback and binary sequence

A1D03
(C)
[97.311g]
Page 1-15

A1D03

What is the maximum transmitter power allowed for spread spectrum transmissions?
A. 5 watts
B. 10 watts
C. 100 watts
D. 1500 watts

A1D04
(D)
[97.3a7]
Page 1-10

A1D04

What is meant by auxiliary station operation?
A. A station operated away from its home location
B. Remote control of model craft
C. A station controlled from a point located other than at the station transmitter
D. Communications sent point-to-point within a system of cooperating amateur stations

A1D05
(A)
[97.3a6,
97.3a7,
97.3a36,
97.201,
97.205,
97.213a]
Page 1-10

A1D05

What is one use for a station in auxiliary operation?
A. Remote control of a station in repeater operation
B. Remote control of model craft
C. Passing of international third-party communications
D. The retransmission of NOAA weather broadcasts

A1D06
Auxiliary stations communicate with which other kind of amateur stations?
A. Those registered with a civil defense organization
B. Those within a system of cooperating amateur stations
C. Those in space station operation
D. Any kind not under manual control

A1D07
On what amateur frequencies above 222.0 MHz (the 1.25-meter band) are auxiliary stations NOT allowed to operate?
A. 222.00 - 223.00 MHz, 432 - 433 MHz and 436 - 438 MHz
B. 222.10 - 223.91 MHz, 431 - 432 MHz and 435 - 437 MHz
C. 222.00 - 222.15 MHz, 431 - 433 MHz and 435 - 438 MHz
D. 222.00 - 222.10 MHz, 430 - 432 MHz and 434 - 437 MHz

A1D08
What class of amateur license must one hold to be the control operator of an auxiliary station?
A. Any class
B. Technician, Technician Plus, General, Advanced or Amateur Extra
C. General, Advanced or Amateur Extra
D. Advanced or Amateur Extra

A1D09
When an auxiliary station is identified in Morse code using an automatic keying device used only for identification, what is the maximum code speed permitted?
A. 13 words per minute
B. 30 words per minute
C. 20 words per minute
D. There is no limitation

A1D10
How often must an auxiliary station be identified?
A. At least once during each transmission
B. Only at the end of a series of transmissions
C. At the beginning of a series of transmissions
D. At least once every ten minutes during and at the end of activity

A1D11
When may an auxiliary station be identified using a digital code?
A. Any time the digital code is used for at least part of the communication
B. Any time
C. Identification by digital code is not allowed
D. No identification is needed for digital transmissions

A1D06
(B)
[97.3a7]
Page 1-10

A1D07
(C)
[97.201b]
Page 1-11

A1D08
(B)
[97.201a]
Page 1-11

A1D09
(C)
[97.119b1]
Page 1-11

A1D10
(D)
[97.119a]
Page 1-11

A1D11
(A)
[97.119b3]
Page 1-11

A1E "Line A"; National Radio Quiet Zone; business communications; restricted operation; antenna structure limitations

A1E01
Which of the following geographic descriptions approximately describes "Line A"?
A. A line roughly parallel to, and south of, the US-Canadian border
B. A line roughly parallel to, and west of, the US Atlantic coastline
C. A line roughly parallel to, and north of, the US-Mexican border and Gulf coastline
D. A line roughly parallel to, and east of, the US Pacific coastline

A1E02
Amateur stations may not transmit in which frequency segment if they are located north of "Line A"?
A. 21.225-21.300 MHz
B. 53-54 MHz
C. 222-223 MHz
D. 420-430 MHz

A1E03
What is the National Radio Quiet Zone?
A. An area in Puerto Rico surrounding the Aricebo Radio Telescope
B. An area in New Mexico surrounding the White Sands Test Area
C. An Area in Maryland, West Virginia and Virginia surrounding the National Radio Astronomy Observatory
D. An area in Florida surrounding Cape Canaveral

A1E04
Which of the following agencies is protected from interference to its operations by the National Radio Quiet Zone?
A. The National Radio Astronomy Observatory at Green Bank, WV
B. NASA's Mission Control Center in Houston, TX
C. The White Sands Test Area in White Sands, NM
D. The space shuttle launch facilities in Cape Canaveral, FL

A1E05
Which communication is NOT a prohibited transmission in the Amateur Service?
A. Sending messages for hire or material compensation
B. Calling a commercial tow truck service for a breakdown on the highway
C. Calling your employer to see if you have any customers to contact
D. Sending a false distress call as a "joke"

A1E06
Under what conditions may you notify other amateurs of the availability of amateur station equipment for sale or trade over the airwaves?
A. You are never allowed to sell or trade equipment on the air
B. Only if this activity does not result in a profit for you
C. Only if this activity is not conducted on a regular basis
D. Only if the equipment is FCC type accepted and has a serial number

A1E06
(C)
[97.113a3]
Page 1-21

A1E07
When may amateurs accept payment for using their own stations (other than a club station) to send messages?
A. When employed by the FCC
B. When passing emergency traffic
C. Under no circumstances
D. When passing international third-party communications

A1E07
(C)
[97.113a2]
Page 1-21

A1E08
When may the control operator of a repeater accept payment for providing communication services to another party?
A. When the repeater is operating under portable power
B. When the repeater is operating under local control
C. During Red Cross or other emergency service drills
D. Under no circumstances

A1E08
(D)
[97.113a2]
Page 1-21

A1E09
When may an amateur station send a message to a business?
A. When the total money involved does not exceed $25
B. When the control operator is employed by the FCC or another government agency
C. When transmitting international third-party communications
D. When neither the amateur nor his or her employer has a pecuniary interest in the communications

A1E09
D)
[97.113a3]
Page 1-21

A1E10
What must an amateur obtain before installing an antenna structure more than 200 feet high?
A. An environmental assessment
B. A Special Temporary Authorization
C. Prior FCC approval
D. An effective radiated power statement

A1E10
(C)
[97.15a]
Page 1-20

A1E11
(A)
[97.15d]
Page 1-20

A1E11
From what government agencies must you obtain permission if you wish to install an antenna structure that exceeds 200 feet above ground level?
A. The Federal Aviation Administration (FAA) and the Federal Communications Commission (FCC)
B. The Environmental Protection Agency (EPA) and the Federal Communications Commission (FCC)
C. The Federal Aviation Administration (FAA) and the Environmental Protection Agency (EPA)
D. The Environmental Protection Agency (EPA) and National Aeronautics and Space Administration (NASA)

A1F Volunteer examinations: when examination is required; exam credit; examination grading; Volunteer Examiner requirements; Volunteer Examiner conduct

A1F01
(B)
[97.505a]
Page 1-24

A1F01
What examination credit must be given to an applicant who holds an unexpired (or expired within the grace period) FCC-issued amateur operator license?
A. No credit
B. Credit for the least elements required for the license
C. Credit for only the telegraphy requirements of the license
D. Credit for only the written element requirements of the license

A1F02
(B)
[97.503a1]
Page 1-23

A1F02
What ability with international Morse code must an applicant demonstrate when taking an Element 1(A) telegraphy examination?
A. To send and receive text at not less than 13 WPM
B. To send and receive text at not less than 5 WPM
C. To send and receive text at not less than 20 WPM
D. To send text at not less than 13 WPM

A1F03
(A)
[97.503a]
Page 1-23

A1F03
Besides all the letters of the alphabet, numerals 0-9 and the period, comma and question mark, what additional characters are used in telegraphy examinations?
A. The slant mark and prosigns AR, BT and SK
B. The slant mark, open and closed parenthesis and prosigns AR, BT and SK
C. The slant mark, dollar sign and prosigns AR, BT and SK
D. No other characters

A1F04

In a telegraphy examination, how many letters of the alphabet are counted as one word?

A. 2
B. 5
C. 8
D. 10

A1F05

What is the minimum age to be a Volunteer Examiner?

A. 16
B. 21
C. 18
D. 13

A1F06

When may a person whose amateur operator or station license has ever been revoked or suspended be a Volunteer Examiner?

A. Under no circumstances
B. After 5 years have elapsed since the revocation or suspension
C. After 3 years have elapsed since the revocation or suspension
D. After review and subsequent approval by a VEC

A1F07

When may an employee of a company engaged in the distribution of equipment used in connection with amateur station transmissions be a Volunteer Examiner?

A. When the employee is employed in the Amateur Radio sales part of the company
B. When the employee does not normally communicate with the manufacturing or distribution part of the company
C. When the employee serves as a Volunteer Examiner for his or her customers
D. When the employee does not normally communicate with the benefits and policies part of the company

A1F08

Who may administer an examination for a Novice license?

A. Three accredited Volunteer Examiners at least 18 years old and holding at least a General class license
B. Three amateur operators at least 18 years old and holding at least a General class license
C. Any accredited Volunteer Examiner at least 21 years old and holding at least a General class license
D. Two amateur operators at least 21 years old and holding at least a Technician class license

| A1F04 |
| (B) |
| [97.507d] |
| Page 1-24 |

| A1F05 |
| (C) |
| [97.509b2] |
| Page 1-27 |

| A1F06 |
| (A) |
| [97.509b4] |
| Page 1-27 |

| A1F07 |
| (B) |
| [97.509b5] |
| Page 1-27 |

| A1F08 |
| (A) |
| [97.509a, |
| b1, b2, b3i] |
| Page 1-27 |

A1F09
(A)
[97.509e]
Page 1-28

A1F09
When may Volunteer Examiners be compensated for their services?
A. Under no circumstances
B. When out-of-pocket expenses exceed $25
C. When traveling over 25 miles to the test site
D. When there are more than 20 applicants attending an examination session

A1F10
(C)
[97.509e]
Page 1-28

A1F10
What are the penalties that may result from fraudulently administering amateur examinations?
A. Suspension of amateur station license for a period not to exceed 3 months
B. A monetary fine not to exceed $500 for each day the offense was committed
C. Revocation of amateur station license and suspension of operator's license
D. Restriction to administering only Novice class license examinations

A1F11
(D)
[97.509e]
Page 1-28

A1F11
What are the penalties that may result from administering examinations for money or other considerations?
A. Suspension of amateur station license for a period not to exceed 3 months
B. A monetary fine not to exceed $500 for each day the offense was committed
C. Restriction to administering only Novice class license examinations
D. Revocation of amateur station license and suspension of operator's license

A1F12
(A)
[97.509h]
Page 1-28

A1F12
How soon must the administering Volunteer Examiners grade an applicant's completed examination element?
A. Immediately
B. Within 48 hours
C. Within 10 days
D. Within 24 hours

A1F13
(B)
[97.509m]
Page 1-28

A1F13
After the successful administration of an examination, within how many days must the Volunteer Examiners submit the application to their coordinating VEC?
A. 7
B. 10
C. 5
D. 30

A1F14
After the successful administration of an examination, where must the
Volunteer Examiners submit the application?
A. To the nearest FCC Field Office
B. To the FCC in Washington, DC
C. To the coordinating VEC
D. To the FCC in Gettysburg, PA

A1F14
(C)
[97.509m]
Page 1-28

A2 — OPERATING PROCEDURES

Subelement
A2

[1 Exam Question — 1 Group]

A2A Facsimile communications; slow-scan TV transmissions; spread-spectrum transmissions; HF digital communications (i.e., PacTOR, CLOVER, HF packet); automatic HF Forwarding

A2A01
What is facsimile?
A. The transmission of characters by radioteletype that form a picture when
   printed
B. The transmission of still pictures by slow-scan television
C. The transmission of video by amateur television
D. The transmission of printed pictures for permanent display on paper

A2A01
(D)
Page 2-2

A2A02
What is the modern standard scan rate for a facsimile picture transmitted by
an amateur station?
A. 240 lines per minute
B. 50 lines per minute
C. 150 lines per second
D. 60 lines per second

A2A02
(A)
Page 2-2

A2A03
What is the approximate transmission time per frame for a facsimile picture
transmitted by an amateur station at 240 lpm?
A. 6 minutes
B. 3.3 minutes
C. 6 seconds
D. 1/60 second

A2A03
(B)
Page 2-2

A2A04
What is the term for the transmission of printed pictures by radio?
A. Television
B. Facsimile
C. Xerography
D. ACSSB

A2A04
(B)
Page 2-2

A2A05
(C)
Page 2-2

A2A06
(D)
Page 2-5

A2A07
(C)
Page 2-6

A2A08
(C)
Page 2-5

A2A09
(D)
Page 2-5

A2A10
(A)
Page 2-11

A2A05
In facsimile, what device converts variations in picture brightness and darkness into voltage variations?
A. An LED
B. A Hall-effect transistor
C. A photodetector
D. An optoisolator

A2A06
What information is sent by slow-scan television transmissions?
A. Baudot or ASCII characters that form a picture when printed
B. Pictures for permanent display on paper
C. Moving pictures
D. Still pictures

A2A07
How many lines are commonly used in each frame on an amateur slow-scan color television picture?
A. 30 or 60
B. 60 or 100
C. 128 or 256
D. 180 or 360

A2A08
What is the audio frequency for black in an amateur slow-scan television picture?
A. 2300 Hz
B. 2000 Hz
C. 1500 Hz
D. 120 Hz

A2A09
What is the audio frequency for white in an amateur slow-scan television picture?
A. 120 Hz
B. 1500 Hz
C. 2000 Hz
D. 2300 Hz

A2A10
Why are received spread-spectrum signals so resistant to interference?
A. Signals not using the spectrum-spreading algorithm are suppressed in the receiver
B. The high power used by a spread-spectrum transmitter keeps its signal from being easily overpowered
C. The receiver is always equipped with a special digital signal processor (DSP) interference filter
D. If interference is detected by the receiver it will signal the transmitter to change frequencies

A2A11
How does the spread-spectrum technique of frequency hopping (FH) work?
A. If interference is detected by the receiver it will signal the transmitter to change frequencies
B. If interference is detected by the receiver it will signal the transmitter to wait until the frequency is clear
C. A pseudo-random binary bit stream is used to shift the phase of an RF carrier very rapidly in a particular sequence
D. The frequency of an RF carrier is changed very rapidly according to a particular pseudo-random sequence

A2A11
(D)
Page 2-11

A2A12
What is the most common data rate used for HF packet communications?
A. 48 bauds
B. 110 bauds
C. 300 bauds
D. 1200 bauds

A2A12
(C)
Page 2-13

A3 — RADIO-WAVE PROPAGATION

Subelement
A3

[2 Exam Questions — 2 Groups]

A3A Sporadic-E; auroral propagation; ground-wave propagation (distances and coverage, and frequency vs. distance in each of these topics)

A3A01
What is a sporadic-E condition?
A. Variations in E-region height caused by sunspot variations
B. A brief decrease in VHF signal levels from meteor trails at E-region height
C. Patches of dense ionization at E-region height
D. Partial tropospheric ducting at E-region height

A3A01
(C)
Page 3-2

A3A02
What is the term for the propagation condition in which scattered patches of relatively dense ionization develop seasonally at E-region heights?
A. Auroral propagation
B. Ducting
C. Scatter
D. Sporadic-E

A3A02
(D)
Page 3-2

A3A03
In what region of the world is sporadic-E most prevalent?
A. The equatorial regions
B. The arctic regions
C. The northern hemisphere
D. The western hemisphere

A3A03
(A)
Page 3-2

A3A04
(B)
Page 3-3

A3A04
On which amateur frequency band is the extended-distance propagation
effect of sporadic-E most often observed?
A. 2 meters
B. 6 meters
C. 20 meters
D. 160 meters

A3A05
(D)
Page 3-6

A3A05
What effect does auroral activity have upon radio communications?
A. The readability of SSB signals increases
B. FM communications are clearer
C. CW signals have a clearer tone
D. CW signals have a fluttery tone

A3A06
(C)
Page 3-5

A3A06
What is the cause of auroral activity?
A. A high sunspot level
B. A low sunspot level
C. The emission of charged particles from the sun
D. Meteor showers concentrated in the northern latitudes

A3A07
(B)
Page 3-7

A3A07
In the northern hemisphere, in which direction should a directional antenna
be pointed to take maximum advantage of auroral propagation?
A. South
B. North
C. East
D. West

A3A08
(D)
Page 3-5

A3A08
Where in the ionosphere does auroral activity occur?
A. At F-region height
B. In the equatorial band
C. At D-region height
D. At E-region height

A3A09
(A)
Page 3-6

A3A09
Which emission modes are best for auroral propagation?
A. CW and SSB
B. SSB and FM
C. FM and CW
D. RTTY and AM

A3A10
As the frequency of a signal is increased, how does its ground-wave propagation distance change?
A. It increases
B. It decreases
C. It stays the same
D. Radio waves don't propagate along the Earth's surface

A3A10
(B)
Page 3-9

A3A11
What typical polarization does ground-wave propagation have?
A. Vertical
B. Horizontal
C. Circular
D. Elliptical

A3A11
(A)
Page 3-10

A3B Selective fading; radio-path horizon; take-off angle over flat or sloping terrain; earth effects on propagation

A3B01
What causes selective fading?
A. Small changes in beam heading at the receiving station
B. Phase differences between radio-wave components of the same transmission, as experienced at the receiving station
C. Large changes in the height of the ionosphere at the receiving station ordinarily occurring shortly after either sunrise or sunset
D. Time differences between the receiving and transmitting stations

A3B01
(B)
Page 3-11

A3B02
What is the propagation effect called that causes selective fading between received wave components of the same transmission?
A. Faraday rotation
B. Diversity reception
C. Phase differences
D. Phase shift

A3B02
(C)
Page 3-11

A3B03
Which emission modes suffer the most from selective fading?
A. CW and SSB
B. FM and double sideband AM
C. SSB and AMTOR
D. SSTV and CW

A3B03
(B)
Page 3-12

A3B04
How does the bandwidth of a transmitted signal affect selective fading?
A. It is more pronounced at wide bandwidths
B. It is more pronounced at narrow bandwidths
C. It is the same for both narrow and wide bandwidths
D. The receiver bandwidth determines the selective fading effect

A3B04
(A)
Page 3-12

A3B05
Why does the radio-path horizon distance exceed the geometric horizon?
A. E-region skip
B. D-region skip
C. Auroral skip
D. Radio waves may be bent

A3B06
How much farther does the VHF/UHF radio-path horizon distance exceed the geometric horizon?
A. By approximately 15% of the distance
B. By approximately twice the distance
C. By approximately one-half the distance
D. By approximately four times the distance

A3B07
For a 3-element Yagi antenna with horizontally mounted elements, how does the main lobe takeoff angle vary with height above flat ground?
A. It increases with increasing height
B. It decreases with increasing height
C. It does not vary with height
D. It depends on E-region height, not antenna height

A3B08
For a 3-element Yagi antenna with horizontally mounted elements, how does the main lobe takeoff angle vary with a downward slope of the ground (moving away from the antenna)?
A. It increases as the slope gets steeper
B. It decreases as the slope gets steeper
C. It does not depend on the ground slope
D. It depends on F-region height, not ground slope

A3B09
What is the name of the high-angle wave in HF propagation that travels for some distance within the F2 region?
A. Oblique-angle ray
B. Pedersen ray
C. Ordinary ray
D. Heaviside ray

A3B10
Excluding enhanced propagation, what is the approximate range of normal VHF propagation?
A. 1000 miles
B. 500 miles
C. 1500 miles
D. 2000 miles

A3B11
What effect is usually responsible for propagating a VHF signal over 500 miles?
A. D-region absorption
B. Faraday rotation
C. Tropospheric ducting
D. Moonbounce

A3B11
(C)
Page 3-15

A3B12
What happens to an electromagnetic wave as it encounters air molecules and other particles?
A. The wave loses kinetic energy
B. The wave gains kinetic energy
C. An aurora is created
D. Nothing happens because the waves have no physical substance

A3B12
(A)
Page 3-15

A4 — AMATEUR RADIO PRACTICE

Subelement
A4

[4 Exam Questions — 4 Groups]

A4A Frequency measurement devices (i.e. frequency counter, oscilloscope Lissajous figures, dip meter); component mounting techniques (i.e. surface, dead bug {raised}, circuit board)

A4A01
What is a frequency standard?
A. A frequency chosen by a net control operator for net operations
B. A device used to produce a highly accurate reference frequency
C. A device for accurately measuring frequency to within 1 Hz
D. A device used to generate wide-band random frequencies

A4A01
(B)
Page 4-2

A4A02
What does a frequency counter do?
A. It makes frequency measurements
B. It produces a reference frequency
C. It measures FM transmitter deviation
D. It generates broad-band white noise

A4A02
(A)
Page 4-4

A4A03
If a 100 Hz signal is fed to the horizontal input of an oscilloscope and a 150 Hz signal is fed to the vertical input, what type of Lissajous figure should be displayed on the screen?
A. A looping pattern with 100 loops horizontally and 150 loops vertically
B. A rectangular pattern 100 mm wide and 150 mm high
C. A looping pattern with 3 loops horizontally and 2 loops vertically
D. An oval pattern 100 mm wide and 150 mm high

A4A03
(C)
Page 4-10

A4A04
What is a dip-meter?
A. A field-strength meter
B. An SWR meter
C. A variable LC oscillator with metered feedback current
D. A marker generator

A4A05
What does a dip-meter do?
A. It accurately indicates signal strength
B. It measures frequency accurately
C. It measures transmitter output power accurately
D. It gives an indication of the resonant frequency of a circuit

A4A06
How does a dip-meter function?
A. Reflected waves at a specific frequency desensitize a detector coil
B. Power coupled from an oscillator causes a decrease in metered current
C. Power from a transmitter cancels feedback current
D. Harmonics from an oscillator cause an increase in resonant circuit Q

A4A07
What two ways could a dip-meter be used in an amateur station?
A. To measure resonant frequency of antenna traps and to measure percentage of modulation
B. To measure antenna resonance and to measure percentage of modulation
C. To measure antenna resonance and to measure antenna impedance
D. To measure resonant frequency of antenna traps and to measure a tuned circuit resonant frequency

A4A08
What types of coupling occur between a dip-meter and a tuned circuit being checked?
A. Resistive and inductive
B. Inductive and capacitive
C. Resistive and capacitive
D. Strong field

A4A09
For best accuracy, how tightly should a dip-meter be coupled with a tuned circuit being checked?
A. As loosely as possible
B. As tightly as possible
C. First loosely, then tightly
D. With a jumper wire between the meter and the circuit to be checked

A4A10
What happens in a dip-meter when it is too tightly coupled with a tuned circuit being checked?
A. Harmonics are generated
B. A less accurate reading results
C. Cross modulation occurs
D. Intermodulation distortion occurs

A4A11
What circuit construction technique uses leadless components mounted between circuit board pads?
A. Raised mounting
B. Integrated circuit mounting
C. Hybrid device mounting
D. Surface mounting

A4B Meter performance limitations; oscilloscope performance limitations; frequency counter performance limitations

A4B01
What factors limit the accuracy, frequency response and stability of a D'Arsonval-type meter?
A. Calibration, coil impedance and meter size
B. Calibration, mechanical tolerance and coil impedance
C. Coil impedance, electromagnet voltage and movement mass
D. Calibration, series resistance and electromagnet current

A4B02
What factors limit the accuracy, frequency response and stability of an oscilloscope?
A. Accuracy and linearity of the time base and the linearity and bandwidth of the deflection amplifiers
B. Tube face voltage increments and deflection amplifier voltage
C. Accuracy and linearity of the time base and tube face voltage increments
D. Deflection amplifier output impedance and tube face frequency increments

A4B03
How can the frequency response of an oscilloscope be improved?
A. By using a triggered sweep and a crystal oscillator as the time base
B. By using a crystal oscillator as the time base and increasing the vertical sweep rate
C. By increasing the vertical sweep rate and the horizontal amplifier frequency response
D. By increasing the horizontal sweep rate and the vertical amplifier frequency response

A4A10
(B)
Page 4-7

A4A11
(D)
Page 4-15

A4B01
(B)
Page 4-14

A4B02
(A)
Page 4-12

A4B03
(D)
Page 4-12

A4B04
What factors limit the accuracy, frequency response and stability of a frequency counter?
A. Number of digits in the readout, speed of the logic and time base stability
B. Time base accuracy, speed of the logic and time base stability
C. Time base accuracy, temperature coefficient of the logic and time base stability
D. Number of digits in the readout, external frequency reference and temperature coefficient of the logic

A4B05
How can the accuracy of a frequency counter be improved?
A. By using slower digital logic
B. By improving the accuracy of the frequency response
C. By increasing the accuracy of the time base
D. By using faster digital logic

A4B06
If a frequency counter with a time base accuracy of ± 1.0 ppm reads 146,520,000 Hz, what is the most the actual frequency being measured could differ from the reading?
A. 165.2 Hz
B. 14.652 kHz
C. 146.52 Hz
D. 1.4652 MHz

A4B07
If a frequency counter with a time base accuracy of ± 0.1 ppm reads 146,520,000 Hz, what is the most the actual frequency being measured could differ from the reading?
A. 14.652 Hz
B. 0.1 MHz
C. 1.4652 Hz
D. 1.4652 kHz

A4B08
If a frequency counter with a time base accuracy of ± 10 ppm reads 146,520,000 Hz, what is the most the actual frequency being measured could differ from the reading?
A. 146.52 Hz
B. 10 Hz
C. 146.52 kHz
D. 1465.20 Hz

A4B09

If a frequency counter with a time base accuracy of ± 1.0 ppm reads 432,100,000 Hz, what is the most the actual frequency being measured could differ from the reading?
A. 43.21 MHz
B. 10 Hz
C. 1.0 MHz
D. 432.1 Hz

A4B09
(D)
Page 4-5

A4B10

If a frequency counter with a time base accuracy of ± 0.1 ppm reads 432,100,000 Hz, what is the most the actual frequency being measured could differ from the reading?
A. 43.21 Hz
B. 0.1 MHz
C. 432.1 Hz
D. 0.2 MHz

A4B10
(A)
Page 4-5

A4B11

If a frequency counter with a time base accuracy of ± 10 ppm reads 432,100,000 Hz, what is the most the actual frequency being measured could differ from the reading?
A. 10 MHz
B. 10 Hz
C. 4321 Hz
D. 432.1 Hz

A4B11
(C)
Page 4-5

A4C Receiver performance characteristics (i.e., phase noise, desensitization, capture effect, intercept point, noise floor, dynamic range {blocking and IMD}, image rejection, MDS, signal-to-noise-ratio)

A4C01
What is the effect of excessive phase noise in a receiver local oscillator?
A. It limits the receiver ability to receive strong signals
B. It reduces the receiver sensitivity
C. It decreases the receiver third-order intermodulation distortion dynamic range
D. It allows strong signals on nearby frequencies to interfere with reception of weak signals

A4C01
(D)
Page 4-18

A4C02
What is the term for the reduction in receiver sensitivity caused by a strong signal near the received frequency?
A. Desensitization
B. Quieting
C. Cross-modulation interference
D. Squelch gain rollback

A4C02
(A)
Page 4-20

A4C03
(B)
Page 4-20

A4C03
What causes receiver desensitization?
A. Audio gain adjusted too low
B. Strong adjacent-channel signals
C. Squelch gain adjusted too high
D. Squelch gain adjusted too low

A4C04
(A)
Page 4-21

A4C04
What is one way receiver desensitization can be reduced?
A. Shield the receiver from the transmitter causing the problem
B. Increase the transmitter audio gain
C. Decrease the receiver squelch gain
D. Increase the receiver bandwidth

A4C05
(C)
Page 4-22

A4C05
What is the capture effect?
A. All signals on a frequency are demodulated by an FM receiver
B. All signals on a frequency are demodulated by an AM receiver
C. The strongest signal received is the only demodulated signal
D. The weakest signal received is the only demodulated signal

A4C06
(C)
Page 4-22

A4C06
What is the term for the blocking of one FM-phone signal by another
stronger FM-phone signal?
A. Desensitization
B. Cross-modulation interference
C. Capture effect
D. Frequency discrimination

A4C07
(A)
Page 4-22

A4C07
With which emission type is capture effect most pronounced?
A. FM
B. SSB
C. AM
D. CW

A4C08
(D)
Page 4-16

A4C08
What is meant by the noise floor of a receiver?
A. The weakest signal that can be detected under noisy atmospheric
conditions
B. The amount of phase noise generated by the receiver local oscillator
C. The minimum level of noise that will overload the receiver RF amplifier
stage
D. The weakest signal that can be detected above the receiver internal
noise

A4C09
What is the blocking dynamic range of a receiver that has an 8-dB noise
figure and an IF bandwidth of 500 Hz if the blocking level (1-dB
compression point) is −20 dBm?
A. −119 dBm
B. 119 dB
C. 146 dB
D. −146 dBm

A4C09
(B)
Page 4-17

A4C10
What part of a superheterodyne receiver determines the image rejection
ratio of the receiver?
A. Product detector
B. RF amplifier
C. AGC loop
D. IF filter

A4C10
(B)
Page 4-20

A4C11
If you measured the MDS of a receiver, what would you be measuring?
A. The meter display sensitivity (MDS), or the responsiveness of the
   receiver S-meter to all signals
B. The minimum discernible signal (MDS), or the weakest signal that the
   receiver can detect
C. The minimum distorting signal (MDS), or the strongest signal the
   receiver can detect without overloading
D. The maximum detectable spectrum (MDS), or the lowest to highest
   frequency range of the receiver

A4C11
(B)
Page 4-15

A4D Intermodulation and cross-modulation interference

A4D01
If the signals of two transmitters mix together in one or both of their final
amplifiers and unwanted signals at the sum and difference frequencies of
the original signals are generated, what is this called?
A. Amplifier desensitization
B. Neutralization
C. Adjacent channel interference
D. Intermodulation interference

A4D01
(D)
Page 4-22

A4D02
How does intermodulation interference between two repeater transmitters usually occur?
A. When the signals from the transmitters are reflected out of phase from airplanes passing overhead
B. When they are in close proximity and the signals mix in one or both of their final amplifiers
C. When they are in close proximity and the signals cause feedback in one or both of their final amplifiers
D. When the signals from the transmitters are reflected in phase from airplanes passing overhead

A4D03
How can intermodulation interference between two repeater transmitters in close proximity often be reduced or eliminated?
A. By using a Class C final amplifier with high driving power
B. By installing a terminated circulator or ferrite isolator in the feed line to the transmitter and duplexer
C. By installing a band-pass filter in the antenna feed line
D. By installing a low-pass filter in the antenna feed line

A4D04
What is cross-modulation interference?
A. Interference between two transmitters of different modulation type
B. Interference caused by audio rectification in the receiver preamp
C. Harmonic distortion of the transmitted signal
D. Modulation from an unwanted signal is heard in addition to the desired signal

A4D05
What is the term used to refer to the condition where the signals from a very strong station are superimposed on other signals being received?
A. Intermodulation distortion
B. Cross-modulation interference
C. Receiver quieting
D. Capture effect

A4D06
How can cross-modulation in a receiver be reduced?
A. By installing a filter at the receiver
B. By using a better antenna
C. By increasing the receiver RF gain while decreasing the AF gain
D. By adjusting the passband tuning

A4D07
What is the result of cross-modulation?
A. A decrease in modulation level of transmitted signals
B. Receiver quieting
C. The modulation of an unwanted signal is heard on the desired signal
D. Inverted sidebands in the final stage of the amplifier

A4D08
What causes intermodulation in an electronic circuit?
A. Too little gain
B. Lack of neutralization
C. Nonlinear circuits or devices
D. Positive feedback

A4D08
(C)
Page 4-22

A4D09
If a receiver tuned to 146.70 MHz receives an intermodulation-product signal whenever a nearby transmitter transmits on 146.52 MHz, what are the two most likely frequencies for the other interfering signal?
A. 146.34 MHz and 146.61 MHz
B. 146.88 MHz and 146.34 MHz
C. 146.10 MHz and 147.30 MHz
D. 73.35 MHz and 239.40 MHz

A4D09
(A)
Page 4-25

A4D10
If a television receiver suffers from cross modulation when a nearby amateur transmitter is operating at 14 MHz, which of the following cures might be effective?
A. A low-pass filter attached to the output of the amateur transmitter
B. A high-pass filter attached to the output of the amateur transmitter
C. A low-pass filter attached to the input of the television receiver
D. A high-pass filter attached to the input of the television receiver

A4D10
(D)
Page 4-26

A4D11
Which of the following is an example of intermodulation distortion?
A. Receiver blocking
B. Splatter from an SSB transmitter
C. Overdeviation of an FM transmitter
D. Excessive 2nd-harmonic output from a transmitter

A4D11
(B)
Page 4-26

A5 — ELECTRICAL PRINCIPLES

Subelement
A5

[10 Exam Questions — 10 Groups]

A5A Characteristics of resonant circuits

A5A01
What can cause the voltage across reactances in series to be larger than the voltage applied to them?
A. Resonance
B. Capacitance
C. Conductance
D. Resistance

A5A01
(A)
Page 5-19

A5A02
(C)
Page 5-18

A5A02
What is resonance in an electrical circuit?
A. The highest frequency that will pass current
B. The lowest frequency that will pass current
C. The frequency at which capacitive reactance equals inductive reactance
D. The frequency at which power factor is at a minimum

A5A03
(B)
Page 5-18

A5A03
What are the conditions for resonance to occur in an electrical circuit?
A. The power factor is at a minimum
B. Inductive and capacitive reactances are equal
C. The square root of the sum of the capacitive and inductive reactance is
   equal to the resonant frequency
D. The square root of the product of the capacitive and inductive reactance
   is equal to the resonant frequency

A5A04
(D)
Page 5-18

A5A04
When the inductive reactance of an electrical circuit equals its capacitive
reactance, what is this condition called?
A. Reactive quiescence
B. High Q
C. Reactive equilibrium
D. Resonance

A5A05
(D)
Page 5-19

A5A05
What is the magnitude of the impedance of a series R-L-C circuit at
resonance?
A. High, as compared to the circuit resistance
B. Approximately equal to capacitive reactance
C. Approximately equal to inductive reactance
D. Approximately equal to circuit resistance

A5A06
(A)
Page 5-21

A5A06
What is the magnitude of the impedance of a circuit with a resistor, an
inductor and a capacitor all in parallel, at resonance?
A. Approximately equal to circuit resistance
B. Approximately equal to inductive reactance
C. Low, as compared to the circuit resistance
D. Approximately equal to capacitive reactance

A5A07
(B)
Page 5-20

A5A07
What is the magnitude of the current at the input of a series R-L-C circuit at
resonance?
A. It is at a minimum
B. It is at a maximum
C. It is DC
D. It is zero

A5A08
What is the magnitude of the circulating current within the components of a parallel L-C circuit at resonance?
A. It is at a minimum
B. It is at a maximum
C. It is DC
D. It is zero

A5A08
(B)
Page 5-21

A5A09
What is the magnitude of the current at the input of a parallel R-L-C circuit at resonance?
A. It is at a minimum
B. It is at a maximum
C. It is DC
D. It is zero

A5A09
(A)
Page 5-21

A5A10
What is the relationship between the current through a resonant circuit and the voltage across the circuit?
A. The voltage leads the current by 90 degrees
B. The current leads the voltage by 90 degrees
C. The voltage and current are in phase
D. The voltage and current are 180 degrees out of phase

A5A10
(C)
Page 5-23

A5A11
What is the relationship between the current into (or out of) a parallel resonant circuit and the voltage across the circuit?
A. The voltage leads the current by 90 degrees
B. The current leads the voltage by 90 degrees
C. The voltage and current are in phase
D. The voltage and current are 180 degrees out of phase

A5A11
(C)
Page 5-23

A5B Series resonance (capacitor and inductor to resonate at a specific frequency)

A5B01
What is the resonant frequency of a series R-L-C circuit if R is 47 ohms, L is 50 microhenrys and C is 40 picofarads?
A. 79.6 MHz
B. 1.78 MHz
C. 3.56 MHz
D. 7.96 MHz

A5B01
(C)
Page 5-20

$$f = \frac{1}{6\sqrt{50 \times 10^{-6} \times 40 \times 10^{-12}}}$$

$$= \frac{10^3}{6\sqrt{2000 \times 10^{-18}}} = \frac{1 \cdot 10^9}{6\sqrt{45}} = \frac{10^9}{240} = \frac{10^7}{2.4}$$

A5B02
What is the resonant frequency of a series R-L-C circuit if R is 47 ohms,
L is 40 microhenrys and C is 200 picofarads?
A. 1.99 kHz
B. 1.78 MHz
C. 1.99 MHz
D. 1.78 kHz

$$\frac{10^9}{6\sqrt{8000}} = \frac{10^9}{540} = 10^?$$

A5B03
What is the resonant frequency of a series R-L-C circuit if R is 47 ohms,
L is 50 microhenrys and C is 10 picofarads?
A. 3.18 MHz
B. 3.18 kHz
C. 7.12 kHz
D. 7.12 MHz

A5B04
What is the resonant frequency of a series R-L-C circuit if R is 47 ohms,
L is 25 microhenrys and C is 10 picofarads?
A. 10.1 MHz
B. 63.7 MHz
C. 10.1 kHz
D. 63.7 kHz

A5B05
What is the resonant frequency of a series R-L-C circuit if R is 47 ohms,
L is 3 microhenrys and C is 40 picofarads?
A. 13.1 MHz
B. 14.5 MHz
C. 14.5 kHz
D. 13.1 kHz

A5B06
What is the resonant frequency of a series R-L-C circuit if R is 47 ohms,
L is 4 microhenrys and C is 20 picofarads?
A. 19.9 kHz
B. 17.8 kHz
C. 19.9 MHz
D. 17.8 MHz

A5B07
What is the resonant frequency of a series R-L-C circuit if R is 47 ohms,
L is 8 microhenrys and C is 7 picofarads?
A. 2.84 MHz
B. 28.4 MHz
C. 21.3 MHz
D. 2.13 MHz

$$\frac{1}{6\sqrt{56 \times 10^{-18}}} = \frac{10^9}{6 \times 7.5} = \frac{1000}{50}$$

10-38   Chapter 10

A5B08
What is the resonant frequency of a series R-L-C circuit if R is 47 ohms,
L is 3 microhenrys and C is 15 picofarads?
A. 23.7 MHz
B. 23.7 kHz
C. 35.4 kHz
D. 35.4 MHz

A5B09
What is the resonant frequency of a series R-L-C circuit if R is 47 ohms,
L is 4 microhenrys and C is 8 picofarads?
A. 28.1 kHz
B. 28.1 MHz
C. 49.7 MHz
D. 49.7 kHz

A5B10
What is the resonant frequency of a series R-L-C circuit if R is 47 ohms,
L is 1 microhenry and C is 9 picofarads?
A. 17.7 MHz
B. 17.7 kHz
C. 53.1 kHz
D. 53.1 MHz

A5B11
What is the value of capacitance (C) in a series R-L-C circuit if the circuit
resonant frequency is 14.25 MHz and L is 2.84 microhenrys?
A. 2.2 microfarads
B. 254 microfarads
C. 44 picofarads
D. 3933 picofarads

A5C Parallel resonance (capacitor and inductor to resonate at a specific frequency)

A5C01
What is the resonant frequency of a parallel R-L-C circuit if R is
4.7 kilohms, L is 1 microhenry and C is 10 picofarads?
A. 50.3 MHz
B. 15.9 MHz
C. 15.9 kHz
D. 50.3 kHz

A5B08
(A)
Page 5-20

A5B09
(B)
Page 5-20

A5B10
(D)
Page 5-20

A5B11
(C)
Page 5-24

A5C01
(A)
Page 5-23

A5C02
(B)
Page 5-23

A5C02
What is the resonant frequency of a parallel R-L-C circuit if R is
4.7 kilohms, L is 2 microhenrys and C is 15 picofarads?
A. 29.1 kHz
B. 29.1 MHz
C. 5.31 MHz
D. 5.31 kHz

A5C03
(C)
Page 5-23

A5C03
What is the resonant frequency of a parallel R-L-C circuit if R is
4.7 kilohms, L is 5 microhenrys and C is 9 picofarads?
A. 23.7 kHz
B. 3.54 kHz
C. 23.7 MHz
D. 3.54 MHz

A5C04
(D)
Page 5-23

A5C04
What is the resonant frequency of a parallel R-L-C circuit if R is
4.7 kilohms, L is 2 microhenrys and C is 30 picofarads?
A. 2.65 kHz
B. 20.5 kHz
C. 2.65 MHz
D. 20.5 MHz

A5C05
(A)
Page 5-23

A5C05
What is the resonant frequency of a parallel R-L-C circuit if R is
4.7 kilohms, L is 15 microhenrys and C is 5 picofarads?
A. 18.4 MHz
B. 2.12 MHz
C. 18.4 kHz
D. 2.12 kHz

A5C06
(B)
Page 5-23

A5C06
What is the resonant frequency of a parallel R-L-C circuit if R is
4.7 kilohms, L is 3 microhenrys and C is 40 picofarads?
A. 1.33 kHz
B. 14.5 MHz
C. 1.33 MHz
D. 14.5 kHz

A5C07
(C)
Page 5-23

A5C07
What is the resonant frequency of a parallel R-L-C circuit if R is
4.7 kilohms, L is 40 microhenrys and C is 6 picofarads?
A. 6.63 MHz
B. 6.63 kHz
C. 10.3 MHz
D. 10.3 kHz

A5C08
What is the resonant frequency of a parallel R-L-C circuit if R is
4.7 kilohms, L is 10 microhenrys and C is 50 picofarads?
A. 3.18 MHz
B. 3.18 kHz
C. 7.12 kHz
D. 7.12 MHz

A5C08
(D)
Page 5-23

A5C09
What is the resonant frequency of a parallel R-L-C circuit if R is
4.7 kilohms, L is 200 microhenrys and C is 10 picofarads?
A. 3.56 MHz
B. 7.96 kHz
C. 3.56 kHz
D. 7.96 MHz

A5C09
(A)
Page 5-23

A5C10
What is the resonant frequency of a parallel R-L-C circuit if R is
4.7 kilohms, L is 90 microhenrys and C is 100 picofarads?
A. 1.77 MHz
B. 1.68 MHz
C. 1.77 kHz
D. 1.68 kHz

A5C10
(B)
Page 5-23

A5C11
What is the value of inductance (L) in a parallel R-L-C circuit if the circuit
resonant frequency is 14.25 MHz and C is 44 picofarads?
A. 253.8 millihenrys
B. 3.9 millihenrys
C. 0.353 microhenrys
D. 2.8 microhenrys

A5C11
(D)
Page 5-24

A5D Skin effect; electrostatic and electromagnetic fields

A5D01
What is the result of skin effect?
A. As frequency increases, RF current flows in a thinner layer of the
conductor, closer to the surface
B. As frequency decreases, RF current flows in a thinner layer of the
conductor, closer to the surface
C. Thermal effects on the surface of the conductor increase the impedance
D. Thermal effects on the surface of the conductor decrease the impedance

A5D01
(A)
Page 5-27

A5D02
What effect causes most of an RF current to flow along the surface of a conductor?
A. Layer effect
B. Seeburg effect
C. Skin effect
D. Resonance effect

A5D03
Where does almost all RF current flow in a conductor?
A. Along the surface of the conductor
B. In the center of the conductor
C. In a magnetic field around the conductor
D. In a magnetic field in the center of the conductor

A5D04
Why does most of an RF current flow within a few thousandths of an inch of its conductor's surface?
A. Because a conductor has AC resistance due to self-inductance
B. Because the RF resistance of a conductor is much less than the DC resistance
C. Because of the heating of the conductor's interior
D. Because of skin effect

A5D05
Why is the resistance of a conductor different for RF currents than for direct currents?
A. Because the insulation conducts current at high frequencies
B. Because of the Heisenburg Effect
C. Because of skin effect
D. Because conductors are non-linear devices

A5D06
What device is used to store electrical energy in an electrostatic field?
A. A battery
B. A transformer
C. A capacitor
D. An inductor

A5D07
What unit measures electrical energy stored in an electrostatic field?
A. Coulomb
B. Joule
C. Watt
D. Volt

A5D08
What is a magnetic field?
A. Current through the space around a permanent magnet
B. The space around a conductor, through which a magnetic force acts
C. The space between the plates of a charged capacitor, through which a
   magnetic force acts
D. The force that drives current through a resistor

A5D09
In what direction is the magnetic field oriented about a conductor in relation
to the direction of electron flow?
A. In the same direction as the current
B. In a direction opposite to the current
C. In all directions; omnidirectional
D. In a direction determined by the left-hand rule

A5D10
What determines the strength of a magnetic field around a conductor?
A. The resistance divided by the current
B. The ratio of the current to the resistance
C. The diameter of the conductor
D. The amount of current

A5D11
What is the term for energy that is stored in an electromagnetic or
electrostatic field?
A. Amperes-joules
B. Potential energy
C. Joules-coulombs
D. Kinetic energy

A5E Half-power bandwidth

A5E01
What is the half-power bandwidth of a parallel resonant circuit that has a
resonant frequency of 1.8 MHz and a Q of 95?
A. 18.9 kHz
B. 1.89 kHz
C. 189 Hz
D. 58.7 kHz

A5E02
What is the half-power bandwidth of a parallel resonant circuit that has a
resonant frequency of 3.6 MHz and a Q of 218?
A. 58.7 kHz
B. 606 kHz
C. 47.3 kHz
D. 16.5 kHz

A5D08
(B)
Page 5-5

A5D09
(D)
Page 5-5

A5D10
(D)
Page 5-5

A5D11
(B)
Page 5-3

A5E01
(A)
Page 5-29

A5E02
(D)
Page 5-29

A5E03
(C)
Page 5-29

A5E03
What is the half-power bandwidth of a parallel resonant circuit that has a resonant frequency of 7.1 MHz and a Q of 150?
A. 211 kHz
B. 16.5 kHz
C. 47.3 kHz
D. 21.1 kHz

A5E04
(D)
Page 5-29

A5E04
What is the half-power bandwidth of a parallel resonant circuit that has a resonant frequency of 12.8 MHz and a Q of 218?
A. 21.1 kHz
B. 27.9 kHz
C. 17 kHz
D. 58.7 kHz

A5E05
(A)
Page 5-29

A5E05
What is the half-power bandwidth of a parallel resonant circuit that has a resonant frequency of 14.25 MHz and a Q of 150?
A. 95 kHz
B. 10.5 kHz
C. 10.5 MHz
D. 17 kHz

A5E06
(D)
Page 5-29

A5E06
What is the half-power bandwidth of a parallel resonant circuit that has a resonant frequency of 21.15 MHz and a Q of 95?
A. 4.49 kHz
B. 44.9 kHz
C. 22.3 kHz
D. 222.6 kHz

A5E07
(B)
Page 5-29

A5E07
What is the half-power bandwidth of a parallel resonant circuit that has a resonant frequency of 10.1 MHz and a Q of 225?
A. 4.49 kHz
B. 44.9 kHz
C. 22.3 kHz
D. 223 kHz

A5E08
(A)
Page 5-29

A5E08
What is the half-power bandwidth of a parallel resonant circuit that has a resonant frequency of 18.1 MHz and a Q of 195?
A. 92.8 kHz
B. 10.8 kHz
C. 22.3 kHz
D. 44.9 kHz

A5E09
What is the half-power bandwidth of a parallel resonant circuit that has a resonant frequency of 3.7 MHz and a Q of 118?
A. 22.3 kHz
B. 76.2 kHz
C. 31.4 kHz
D. 10.8 kHz

A5E09
(C)
Page 5-29

A5E10
What is the half-power bandwidth of a parallel resonant circuit that has a resonant frequency of 14.25 MHz and a Q of 187?
A. 22.3 kHz
B. 10.8 kHz
C. 76.2 kHz
D. 13.1 kHz

$$Q = \frac{fc}{\Delta f} \qquad \Delta f = \frac{14.25}{187} =$$

A5E10
(C)
Page 5-29

A5E11
What term describes the frequency range over which the circuit response is no more than 3 dB below the peak response?
A. Resonance
B. Half-power bandwidth
C. Circuit Q
D. 2:1 bandwidth

A5E11
(B)
Page 5-28

A5F Circuit Q

A5F01
What is the Q of a parallel R-L-C circuit if the resonant frequency is 14.128 MHz, L is 2.7 microhenrys and R is 18 kilohms?
A. 75.1
B. 7.51
C. 71.5
D. 0.013

$$Q = \frac{R}{X_c}$$

$$\frac{6.2 \times 14.1 \times 10^{?} \times 2.7 \times 10^{-6}}{18 \times 10^{3}}$$

A5F01
(A)
Page 5-28

A5F02
What is the Q of a parallel R-L-C circuit if the resonant frequency is 14.128 MHz, L is 4.7 microhenrys and R is 18 kilohms?
A. 4.31
B. 43.1
C. 13.3
D. 0.023

A5F02
(B)
Page 5-28

A5F03
What is the Q of a parallel R-L-C circuit if the resonant frequency is
4.468 MHz, L is 47 microhenrys and R is 180 ohms?
A. 0.00735
B. 7.35
C. 0.136
D. 13.3

A5F04
What is the Q of a parallel R-L-C circuit if the resonant frequency is
14.225 MHz, L is 3.5 microhenrys and R is 10 kilohms?
A. 7.35
B. 0.0319
C. 71.5
D. 31.9

A5F05
What is the Q of a parallel R-L-C circuit if the resonant frequency is
7.125 MHz, L is 8.2 microhenrys and R is 1 kilohm?
A. 36.8
B. 0.273
C. 0.368
D. 2.73

A5F06
What is the Q of a parallel R-L-C circuit if the resonant frequency is
7.125 MHz, L is 10.1 microhenrys and R is 100 ohms?
A. 0.221
B. 4.52
C. 0.00452
D. 22.1

A5F07
What is the Q of a parallel R-L-C circuit if the resonant frequency is
7.125 MHz, L is 12.6 microhenrys and R is 22 kilohms?
A. 22.1
B. 39
C. 25.6
D. 0.0256

A5F08
What is the Q of a parallel R-L-C circuit if the resonant frequency is
3.625 MHz, L is 3 microhenrys and R is 2.2 kilohms?
A. 0.031
B. 32.2
C. 31.1
D. 25.6

A5F09
What is the Q of a parallel R-L-C circuit if the resonant frequency is
3.625 MHz, L is 42 microhenrys and R is 220 ohms?
A. 23
B. 0.00435
C. 4.35
D. 0.23

A5F09
(D)
Page 5-28

A5F10
What is the Q of a parallel R-L-C circuit if the resonant frequency is
3.625 MHz, L is 43 microhenrys and R is 1.8 kilohms?
A. 1.84
B. 0.543
C. 54.3
D. 23

A5F10
(A)
Page 5-28

A5F11
Why is a resistor often included in a parallel resonant circuit?
A. To increase the Q and decrease the skin effect
B. To decrease the Q and increase the resonant frequency
C. To decrease the Q and increase the bandwidth
D. To increase the Q and decrease the bandwidth

A5F11
(C)
Page 5-28

A5G Phase angle between voltage and current

A5G01
What is the phase angle between the voltage across and the current
through a series R-L-C circuit if XC is 25 ohms, R is 100 ohms, and XL is
100 ohms?
A. 36.9 degrees with the voltage leading the current
B. 53.1 degrees with the voltage lagging the current
C. 36.9 degrees with the voltage lagging the current
D. 53.1 degrees with the voltage leading the current

A5G01
(A)
Page 5-15

A5G02
What is the phase angle between the voltage across and the current
through a series R-L-C circuit if XC is 25 ohms, R is 100 ohms, and XL is
50 ohms?
A. 14 degrees with the voltage lagging the current
B. 14 degrees with the voltage leading the current
C. 76 degrees with the voltage lagging the current
D. 76 degrees with the voltage leading the current

A5G02
(B)
Page 5-15

A5G03
What is the phase angle between the voltage across and the current
through a series R-L-C circuit if XC is 500 ohms, R is 1 kilohm, and XL is
250 ohms?
A. 68.2 degrees with the voltage leading the current
B. 14.1 degrees with the voltage leading the current
C. 14.1 degrees with the voltage lagging the current
D. 68.2 degrees with the voltage lagging the current

A5G04
What is the phase angle between the voltage across and the current
through a series R-L-C circuit if XC is 75 ohms, R is 100 ohms, and XL is
100 ohms?
A. 76 degrees with the voltage leading the current
B. 14 degrees with the voltage leading the current
C. 14 degrees with the voltage lagging the current
D. 76 degrees with the voltage lagging the current

A5G05
What is the phase angle between the voltage across and the current
through a series R-L-C circuit if XC is 50 ohms, R is 100 ohms, and XL is
25 ohms?
A. 76 degrees with the voltage lagging the current
B. 14 degrees with the voltage leading the current
C. 76 degrees with the voltage leading the current
D. 14 degrees with the voltage lagging the current

A5G06
What is the phase angle between the voltage across and the current
through a series R-L-C circuit if XC is 75 ohms, R is 100 ohms, and XL is
50 ohms?
A. 76 degrees with the voltage lagging the current
B. 14 degrees with the voltage leading the current
C. 14 degrees with the voltage lagging the current
D. 76 degrees with the voltage leading the current

A5G07
What is the phase angle between the voltage across and the current
through a series R-L-C circuit if XC is 100 ohms, R is 100 ohms, and XL is
75 ohms?
A. 14 degrees with the voltage lagging the current
B. 14 degrees with the voltage leading the current
C. 76 degrees with the voltage leading the current
D. 76 degrees with the voltage lagging the current

A5G08
What is the phase angle between the voltage across and the current through a series R-L-C circuit if XC is 250 ohms, R is 1 kilohm, and XL is 500 ohms?
A. 81.47 degrees with the voltage lagging the current
B. 81.47 degrees with the voltage leading the current
C. 14.04 degrees with the voltage lagging the current
D. 14.04 degrees with the voltage leading the current

A5G08
(D)
Page 5-15

A5G09
What is the phase angle between the voltage across and the current through a series R-L-C circuit if XC is 50 ohms, R is 100 ohms, and XL is 75 ohms?
A. 76 degrees with the voltage leading the current
B. 76 degrees with the voltage lagging the current
C. 14 degrees with the voltage lagging the current
D. 14 degrees with the voltage leading the current

A5G09
(D)
Page 5-15

A5G10
What is the relationship between the current through and the voltage across a capacitor?
A. Voltage and current are in phase
B. Voltage and current are 180 degrees out of phase
C. Voltage leads current by 90 degrees
D. Current leads voltage by 90 degrees

A5G10
(D)
Page 5-8

A5G11
What is the relationship between the current through an inductor and the voltage across an inductor?
A. Voltage leads current by 90 degrees
B. Current leads voltage by 90 degrees
C. Voltage and current are 180 degrees out of phase
D. Voltage and current are in phase

A5G11
(A)
Page 5-10

A5H Reactive power; power factor
A5H01
What is reactive power?
A. Wattless, nonproductive power
B. Power consumed in wire resistance in an inductor
C. Power lost because of capacitor leakage
D. Power consumed in circuit Q

A5H01
(A)
Page 5-31

A5H02
(D)
Page 5-31

A5H02
What is the term for an out-of-phase, nonproductive power associated with inductors and capacitors?
A. Effective power
B. True power
C. Peak envelope power
D. Reactive power

A5H03
(B)
Page 5-31

A5H03
In a circuit that has both inductors and capacitors, what happens to reactive power?
A. It is dissipated as heat in the circuit
B. It goes back and forth between magnetic and electric fields, but is not dissipated
C. It is dissipated as kinetic energy in the circuit
D. It is dissipated in the formation of inductive and capacitive fields

A5H04
(A)
Page 5-32

A5H04
In a circuit where the AC voltage and current are out of phase, how can the true power be determined?
A. By multiplying the apparent power times the power factor
B. By subtracting the apparent power from the power factor
C. By dividing the apparent power by the power factor
D. By multiplying the RMS voltage times the RMS current

A5H05
(C)
Page 5-32

A5H05
What is the power factor of an R-L circuit having a 60 degree phase angle between the voltage and the current?
A. 1.414
B. 0.866
C. 0.5
D. 1.73

A5H06
(D)
Page 5-32

A5H06
What is the power factor of an R-L circuit having a 45 degree phase angle between the voltage and the current?
A. 0.866
B. 1.0
C. 0.5
D. 0.707

A5H07
(C)
Page 5-32

A5H07
What is the power factor of an R-L circuit having a 30 degree phase angle between the voltage and the current?
A. 1.73
B. 0.5
C. 0.866
D. 0.577

A5H08
How many watts are consumed in a circuit having a power factor of 0.2 if
the input is 100-V AC at 4 amperes?
A. 400 watts
B. 80 watts
C. 2000 watts
D. 50 watts

A5H09
How many watts are consumed in a circuit having a power factor of 0.6 if
the input is 200-V AC at 5 amperes?
A. 200 watts
B. 1000 watts
C. 1600 watts
D. 600 watts

A5H10
How many watts are consumed in a circuit having a power factor of 0.71 if
the apparent power is 500 watts?
A. 704 W
B. 355 W
C. 252 W
D. 1.42 mW

A5H11
Why would the power used in a circuit be less than the product of the
magnitudes of the AC voltage and current?
A. Because there is a phase angle greater than zero between the current
and voltage
B. Because there are only resistances in the circuit
C. Because there are no reactances in the circuit
D. Because there is a phase angle equal to zero between the current and
voltage

A5I Effective radiated power, system gains and losses

A5I01
What is the effective radiated power of a repeater station with 50 watts
transmitter power output, 4-dB feed line loss, 2-dB duplexer loss, 1-dB
circulator loss and 6-dBd antenna gain?
A. 199 watts
B. 39.7 watts
C. 45 watts
D. 62.9 watts

A5H08
(B)
Page 5-33

A5H09
(D)
Page 5-33

A5H10
(B)
Page 5-33

A5H11
(A)
Page 5-31

A5I01
(B)
Page 5-34

A5I02
What is the effective radiated power of a repeater station with 50 watts transmitter power output, 5-dB feed line loss, 3-dB duplexer loss, 1-dB circulator loss and 7-dBd antenna gain?
A. 79.2 watts
B. 315 watts
C. 31.5 watts
D. 40.5 watts

A5I03
What is the effective radiated power of a station with 75 watts transmitter power output, 4-dB feed line loss and 10-dBd antenna gain?
A. 600 watts
B. 75 watts
C. 150 watts
D. 299 watts

A5I04
What is the effective radiated power of a repeater station with 75 watts transmitter power output, 5-dB feed line loss, 3-dB duplexer loss, 1-dB circulator loss and 6-dBd antenna gain?
A. 37.6 watts
B. 237 watts
C. 150 watts
D. 23.7 watts

A5I05
What is the effective radiated power of a station with 100 watts transmitter power output, 1-dB feed line loss and 6-dBd antenna gain?
A. 350 watts
B. 500 watts
C. 20 watts
D. 316 watts

A5I06
What is the effective radiated power of a repeater station with 100 watts transmitter power output, 5-dB feed line loss, 3-dB duplexer loss, 1-dB circulator loss and 10-dBd antenna gain?
A. 794 watts
B. 126 watts
C. 79.4 watts
D. 1260 watts

A5I07
What is the effective radiated power of a repeater station with 120 watts
transmitter power output, 5-dB feed line loss, 3-dB duplexer loss, 1-dB
circulator loss and 6-dBd antenna gain?
A. 601 watts
B. 240 watts
C. 60 watts
D. 79 watts

$-3 \, dL = 10 \, \dfrac{P?}{P_{ref}}$

$antilog \left(\dfrac{-3}{10} \right) \times 120 = P_?$

A5I08
What is the effective radiated power of a repeater station with 150 watts
transmitter power output, 2-dB feed line loss, 2.2-dB duplexer loss and
7-dBd antenna gain?
A. 1977 watts
B. 78.7 watts
C. 420 watts
D. 286 watts

$\begin{array}{r} 4.2 \\ + \overline{2.8} \end{array}$

A5I09
What is the effective radiated power of a repeater station with 200 watts
transmitter power output, 4-dB feed line loss, 3.2-dB duplexer loss, 0.8-dB
circulator loss and 10-dBd antenna gain?
A. 317 watts
B. 2000 watts
C. 126 watts
D. 300 watts

A5I10
What is the effective radiated power of a repeater station with 200 watts
transmitter power output, 2-dB feed line loss, 2.8-dB duplexer loss, 1.2-dB
circulator loss and 7-dBd antenna gain?
A. 159 watts
B. 252 watts
C. 632 watts
D. 63.2 watts

A5I11
What term describes station output (including the transmitter, antenna and
everything in between), when considering transmitter power and system
gains and losses?
A. Power factor
B. Half-power bandwidth
C. Effective radiated power
D. Apparent power

A5I07
(C)
Page 5-34

A5I08
(D)
Page 5-34

A5I09
(A)
Page 5-34

A5I10
(B)
Page 5-34

A5I11
(C)
Page 5-33

A5J Replacement of voltage source and resistive voltage divider with equivalent voltage source and one resistor (Thevenin's Theorem)

A5J01
In Figure A5-1, what values of V2 and R3 result in the same voltage and current as when V1 is 8 volts, R1 is 8 kilohms, and R2 is 8 kilohms?
A. R3 = 4 kilohms and V2 = 8 volts
B. R3 = 4 kilohms and V2 = 4 volts
C. R3 = 16 kilohms and V2 = 8 volts
D. R3 = 16 kilohms and V2 = 4 volts

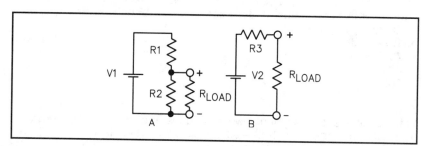

Figure A5-1 — Refer to questions A5J01 through A5J10.

A5J02
In Figure A5-1, what values of V2 and R3 result in the same voltage and current as when V1 is 8 volts, R1 is 16 kilohms, and R2 is 8 kilohms?
A. R3 = 24 kilohms and V2 = 5.33 volts
B. R3 = 5.33 kilohms and V2 = 8 volts
C. R3 = 5.33 kilohms and V2 = 2.67 volts
D. R3 = 24 kilohms and V2 = 8 volts

A5J03
In Figure A5-1, what values of V2 and R3 result in the same voltage and current as when V1 is 8 volts, R1 is 8 kilohms, and R2 is 16 kilohms?
A. R3 = 5.33 kilohms and V2 = 5.33 volts
B. R3 = 8 kilohms and V2 = 4 volts
C. R3 = 24 kilohms and V2 = 8 volts
D. R3 = 5.33 kilohms and V2 = 8 volts

A5J04
In Figure A5-1, what values of V2 and R3 result in the same voltage and current as when V1 is 10 volts, R1 is 10 kilohms, and R2 is 10 kilohms?
A. R3 = 10 kilohms and V2 = 5 volts
B. R3 = 20 kilohms and V2 = 5 volts
C. R3 = 20 kilohms and V2 = 10 volts
D. R3 = 5 kilohms and V2 = 5 volts

A5J05
In Figure A5-1, what values of V2 and R3 result in the same voltage and current as when V1 is 10 volts, R1 is 20 kilohms, and R2 is 10 kilohms?
A. R3 = 30 kilohms and V2 = 10 volts
B. R3 = 6.67 kilohms and V2 = 10 volts
C. R3 = 6.67 kilohms and V2 = 3.33 volts
D. R3 = 30 kilohms and V2 = 3.33 volts

A5J06
In Figure A5-1, what values of V2 and R3 result in the same voltage and current as when V1 is 10 volts, R1 is 10 kilohms, and R2 is 20 kilohms?
A. R3 = 6.67 kilohms and V2 = 6.67 volts
B. R3 = 6.67 kilohms and V2 = 10 volts
C. R3 = 30 kilohms and V2 = 6.67 volts
D. R3 = 30 kilohms and V2 = 10 volts

A5J07
In Figure A5-1, what values of V2 and R3 result in the same voltage and current as when V1 is 12 volts, R1 is 10 kilohms, and R2 is 10 kilohms?
A. R3 = 20 kilohms and V2 = 12 volts
B. R3 = 5 kilohms and V2 = 6 volts
C. R3 = 5 kilohms and V2 = 12 volts
D. R3 = 30 kilohms and V2 = 6 volts

A5J08
In Figure A5-1, what values of V2 and R3 result in the same voltage and current as when V1 is 12 volts, R1 is 20 kilohms, and R2 is 10 kilohms?
A. R3 = 30 kilohms and V2 = 4 volts
B. R3 = 6.67 kilohms and V2 = 4 volts
C. R3 = 30 kilohms and V2 = 12 volts
D. R3 = 6.67 kilohms and V2 = 12 volts

A5J09
In Figure A5-1, what values of V2 and R3 result in the same voltage and current as when V1 is 12 volts, R1 is 10 kilohms, and R2 is 20 kilohms?
A. R3 = 6.67 kilohms and V2 = 12 volts
B. R3 = 30 kilohms and V2 = 12 volts
C. R3 = 6.67 kilohms and V2 = 8 volts
D. R3 = 30 kilohms and V2 = 8 volts

A5J10
In Figure A5-1, what values of V2 and R3 result in the same voltage and current as when V1 is 12 volts, R1 is 20 kilohms, and R2 is 20 kilohms?
A. R3 = 10 kilohms and V2 = 6 volts
B. R3 = 40 kilohms and V2 = 6 volts
C. R3 = 40 kilohms and V2 = 12 volts
D. R3 = 10 kilohms and V2 = 12 volts

A5J05
(C)
Page 5-36

A5J06
(A)
Page 5-36

A5J07
(B)
Page 5-36

A5J08
(B)
Page 5-36

A5J09
(C)
Page 5-36

A5J10
(A)
Page 5-36

A5J11
(D)
Page 5-36

A5J11
What circuit principle describes the replacement of any complex two-terminal network of voltage sources and resistances with a single voltage source and a single resistor?
A. Ohm's Law
B. Kirchhoff's Law
C. Laplace's Theorem
D. Thevenin's Theorem

Subelement A6

A6 — CIRCUIT COMPONENTS

[6 Exam Questions — 6 Groups]

A6A Semiconductor material: Germanium, Silicon, P-type, N-type

A6A01
(B)
Page 6-2

A6A01
What two elements widely used in semiconductor devices exhibit both metallic and nonmetallic characteristics?
A. Silicon and gold
B. Silicon and germanium
C. Galena and germanium
D. Galena and bismuth

A6A02
(C)
Page 6-5

A6A02
In what application is gallium arsenide used as a semiconductor material in preference to germanium or silicon?
A. In bipolar transistors
B. In high-power circuits
C. At microwave frequencies
D. At very low frequencies

A6A03
(C)
Page 6-2

A6A03
What type of semiconductor material might be produced by adding some antimony atoms to germanium crystals?
A. J-type
B. MOS-type
C. N-type
D. P-type

A6A04
(B)
Page 6-4

A6A04
What type of semiconductor material might be produced by adding some gallium atoms to silicon crystals?
A. N-type
B. P-type
C. MOS-type
D. J-type

A6A05
What type of semiconductor material contains more free electrons than
pure germanium or silicon crystals?
A. N-type
B. P-type
C. Bipolar
D. Insulated gate

A6A05
(A)
Page 6-2

A6A06
What type of semiconductor material might be produced by adding some
arsenic atoms to silicon crystals?
A. N-type
B. P-type
C. MOS-type
D. J-type

A6A06
(A)
Page 6-2

A6A07
What type of semiconductor material might be produced by adding some
indium atoms to germanium crystals?
A. J-type
B. MOS-type
C. N-type
D. P-type

A6A07
(D)
Page 6-4

A6A08
What type of semiconductor material contains fewer free electrons than
pure germanium or silicon crystals?
A. N-type
B. P-type
C. Superconductor-type
D. Bipolar-type

A6A08
(B)
Page 6-4

A6A09
What are the majority charge carriers in P-type semiconductor material?
A. Free neutrons
B. Free protons
C. Holes
D. Free electrons

A6A09
(C)
Page 6-5

A6A10
What are the majority charge carriers in N-type semiconductor material?
A. Holes
B. Free electrons
C. Free protons
D. Free neutrons

A6A10
(B)
Page 6-5

A6A11
(B)
Page 6-4

A6A11
What is the name given to an impurity atom that provides excess electrons
to a semiconductor crystal structure?
A. Acceptor impurity
B. Donor impurity
C. P-type impurity
D. Conductor impurity

A6A12
(C)
Page 6-4

A6A12
What is the name given to an impurity atom that adds holes to a
semiconductor crystal structure?
A. Insulator impurity
B. N-type impurity
C. Acceptor impurity
D. Donor impurity

A6B Diodes: Zener, Tunnel, Varactor, Hot-carrier, Junction, Point contact, PIN and Light-emitting

A6B01
(B)
Page 6-11

A6B01
What is the principal characteristic of a Zener diode?
A. A constant current under conditions of varying voltage
B. A constant voltage under conditions of varying current
C. A negative resistance region
D. An internal capacitance that varies with the applied voltage

A6B02
(D)
Page 6-12

A6B02
In Figure A6-1, what is the schematic symbol for a Zener diode?
A. 7
B. 6
C. 4
D. 3

Figure A6-1 — Refer to questions A6B02, A6B05, A6B07, A6B13 and
A6B15.

A6B03
What is the principal characteristic of a tunnel diode?
A. A high forward resistance
B. A very high PIV
C. A negative resistance region
D. A high forward current rating

A6B04
What special type of diode is capable of both amplification and oscillation?
A. Point contact
B. Zener
C. Tunnel
D. Junction

A6B05
In Figure A6-1, what is the schematic symbol for a tunnel diode?
A. 8
B. 6
C. 2
D. 1

A6B06
What type of semiconductor diode varies its internal capacitance as the voltage applied to its terminals varies?
A. Varactor
B. Tunnel
C. Silicon-controlled rectifier
D. Zener

A6B07
In Figure A6-1, what is the schematic symbol for a varactor diode?
A. 8
B. 6
C. 2
D. 1

A6B08
What is a common use of a hot-carrier diode?
A. As balanced mixers in FM generation
B. As a variable capacitance in an automatic frequency control circuit
C. As a constant voltage reference in a power supply
D. As VHF and UHF mixers and detectors

A6B09
What limits the maximum forward current in a junction diode?
A. Peak inverse voltage
B. Junction temperature
C. Forward voltage
D. Back EMF

A6B03
(C)
Page 6-12

A6B04
(C)
Page 6-12

A6B05
(C)
Page 6-12

A6B06
(A)
Page 6-9

A6B07
(D)
Page 6-9

A6B08
(D)
Page 6-14

A6B09
(B)
Page 6-8

A6B10
(D)
Page 6-7

A6B11
(A)
Page 6-5

A6B12
(C)
Page 6-1

A6B13
(D)
Page 6-6

A6B14
(C)
Page 6-10

A6B15
(B)
Page 6-13

A6B16
(B)
Page 6-13

A6B10
How are junction diodes rated?
A. Maximum forward current and capacitance
B. Maximum reverse current and PIV
C. Maximum reverse current and capacitance
D. Maximum forward current and PIV

A6B11
Structurally, what are the two main categories of semiconductor diodes?
A. Junction and point contact
B. Electrolytic and junction
C. Electrolytic and point contact
D. Vacuum and point contact

A6B12
What is a common use for point contact diodes?
A. As a constant current source
B. As a constant voltage source
C. As an RF detector
D. As a high voltage rectifier

A6B13
In Figure A6-1, what is the schematic symbol for a semiconductor diode/
rectifier?
A. 1
B. 2
C. 3
D. 4

A6B14
What is one common use for PIN diodes?
A. As a constant current source
B. As a constant voltage source
C. As an RF switch
D. As a high voltage rectifier

A6B15
In Figure A6-1, what is the schematic symbol for a light-emitting diode?
A. 1
B. 5
C. 6
D. 7

A6B16
What type of bias is required for an LED to produce luminescence?
A. Reverse bias
B. Forward bias
C. Zero bias
D. Inductive bias

A6C Toroids: Permeability, core material, selecting, winding

A6C01
What material property determines the inductance of a toroidal inductor with a 10-turn winding?
A. Core load current
B. Core resistance
C. Core reactivity
D. Core permeability

A6C02
By careful selection of core material, over what frequency range can toroidal cores produce useful inductors?
A. From a few kHz to no more than several MHz
B. From DC to at least 1000 MHz
C. From DC to no more than 3000 kHz
D. From a few hundred MHz to at least 1000 GHz

A6C03
What materials are used to make ferromagnetic inductors and transformers?
A. Ferrite and powdered-iron toroids
B. Silicon-ferrite toroids and shellac
C. Powdered-ferrite and silicon toroids
D. Ferrite and silicon-epoxy toroids

A6C04
What is one important reason for using powdered-iron toroids rather than ferrite toroids in an inductor?
A. Powdered-iron toroids generally have greater initial permeabilities
B. Powdered-iron toroids generally have better temperature stability
C. Powdered-iron toroids generally require fewer turns to produce a given inductance value
D. Powdered-iron toroids are easier to use with surface-mount technology

A6C05
What is one important reason for using ferrite toroids rather than powdered-iron toroids in an inductor?
A. Ferrite toroids generally have lower initial permeabilities
B. Ferrite toroids generally have better temperature stability
C. Ferrite toroids generally require fewer turns to produce a given inductance value
D. Ferrite toroids are easier to use with surface-mount technology

A6C01
(D)
Page 6-25

A6C02
(B)
Page 6-25

A6C03
(A)
Page 6-24

A6C04
(B)
Page 6-25

A6C05
(C)
Page 6-25

A6C06
(B)
Page 6-28

A6C06
What would be a good choice of toroid core material to make a common-mode choke (such as winding telephone wires or stereo speaker leads on a core) to cure an HF RFI problem?
A. Type 61 mix ferrite (initial permeability of 125)
B. Type 43 mix ferrite (initial permeability of 850)
C. Type 6 mix powdered iron (initial permeability of 8)
D. Type 12 mix powdered iron (initial permeability of 3)

A6C07
(C)
Page 6-28

A6C07
What devices are commonly used as parasitic suppressors at the input and output terminals of VHF and UHF amplifiers?
A. Electrolytic capacitors
B. Butterworth filters
C. Ferrite beads
D. Steel-core toroids

A6C08
(A)
Page 6-24

A6C08
What is a primary advantage of using a toroidal core instead of a linear core in an inductor?
A. Toroidal cores contain most of the magnetic field within the core material
B. Toroidal cores make it easier to couple the magnetic energy into other components
C. Toroidal cores exhibit greater hysteresis
D. Toroidal cores have lower Q characteristics

A6C09
(D)
Page 6-28

A6C09
What is a bifilar-wound toroid?
A. An inductor that has two cores taped together to double the inductance value
B. An inductor wound on a core with two holes (binocular core)
C. A transformer designed to provide a 2-to-1 impedance transformation
D. An inductor that uses a pair of wires to place two windings on the core

A6C10
(C)
Page 6-27

A6C10
How many turns will be required to produce a 1-mH inductor using a ferrite toroidal core that has an inductance index (A sub L) value of 523?
A. 2 turns
B. 4 turns
C. 43 turns
D. 229 turns

A6C11
How many turns will be required to produce a 5-microhenry inductor using a powdered-iron toroidal core that has an inductance index (A sub L) value of 40?
A. 35 turns
B. 13 turns
C. 79 turns
D. 141 turns

A6D Transistor types: NPN, PNP, Junction, Unijunction

A6D01
What are the three terminals of a bipolar transistor?
A. Cathode, plate and grid
B. Base, collector and emitter
C. Gate, source and sink
D. Input, output and ground

A6D02
What is the alpha of a bipolar transistor?
A. The change of collector current with respect to base current
B. The change of base current with respect to collector current
C. The change of collector current with respect to emitter current
D. The change of collector current with respect to gate current

A6D03
What is the beta of a bipolar transistor?
A. The change of collector current with respect to base current
B. The change of base current with respect to emitter current
C. The change of collector current with respect to emitter current
D. The change of base current with respect to gate current

A6D04
What is the alpha cutoff frequency of a bipolar transistor?
A. The practical lower frequency limit of a transistor in common emitter configuration
B. The practical upper frequency limit of a transistor in common emitter configuration
C. The practical lower frequency limit of a transistor in common base configuration
D. The practical upper frequency limit of a transistor in common base configuration

A6C11
(A)
Page 6-26

A6D01
(B)
Page 6-15

A6D02
(C)
Page 6-17

A6D03
(A)
Page 6-16

A6D04
(D)
Page 6-17

A6D05
In Figure A6-2, what is the schematic symbol for an NPN transistor?
A. 1
B. 2
C. 4
D. 5

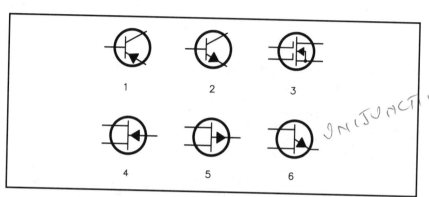

UNIJUNCTIO

Figure A6-2 — Refer to questions A6D05, A6D06 and A6D12.

A6D06
In Figure A6-2, what is the schematic symbol for a PNP transistor?
A. 1
B. 2
C. 4
D. 5

A6D07
What term indicates the frequency at which a transistor grounded base current gain has decreased to 0.7 of the gain obtainable at 1 kHz?
A. Corner frequency
B. Alpha rejection frequency
C. Beta cutoff frequency
D. Alpha cutoff frequency

A6D08
What does the beta cutoff of a bipolar transistor indicate?
A. The frequency at which the grounded base current gain has decreased to 0.7 of that obtainable at 1 kHz
B. The frequency at which the grounded emitter current gain has decreased to 0.7 of that obtainable at 1 kHz
C. The frequency at which the grounded collector current gain has decreased to 0.7 of that obtainable at 1 kHz
D. The frequency at which the grounded gate current gain has decreased to 0.7 of that obtainable at 1 kHz

A6D09
What is the transition region of a transistor?
A. An area of low charge density around the P-N junction
B. The area of maximum P-type charge
C. The area of maximum N-type charge
D. The point where wire leads are connected to the P- or N-type material

A6D10
What does it mean for a transistor to be fully saturated?
A. The collector current is at its maximum value
B. The collector current is at its minimum value
C. The transistor alpha is at its maximum value
D. The transistor beta is at its maximum value

A6D11
What does it mean for a transistor to be cut off?
A. There is no base current
B. The transistor is at its operating point
C. No current flows from emitter to collector
D. Maximum current flows from emitter to collector

A6D12
In Figure A6-2, what is the schematic symbol for a unijunction transistor?
A. 3
B. 4
C. 5
D. 6

A6D13
What are the elements of a unijunction transistor?
A. Gate, base 1 and base 2
B. Gate, cathode and anode
C. Base 1, base 2 and emitter
D. Gate, source and sink

A6E Silicon controlled rectifier (SCR); Triac; neon lamp

A6E01
What are the three terminals of a silicon controlled rectifier (SCR)?
A. Gate, source and sink
B. Anode, cathode and gate
C. Base, collector and emitter
D. Gate, base 1 and base 2

A6D09
(A)
Page 6-16

A6D10
(A)
Page 6-16

A6D11
(C)
Page 6-16

A6D12
(D)
Page 6-18

A6D13
(C)
Page 6-18

A6E01
(B)
Page 6-18

A6E02
(A)
Page 6-19

A6E02
What are the two stable operating conditions of a silicon controlled rectifier (SCR)?
A. Conducting and nonconducting
B. Oscillating and quiescent
C. Forward conducting and reverse conducting
D. NPN conduction and PNP conduction

A6E03
(A)
Page 6-18

A6E03
When a silicon controlled rectifier (SCR) is triggered, to what other solid-state device are its electrical characteristics similar (as measured between its cathode and anode)?
A. The junction diode
B. The tunnel diode
C. The hot-carrier diode
D. The varactor diode

A6E04
(D)
Page 6-19

A6E04
Under what operating conditions does a silicon controlled rectifier (SCR) exhibit electrical characteristics similar to a forward-biased silicon rectifier?
A. During a switching transition
B. When it is used as a detector
C. When it is gated "off"
D. When it is gated "on"

A6E05
(C)
Page 6-19

A6E05
In Figure A6-3, what is the schematic symbol for a silicon controlled rectifier (SCR)?
A. 1
B. 2
C. 5
D. 6

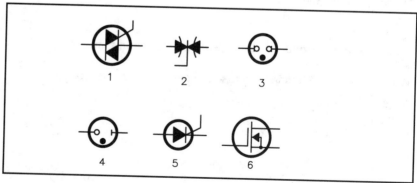

Figure A6-3 — Refer to questions A6E05, A6E08 and A6E11.

A6E06
What is the name of the device that is fabricated as two complementary silicon controlled rectifiers (SCRs) in parallel with a common gate terminal?
A. Bilateral SCR
B. TRIAC
C. Unijunction transistor
D. Field-effect transistor

A6E07
What are the three terminals of a TRIAC?
A. Emitter, base 1 and base 2
B. Gate, anode 1 and anode 2
C. Base, emitter and collector
D. Gate, source and sink

A6E08
In Figure A6-3, what is the schematic symbol for a TRIAC?
A. 1
B. 2
C. 3
D. 5

A6E09
What will happen to a neon lamp in the presence of RF?
A. It will glow only in the presence of very high frequency radio energy
B. It will change color
C. It will glow only in the presence of very low frequency radio energy
D. It will glow

A6E10
If an NE-2 neon bulb is to be used as a dial lamp with a 120 V AC line, what additional component must be connected to it?
A. A 150-pF capacitor in parallel with the bulb
B. A 10-mH inductor in series with the bulb
C. A 150-kilohm resistor in series with the bulb
D. A 10-kilohm resistor in parallel with the bulb

A6E11
In Figure A6-3, what is the schematic symbol for a neon lamp?
A. 1
B. 2
C. 3
D. 4

A6E06
(B)
Page 6-19

A6E07
(B)
Page 6-19

A6E08
(A)
Page 6-19

A6E09
(D)
Page 6-20

A6E10
(C)
Page 6-20

A6E11
(C)
Page 6-20

A6F01
(B)
Page 6-22

A6F02
(C)
Page 6-22

A6F03
(D)
Page 6-20

A6F04
(D)
Page 6-22

A6F05
(A)
Page 6-22

A6F Quartz crystal (frequency determining properties as used in oscillators and filters); monolithic amplifiers (MMICs)

A6F01
For single-sideband phone emissions, what would be the bandwidth of a good crystal lattice band-pass filter?
A. 6 kHz at –6 dB
B. 2.1 kHz at –6 dB
C. 500 Hz at –6 dB
D. 15 kHz at –6 dB

A6F02
For double-sideband phone emissions, what would be the bandwidth of a good crystal lattice band-pass filter?
A. 1 kHz at –6 dB
B. 500 Hz at –6 dB
C. 6 kHz at –6 dB
D. 15 kHz at –6 dB

A6F03
What is a crystal lattice filter?
A. A power supply filter made with interlaced quartz crystals
B. An audio filter made with four quartz crystals that resonate at 1-kHz intervals
C. A filter with wide bandwidth and shallow skirts made using quartz crystals
D. A filter with narrow bandwidth and steep skirts made using quartz crystals

A6F04
What technique is used to construct low-cost, high-performance crystal filters?
A. Choose a center frequency that matches the available crystals
B. Choose a crystal with the desired bandwidth and operating frequency to match a desired center frequency
C. Measure crystal bandwidth to ensure at least 20% coupling
D. Measure crystal frequencies and carefully select units with less than 10% frequency difference

A6F05
Which factor helps determine the bandwidth and response shape of a crystal filter?
A. The relative frequencies of the individual crystals
B. The center frequency chosen for the filter
C. The gain of the RF stage preceding the filter
D. The amplitude of the signals passing through the filter

A6F06
What is the piezoelectric effect?
A. Physical deformation of a crystal by the application of a voltage
B. Mechanical deformation of a crystal by the application of a magnetic field
C. The generation of electrical energy by the application of light
D. Reversed conduction states when a P-N junction is exposed to light

A6F07
Which of the following devices would be most suitable for constructing a receive preamplifier for 1296 MHz?
A. A 2N2222 bipolar transistor
B. An MRF901 bipolar transistor
C. An MSA-0135 monolithic microwave integrated circuit (MMIC)
D. An MPF102 N-junction field-effect transistor (JFET)

A6F08
Which device might be used to simplify the design and construction of a 3456-MHz receiver?
A. An MSA-0735 monolithic microwave integrated circuit (MMIC).
B. An MRF901 bipolar transistor
C. An MGF1402 gallium arsenide field-effect transistor (GaAsFET)
D. An MPF102 N-junction field-effect transistor (JFET)

A6F09
What type of amplifier device consists of a small "pill sized" package with an input lead, an output lead and 2 ground leads?
A. A gallium arsenide field-effect transistor (GaAsFET)
B. An operational amplifier integrated circuit (OAIC)
C. An indium arsenide integrated circuit (IAIC)
D. A monolithic microwave integrated circuit (MMIC)

A6F10
What typical construction technique do amateurs use when building an amplifier containing a monolithic microwave integrated circuit (MMIC)?
A. Ground-plane "ugly" construction
B. Microstrip construction
C. Point-to-point construction
D. Wave-soldering construction

A6F11
How is the operating bias voltage supplied to a monolithic microwave integrated circuit (MMIC)?
A. Through a resistor and RF choke connected to the amplifier output lead
B. MMICs require no operating bias
C. Through a capacitor and RF choke connected to the amplifier input lead
D. Directly to the bias-voltage (VCC IN) lead

A7 — PRACTICAL CIRCUITS

[10 Exam Questions — 10 Groups]

A7A Amplifier circuits: Class A, Class AB, Class B, Class C, amplifier operating efficiency (i.e., DC input vs. PEP); transmitter final amplifiers

A7A01
(B)
Page 7-3

A7A01
For what portion of a signal cycle does a Class A amplifier operate?
A. Less than 180 degrees
B. The entire cycle
C. More than 180 degrees and less than 360 degrees
D. Exactly 180 degrees

A7A02
(A)
Page 7-3

A7A02
Which class of amplifier has the highest linearity and least distortion?
A. Class A
B. Class B
C. Class C
D. Class AB

A7A03
(A)
Page 7-4

A7A03
For what portion of a signal cycle does a Class AB amplifier operate?
A. More than 180 degrees but less than 360 degrees
B. Exactly 180 degrees
C. The entire cycle
D. Less than 180 degrees

A7A04
(D)
Page 7-4

A7A04
For what portion of a signal cycle does a Class B amplifier operate?
A. The entire cycle
B. Greater than 180 degrees and less than 360 degrees
C. Less than 180 degrees
D. 180 degrees

A7A05
(A)
Page 7-4

A7A05
For what portion of a signal cycle does a Class C amplifier operate?
A. Less than 180 degrees
B. Exactly 180 degrees
C. The entire cycle
D. More than 180 degrees but less than 360 degrees

A7A06
(C)
Page 7-4

A7A06
Which class of amplifier provides the highest efficiency?
A. Class A
B. Class B
C. Class C
D. Class AB

A7A07
Where on the load line should a solid-state power amplifier be operated for best efficiency and stability?
A. Just below the saturation point
B. Just above the saturation point
C. At the saturation point
D. At 1.414 times the saturation point

A7A08
What is the formula for the efficiency of a power amplifier?
A. Efficiency = (RF power out / DC power in) x 100%
B. Efficiency = (RF power in / RF power out) x 100%
C. Efficiency = (RF power in / DC power in) x 100%
D. Efficiency = (DC power in / RF power in) x 100%

A7A09
How can parasitic oscillations be eliminated from a power amplifier?
A. By tuning for maximum SWR
B. By tuning for maximum power output
C. By neutralization
D. By tuning the output

A7A10
What is the procedure for tuning a vacuum-tube power amplifier having an output pi-network?
A. Adjust the loading capacitor to maximum capacitance and then dip the plate current with the tuning capacitor
B. Alternately increase the plate current with the tuning capacitor and dip the plate current with the loading capacitor
C. Adjust the tuning capacitor to maximum capacitance and then dip the plate current with the loading capacitor
D. Alternately increase the plate current with the loading capacitor and dip the plate current with the tuning capacitor

A7A11
How can even-order harmonics be reduced or prevented in transmitter amplifiers?
A. By using a push-push amplifier
B. By using a push-pull amplifier
C. By operating Class C
D. By operating Class AB

A7A12
What can occur when a nonlinear amplifier is used with a single-sideband phone transmitter?
A. Reduced amplifier efficiency
B. Increased intelligibility
C. Sideband inversion
D. Distortion

A7A07
(A)
Page 7-11

A7A08
(A)
Page 7-14

A7A09
(C)
Page 7-19

A7A10
(D)
Page 7-21

A7A11
(B)
Page 7-6

A7A12
(D)
Page 7-4

A7B Amplifier circuits: tube, bipolar transistor, FET

A7B01
How can a vacuum-tube power amplifier be neutralized?
A. By increasing the grid drive
B. By feeding back an in-phase component of the output to the input
C. By feeding back an out-of-phase component of the output to the input
D. By feeding back an out-of-phase component of the input to the output

A7B02
What is the flywheel effect?
A. The continued motion of a radio wave through space when the transmitter is turned off
B. The back and forth oscillation of electrons in an LC circuit
C. The use of a capacitor in a power supply to filter rectified AC
D. The transmission of a radio signal to a distant station by several hops through the ionosphere

A7B03
What tank-circuit Q is required to reduce harmonics to an acceptable level?
A. Approximately 120
B. Approximately 12
C. Approximately 1200
D. Approximately 1.2

A7B04
What type of circuit is shown in Figure A7-1?
A. Switching voltage regulator
B. Linear voltage regulator
C. Common emitter amplifier
D. Emitter follower amplifier

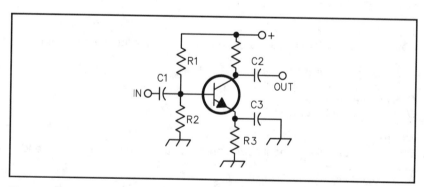

Figure A7-1 — Refer to questions A7B04 through A7B08.

A7B05
In Figure A7-1, what is the purpose of R1 and R2?
A. Load resistors
B. Fixed bias
C. Self bias
D. Feedback

A7B06
In Figure A7-1, what is the purpose of C1?
A. Decoupling
B. Output coupling
C. Self bias
D. Input coupling

A7B07
In Figure A7-1, what is the purpose of C3?
A. AC feedback
B. Input coupling
C. Power supply decoupling
D. Emitter bypass

A7B08
In Figure A7-1, what is the purpose of R3?
A. Fixed bias
B. Emitter bypass
C. Output load resistor
D. Self bias

A7B09
What type of circuit is shown in Figure A7-2?
A. High-gain amplifier
B. Common-collector amplifier
C. Linear voltage regulator
D. Grounded-emitter amplifier

A7B05
(B)
Page 7-8

A7B06
(D)
Page 7-8

A7B07
(D)
Page 7-8

A7B08
(D)
Page 7-8

A7B09
(B)
Page 7-11

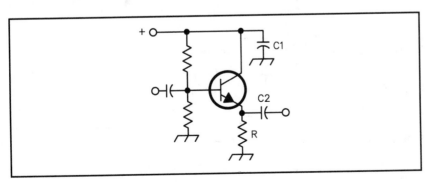

Figure A7-2 — Refer to questions A7B09 through A7B12.

A7B10
In Figure A7-2, what is the purpose of R?
A. Emitter load
B. Fixed bias
C. Collector load
D. Voltage regulation

A7B11
In Figure A7-2, what is the purpose of C1?
A. Input coupling
B. Output coupling
C. Emitter bypass
D. Collector bypass

A7B12
In Figure A7-2, what is the purpose of C2?
A. Output coupling
B. Emitter bypass
C. Input coupling
D. Hum filtering

A7B13
What type of circuit is shown in Figure A7-3?
A. Switching voltage regulator
B. Grounded emitter amplifier
C. Linear voltage regulator
D. Emitter follower

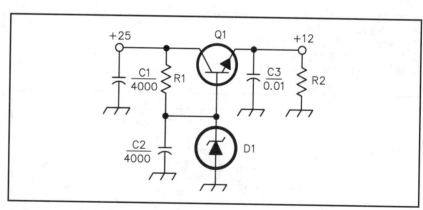

Figure A7-3 — Refer to questions A7B13 through A7B20.

A7B14
What is the purpose of D1 in the circuit shown in Figure A7-3?
A. Line voltage stabilization
B. Voltage reference
C. Peak clipping
D. Hum filtering

A7B15
What is the purpose of Q1 in the circuit shown in Figure A7-3?
A. It increases the output ripple
B. It provides a constant load for the voltage source
C. It increases the current-handling capability
D. It provides D1 with current

A7B16
What is the purpose of C1 in the circuit shown in Figure A7-3?
A. It resonates at the ripple frequency
B. It provides fixed bias for Q1
C. It decouples the output
D. It filters the supply voltage

A7B17
What is the purpose of C2 in the circuit shown in Figure A7-3?
A. It bypasses hum around D1
B. It is a brute force filter for the output
C. To self resonate at the hum frequency
D. To provide fixed DC bias for Q1

A7B18
What is the purpose of C3 in the circuit shown in Figure A7-3?
A. It prevents self-oscillation
B. It provides brute force filtering of the output
C. It provides fixed bias for Q1
D. It clips the peaks of the ripple

A7B19
What is the purpose of R1 in the circuit shown in Figure A7-3?
A. It provides a constant load to the voltage source
B. It couples hum to D1
C. It supplies current to D1
D. It bypasses hum around D1

A7B20
What is the purpose of R2 in the circuit shown in Figure A7-3?
A. It provides fixed bias for Q1
B. It provides fixed bias for D1
C. It decouples hum from D1
D. It provides a constant minimum load for Q1

A7B14
(B)
Page 7-37

A7B15
(C)
Page 7-37

A7B16
(D)
Page 7-37

A7B17
(A)
Page 7-37

A7B18
(A)
Page 7-37

A7B19
(C)
Page 7-37

A7B20
(D)
Page 7-37

A7C Impedance-matching networks: Pi, L, Pi-L

A7C01
(D)
Page 7-22

A7C01
What is a pi-network?
A. A network consisting entirely of four inductors or four capacitors
B. A Power Incidence network
C. An antenna matching network that is isolated from ground
D. A network consisting of one inductor and two capacitors or two inductors and one capacitor

A7C02
(B)
Page 7-22

A7C02
Which type of network offers the greater transformation ratio?
A. L-network
B. Pi-network
C. Constant-K
D. Constant-M

A7C03
(D)
Page 7-22

A7C03
How are the capacitors and inductors of a pi-network arranged between the network's input and output?
A. Two inductors are in series between the input and output and a capacitor is connected between the two inductors and ground
B. Two capacitors are in series between the input and output and an inductor is connected between the two capacitors and ground
C. An inductor is in parallel with the input, another inductor is in parallel with the output, and a capacitor is in series between the two
D. A capacitor is in parallel with the input, another capacitor is in parallel with the output, and an inductor is in series between the two

A7C04
(B)
Page 7-24

A7C04
What is an L-network?
A. A network consisting entirely of four inductors
B. A network consisting of an inductor and a capacitor
C. A network used to generate a leading phase angle
D. A network used to generate a lagging phase angle

A7C05
(A)
Page 7-24

A7C05
Why is an L-network of limited utility in impedance matching?
A. It matches a small impedance range
B. It has limited power-handling capabilities
C. It is thermally unstable
D. It is prone to self resonance

A7C06
(B)
Page 7-25

A7C06
What is a pi-L-network?
A. A Phase Inverter Load network
B. A network consisting of two inductors and two capacitors
C. A network with only three discrete parts
D. A matching network in which all components are isolated from ground

A7C07
A T-network with series capacitors and a parallel (shunt) inductor has which
of the following properties?
A. It transforms impedances and is a low-pass filter
B. It transforms reactances and is a low-pass filter
C. It transforms impedances and is a high-pass filter
D. It transforms reactances and is a high-pass filter

A7C08
What advantage does a pi-L-network have over a pi-network for impedance
matching between the final amplifier of a vacuum-tube type transmitter and
a multiband antenna?
A. Greater harmonic suppression
B. Higher efficiency
C. Lower losses
D. Greater transformation range

A7C09
Which type of network provides the greatest harmonic suppression?
A. L-network
B. Pi-network
C. Pi-L-network
D. Inverse-Pi network

A7C10
Which three types of networks are most commonly used to match an
amplifying device and a transmission line?
A. M, pi and T
B. T, M and Q
C. L, pi and pi-L
D. L, M and C

A7C11
How does a network transform one impedance to another?
A. It introduces negative resistance to cancel the resistive part of an
   impedance
B. It introduces transconductance to cancel the reactive part of an
   impedance
C. It cancels the reactive part of an impedance and changes the resistive
   part
D. Network resistances substitute for load resistances

| A7C07
| (C)
| Page 7-25

| A7C08
| (A)
| Page 7-25

| A7C09
| (C)
| Page 7-25

| A7C10
| (C)
| Page 7-25

| A7C11
| (C)
| Page 7-24

A7D01
(A)
Page 7-26

A7D02
(B)
Page 5-24

A7D03
(D)
Page 5-24

A7D04
(A)
Page 5-24

A7D05
(C)
Page 5-24

A7D06
(A)
Page 5-24

A7D Filter circuits: constant K, M-derived, band-stop, notch, crystal lattice, Pi-section, T-section, L-section, Butterworth, Chebyshev, elliptical

A7D01
What are the three general groupings of filters?
A. High-pass, low-pass and band-pass
B. Inductive, capacitive and resistive
C. Audio, radio and capacitive
D. Hartley, Colpitts and Pierce

A7D02
What value capacitor would be required to tune a 20-microhenry inductor to resonate in the 80-meter band?
A. 150 picofarads
B. 100 picofarads
C. 200 picofarads
D. 100 microfarads

A7D03
What value inductor would be required to tune a 100-picofarad capacitor to resonate in the 40-meter band?
A. 200 microhenrys
B. 150 microhenrys
C. 5 millihenrys
D. 5 microhenrys

A7D04
What value capacitor would be required to tune a 2-microhenry inductor to resonate in the 20-meter band?
A. 64 picofarads
B. 6 picofarads
C. 12 picofarads
D. 88 microfarads

A7D05
What value inductor would be required to tune a 15-picofarad capacitor to resonate in the 15-meter band?
A. 2 microhenrys
B. 30 microhenrys
C. 4 microhenrys
D. 15 microhenrys

A7D06
What value capacitor would be required to tune a 100-microhenry inductor to resonate in the 160-meter band?
A. 78 picofarads
B. 25 picofarads
C. 405 picofarads
D. 40.5 microfarads

A7D07
What are the distinguishing features of a Butterworth filter?
A. The product of its series- and shunt-element impedances is a constant for all frequencies
B. It only requires capacitors
C. It has a maximally flat response over its passband
D. It requires only inductors

A7D07
(C)
Page 7-30

A7D08
What are the distinguishing features of a Chebyshev filter?
A. It has a maximally flat response over its passband
B. It allows ripple in the passband
C. It only requires inductors
D. The product of its series- and shunt-element impedances is a constant for all frequencies

A7D08
(B)
Page 7-30

A7D09
Which filter type is described as having ripple in the passband and a sharp cutoff?
A. A Butterworth filter
B. An active LC filter
C. A passive op-amp filter
D. A Chebyshev filter

A7D09
(D)
Page 7-30

A7D10
What are the distinguishing features of an elliptical filter?
A. Gradual passband rolloff with minimal stop-band ripple
B. Extremely flat response over its passband, with gradually rounded stop-band corners
C. Extremely sharp cutoff, with one or more infinitely deep notches in the stop band
D. Gradual passband rolloff with extreme stop-band ripple

A7D10
(C)
Page 7-30

A7D11
Which filter type has an extremely sharp cutoff, with one or more infinitely deep notches in the stop band?
A. Chebyshev
B. Elliptical
C. Butterworth
D. Crystal lattice

A7D11
(B)
Page 7-30

A7E Voltage-regulator circuits: discrete, integrated and switched mode

A7E01
What is one characteristic of a linear electronic voltage regulator?
A. It has a ramp voltage as its output
B. The pass transistor switches from the "off" state to the "on" state
C. The control device is switched on or off, with the duty cycle proportional to the line or load conditions
D. The conduction of a control element is varied in direct proportion to the line voltage or load current

A7E02
What is one characteristic of a switching electronic voltage regulator?
A. The conduction of a control element is varied in direct proportion to the line voltage or load current
B. It provides more than one output voltage
C. The control device is switched on or off, with the duty cycle proportional to the line or load conditions
D. It gives a ramp voltage at its output

A7E03
What device is typically used as a stable reference voltage in a linear voltage regulator?
A. A Zener diode
B. A tunnel diode
C. An SCR
D. A varactor diode

A7E04
What type of linear regulator is used in applications requiring efficient utilization of the primary power source?
A. A constant current source
B. A series regulator
C. A shunt regulator
D. A shunt current source

A7E05
What type of linear voltage regulator is used in applications requiring a constant load on the unregulated voltage source?
A. A constant current source
B. A series regulator
C. A shunt current source
D. A shunt regulator

A7E06
To obtain the best temperature stability, approximately what operating voltage should be used for the reference diode in a linear voltage regulator?
A. 2 volts
B. 3 volts
C. 6 volts
D. 10 volts

A7E06
(C)
Page 7-35

A7E07
How is remote sensing accomplished in a linear voltage regulator?
A. A feedback connection to an error amplifier is made directly to the load
B. By wireless inductive loops
C. A load connection is made outside the feedback loop
D. An error amplifier compares the input voltage to the reference voltage

A7E07
(A)
Page 7-37

A7E08
What is a three-terminal regulator?
A. A regulator that supplies three voltages with variable current
B. A regulator that supplies three voltages at a constant current
C. A regulator containing three error amplifiers and sensing transistors
D. A regulator containing a voltage reference, error amplifier, sensing resistors and transistors, and a pass element

A7E08
(D)
Page 7-39

A7E09
What are the important characteristics of a three-terminal regulator?
A. Maximum and minimum input voltage, minimum output current and voltage
B. Maximum and minimum input voltage, maximum output current and voltage
C. Maximum and minimum input voltage, minimum output current and maximum output voltage
D. Maximum and minimum input voltage, minimum output voltage and maximum output current

A7E09
(B)
Page 7-40

A7E10
What type of voltage regulator limits the voltage drop across its junction when a specified current passes through it in the reverse-breakdown direction?
A. A Zener diode
B. A three-terminal regulator
C. A bipolar regulator
D. A pass-transistor regulator

A7E10
(A)
Page 7-34

A7E11
(C)
Page 7-39

A7E11
What type of voltage regulator contains a voltage reference, error amplifier, sensing resistors and transistors, and a pass element in one package?
A. A switching regulator
B. A Zener regulator
C. A three-terminal regulator
D. An op-amp regulator

A7F Oscillators: types, applications, stability

A7F01
(D)
Page 7-42

A7F01
What are three major oscillator circuits often used in Amateur Radio equipment?
A. Taft, Pierce and negative feedback
B. Colpitts, Hartley and Taft
C. Taft, Hartley and Pierce
D. Colpitts, Hartley and Pierce

A7F02
(C)
Page 7-42

A7F02
What condition must exist for a circuit to oscillate?
A. It must have a gain of less than 1
B. It must be neutralized
C. It must have positive feedback sufficient to overcome losses
D. It must have negative feedback sufficient to cancel the input

A7F03
(A)
Page 7-42

A7F03
How is the positive feedback coupled to the input in a Hartley oscillator?
A. Through a tapped coil
B. Through a capacitive divider
C. Through link coupling
D. Through a neutralizing capacitor

A7F04
(C)
Page 7-42

A7F04
How is the positive feedback coupled to the input in a Colpitts oscillator?
A. Through a tapped coil
B. Through link coupling
C. Through a capacitive divider
D. Through a neutralizing capacitor

A7F05
(D)
Page 7-42

A7F05
How is the positive feedback coupled to the input in a Pierce oscillator?
A. Through a tapped coil
B. Through link coupling
C. Through a neutralizing capacitor
D. Through capacitive coupling

A7F06
Which of the three major oscillator circuits used in Amateur Radio equipment uses a quartz crystal?
A. Negative feedback
B. Hartley
C. Colpitts
D. Pierce

A7F07
What is the major advantage of a Pierce oscillator?
A. It is easy to neutralize
B. It doesn't require an LC tank circuit
C. It can be tuned over a wide range
D. It has a high output power

A7F08
Which type of oscillator circuits are commonly used in a VFO?
A. Pierce and Zener
B. Colpitts and Hartley
C. Armstrong and deForest
D. Negative feedback and Balanced feedback

A7F09
Why is the Colpitts oscillator circuit commonly used in a VFO?
A. The frequency is a linear function of the load impedance
B. It can be used with or without crystal lock-in
C. It is stable
D. It has high output power

A7F10
What component is often used to control an oscillator frequency by varying a control voltage?
A. A varactor diode
B. A piezoelectric crystal
C. A Zener diode
D. A Pierce crystal

A7F11
Why must a very stable reference oscillator be used as part of a phase-locked loop (PLL) frequency synthesizer?
A. Any amplitude variations in the reference oscillator signal will prevent the loop from locking to the desired signal
B. Any phase variations in the reference oscillator signal will produce phase noise in the synthesizer output
C. Any phase variations in the reference oscillator signal will produce harmonic distortion in the modulating signal
D. Any amplitude variations in the reference oscillator signal will prevent the loop from changing frequency

A7F06
(D)
Page 7-42

A7F07
(B)
Page 7-42

A7F08
(B)
Page 7-45

A7F09
(C)
Page 7-45

A7F10
(A)
Page 7-46

A7F11
(B)
Page 7-48

A7G01
(D)
Page 7-55

A7G02
(B)
Page 7-63

A7G03
(C)
Page 7-63

A7G04
(B)
Page 7-63

A7G05
(C)
Page 7-66

A7G06
(C)
Page 7-66

A7G Modulators: Reactance, Phase, Balanced

A7G01
What is meant by modulation?
A. The squelching of a signal until a critical signal-to-noise ratio is reached
B. Carrier rejection through phase nulling
C. A linear amplification mode
D. A mixing process whereby information is imposed upon a carrier

A7G02
How is an F3E FM-phone emission produced?
A. With a balanced modulator on the audio amplifier
B. With a reactance modulator on the oscillator
C. With a reactance modulator on the final amplifier
D. With a balanced modulator on the oscillator

A7G03
How does a reactance modulator work?
A. It acts as a variable resistance or capacitance to produce FM signals
B. It acts as a variable resistance or capacitance to produce AM signals
C. It acts as a variable inductance or capacitance to produce FM signals
D. It acts as a variable inductance or capacitance to produce AM signals

A7G04
What type of circuit varies the tuning of an oscillator circuit to produce FM signals?
A. A balanced modulator
B. A reactance modulator
C. A double balanced mixer
D. An audio modulator

A7G05
How does a phase modulator work?
A. It varies the tuning of a microphone preamplifier to produce FM signals
B. It varies the tuning of an amplifier tank circuit to produce AM signals
C. It varies the tuning of an amplifier tank circuit to produce FM signals
D. It varies the tuning of a microphone preamplifier to produce AM signals

A7G06
What type of circuit varies the tuning of an amplifier tank circuit to produce FM signals?
A. A balanced modulator
B. A double balanced mixer
C. A phase modulator
D. An audio modulator

A7G07
What type of signal does a balanced modulator produce?
A. FM with balanced deviation
B. Double sideband, suppressed carrier
C. Single sideband, suppressed carrier
D. Full carrier

A7G08
How can a single-sideband phone signal be generated?
A. By using a balanced modulator followed by a filter
B. By using a reactance modulator followed by a mixer
C. By using a loop modulator followed by a mixer
D. By driving a product detector with a DSB signal

A7G09
How can a double-sideband phone signal be generated?
A. By feeding a phase modulated signal into a low-pass filter
B. By using a balanced modulator followed by a filter
C. By detuning a Hartley oscillator
D. By modulating the plate voltage of a Class C amplifier

A7G10
What audio shaping network is added at a transmitter to proportionally attenuate the lower audio frequencies, giving an even spread to the energy in the audio band?
A. A de-emphasis network
B. A heterodyne suppressor
C. An audio prescaler
D. A pre-emphasis network

A7G11
What audio shaping network is added at a receiver to restore proportionally attenuated lower audio frequencies?
A. A de-emphasis network
B. A heterodyne suppressor
C. An audio prescaler
D. A pre-emphasis network

A7H Detectors; filter applications (audio, IF, Digital signal processing {DSP})

A7H01
What is the process of detection?
A. The masking of the intelligence on a received carrier
B. The recovery of the intelligence from a modulated RF signal
C. The modulation of a carrier
D. The mixing of noise with a received signal

A7G07
(B)
Page 7-57

A7G08
(A)
Page 7-57

A7G09
(D)
Page 7-56

A7G10
(D)
Page 7-67

A7G11
(A)
Page 7-67

A7H01
(B)
Page 7-52

A7H02
What is the principle of detection in a diode detector?
A. Rectification and filtering of RF
B. Breakdown of the Zener voltage
C. Mixing with noise in the transition region of the diode
D. The change of reactance in the diode with respect to frequency

A7H03
What does a product detector do?
A. It provides local oscillations for input to a mixer
B. It amplifies and narrows band-pass frequencies
C. It mixes an incoming signal with a locally generated carrier
D. It detects cross-modulation products

A7H04
How are FM-phone signals detected?
A. With a balanced modulator
B. With a frequency discriminator
C. With a product detector
D. With a phase splitter

A7H05
What is a frequency discriminator?
A. An FM generator
B. A circuit for filtering two closely adjacent signals
C. An automatic band-switching circuit
D. A circuit for detecting FM signals

A7H06
Which of the following is NOT an advantage of using active filters rather than L-C filters at audio frequencies?
A. Active filters have higher signal-to-noise ratios
B. Active filters can provide gain as well as frequency selection
C. Active filters do not require the use of inductors
D. Active filters can use potentiometers for tuning

A7H07
What kind of audio filter would you use to attenuate an interfering carrier signal while receiving an SSB transmission?
A. A band-pass filter
B. A notch filter
C. A pi-network filter
D. An all-pass filter

A7H08
What characteristic do typical SSB receiver IF filters lack that is important to digital communications?
A. Steep amplitude-response skirts
B. Passband ripple
C. High input impedance
D. Linear phase response

A7H09
What kind of digital signal processing audio filter might be used to remove unwanted noise from a received SSB signal?
A. An adaptive filter
B. A notch filter
C. A Hilbert-transform filter
D. A phase-inverting filter

A7H10
What kind of digital signal processing filter might be used in generating an SSB signal?
A. An adaptive filter
B. A notch filter
C. A Hilbert-transform filter
D. An elliptical filter

A7H11
Which type of filter would be the best to use in a 2-meter repeater duplexer?
A. A crystal filter
B. A cavity filter
C. A DSP filter
D. An L-C filter

A7I Mixer stages; Frequency synthesizers

A7I01
What is the mixing process?
A. The elimination of noise in a wideband receiver by phase comparison
B. The elimination of noise in a wideband receiver by phase differentiation
C. The recovery of the intelligence from a modulated RF signal
D. The combination of two signals to produce sum and difference frequencies

A7H08
(D)
Page 7-32

A7H09
(A)
Page 7-33

A7H10
(C)
Page 7-33

A7H11
(B)
Page 7-26

A7I01
(D)
Page 7-50

A7I02
(C)
Page 7-50

A7I02
What are the principal frequencies that appear at the output of a mixer circuit?
A. Two and four times the original frequency
B. The sum, difference and square root of the input frequencies
C. The original frequencies and the sum and difference frequencies
D. 1.414 and 0.707 times the input frequency

A7I03
(B)
Page 7-49

A7I03
What are the advantages of the frequency-conversion process?
A. Automatic squelching and increased selectivity
B. Increased selectivity and optimal tuned-circuit design
C. Automatic soft limiting and automatic squelching
D. Automatic detection in the RF amplifier and increased selectivity

A7I04
(A)
Page 7-50

A7I04
What occurs in a receiver when an excessive amount of signal energy reaches the mixer circuit?
A. Spurious mixer products are generated
B. Mixer blanking occurs
C. Automatic limiting occurs
D. A beat frequency is generated

A7I05
(C)
Page 7-47

A7I05
What type of frequency synthesizer circuit uses a stable voltage-controlled oscillator, programmable divider, phase detector, loop filter and a reference frequency source?
A. A direct digital synthesizer
B. A hybrid synthesizer
C. A phase-locked loop synthesizer
D. A diode-switching matrix synthesizer

A7I06
(A)
Page 7-48

A7I06
What type of frequency synthesizer circuit uses a phase accumulator, lookup table, digital to analog converter and a low-pass antialias filter?
A. A direct digital synthesizer
B. A hybrid synthesizer
C. A phase-locked loop synthesizer
D. A diode-switching matrix synthesizer

A7I07
(B)
Page 7-47

A7I07
What are the main blocks of a phase-locked loop frequency synthesizer?
A. A variable-frequency crystal oscillator, programmable divider, digital to analog converter and a loop filter
B. A stable voltage-controlled oscillator, programmable divider, phase detector, loop filter and a reference frequency source
C. A phase accumulator, lookup table, digital to analog converter and a low-pass antialias filter
D. A variable-frequency oscillator, programmable divider, phase detector and a low-pass antialias filter

A7I08
What are the main blocks of a direct digital frequency synthesizer?
A. A variable-frequency crystal oscillator, phase accumulator, digital to analog converter and a loop filter
B. A stable voltage-controlled oscillator, programmable divider, phase detector, loop filter and a digital to analog converter
C. A variable-frequency oscillator, programmable divider, phase detector and a low-pass antialias filter
D. A phase accumulator, lookup table, digital to analog converter and a low-pass antialias filter

A7I08
(D)
Page 7-48

A7I09
What information is contained in the lookup table of a direct digital frequency synthesizer?
A. The phase relationship between a reference oscillator and the output waveform
B. The amplitude values that represent a sine-wave output
C. The phase relationship between a voltage-controlled oscillator and the output waveform
D. The synthesizer frequency limits and frequency values stored in the radio memories

A7I09
(B)
Page 7-49

A7I10
What are the major spectral impurity components of direct digital synthesizers?
A. Broadband noise
B. Digital conversion noise
C. Spurs at discrete frequencies
D. Nyquist limit noise

A7I10
(C)
Page 7-49

A7I11
What are the major spectral impurity components of phase-locked loop synthesizers?
A. Broadband noise
B. Digital conversion noise
C. Spurs at discrete frequencies
D. Nyquist limit noise

A7I11
(A)
Page 7-48

A7J Amplifier applications: AF, IF, RF

A7J01
For most amateur phone communications, what should be the upper frequency limit of an audio amplifier?
A. No more than 1000 Hz
B. About 3000 Hz
C. At least 10,000 Hz
D. More than 20,000 Hz

A7J01
(B)
Page 7-17

A7J02
What is the term for the ratio of the RMS voltage for all harmonics in an audio-amplifier output to the total RMS voltage of the output for a pure sine-wave input?
A. Total harmonic distortion
B. Maximum frequency deviation
C. Full quieting ratio
D. Harmonic signal ratio

A7J03
What are the advantages of a Darlington pair audio amplifier?
A. Mutual gain, low input impedance and low output impedance
B. Low output impedance, high mutual inductance and low output current
C. Mutual gain, high stability and low mutual inductance
D. High gain, high input impedance and low output impedance

A7J04
What is the purpose of a speech amplifier in an amateur phone transmitter?
A. To increase the dynamic range of the audio
B. To raise the microphone audio output to the level required by the modulator
C. To match the microphone impedance to the transmitter input impedance
D. To provide adequate AGC drive to the transmitter

A7J05
What is an IF amplifier stage?
A. A fixed-tuned pass-band amplifier
B. A receiver demodulator
C. A receiver filter
D. A buffer oscillator

A7J06
What factors should be considered when selecting an intermediate frequency?
A. Cross-modulation distortion and interference
B. Interference to other services
C. Image rejection and selectivity
D. Noise figure and distortion

A7J07
Which of the following is a purpose of the first IF amplifier stage in a receiver?
A. To improve noise figure performance
B. To tune out cross-modulation distortion
C. To increase the dynamic response
D. To provide selectivity

A7J08
Which of the following is an important reason for using a VHF intermediate frequency in an HF receiver?
A. To provide a greater tuning range
B. To move the image response far away from the filter passband
C. To tune out cross-modulation distortion
D. To prevent the generation of spurious mixer products

A7J09
How much gain should be used in the RF amplifier stage of a receiver?
A. As much gain as possible, short of self oscillation
B. Sufficient gain to allow weak signals to overcome noise generated in the first mixer stage
C. Sufficient gain to keep weak signals below the noise of the first mixer stage
D. It depends on the amplification factor of the first IF stage

A7J10
Why should the RF amplifier stage of a receiver have only sufficient gain to allow weak signals to overcome noise generated in the first mixer stage?
A. To prevent the sum and difference frequencies from being generated
B. To prevent bleed-through of the desired signal
C. To prevent the generation of spurious mixer products
D. To prevent bleed-through of the local oscillator

A7J11
What is the primary purpose of an RF amplifier in a receiver?
A. To improve the receiver noise figure
B. To vary the receiver image rejection by using the AGC
C. To provide most of the receiver gain
D. To develop the AGC voltage

A8 — SIGNALS AND EMISSIONS

[6 Exam Questions — 6 Groups]

A8A FCC emission designators vs. emission types

A8A01
What is emission A3C?
A. Facsimile
B. RTTY
C. ATV
D. Slow Scan TV

A7J08
(B)
Page 7-17

A7J09
(B)
Page 7-12

A7J10
(C)
Page 7-12

A7J11
(A)
Page 7-12

Subelement
A8

A8A01
(A)
Page 8-4

A8A02
(B)
Page 8-4

A8A02
What type of emission is produced when an AM transmitter is modulated by a facsimile signal?
A. A3F
B. A3C
C. F3F
D. F3C

A8A03
(C)
Page 8-4

A8A03
What does a facsimile transmission produce?
A. Tone-modulated telegraphy
B. A pattern of printed characters designed to form a picture
C. Printed pictures by electrical means
D. Moving pictures by electrical means

A8A04
(D)
Page 8-4

A8A04
What is emission F3C?
A. Voice transmission
B. Slow Scan TV
C. RTTY
D. Facsimile

A8A05
(A)
Page 8-4

A8A05
What type of emission is produced when an FM transmitter is modulated by a facsimile signal?
A. F3C
B. A3C
C. F3F
D. A3F

A8A06
(B)
Page 8-4

A8A06
What is emission A3F?
A. RTTY
B. Television
C. SSB
D. Modulated CW

A8A07
(B)
Page 8-4

A8A07
What type of emission is produced when an AM transmitter is modulated by a television signal?
A. F3F
B. A3F
C. A3C
D. F3C

A8A08
What is emission F3F?
A. Modulated CW
B. Facsimile
C. RTTY
D. Television

A8A08
(D)
Page 8-4

A8A09
What type of emission is produced when an FM transmitter is modulated by a television signal?
A. A3F
B. A3C
C. F3F
D. F3C

A8A09
(C)
Page 8-4

A8A10
What type of emission is produced when an SSB transmitter is modulated by a slow-scan television signal?
A. J3A
B. F3F
C. A3F
D. J3F

A8A10
(D)
Page 8-4

A8A11
What emission is produced when an AM transmitter is modulated by a single-channel signal containing digital information without the use of a modulating subcarrier, resulting in telegraphy for aural reception?
A. CW
B. RTTY
C. Data
D. MCW

A8A11
(A)
Page 8-3

A8B Modulation symbols and transmission characteristics

A8B01
What International Telecommunication Union (ITU) system describes the characteristics and necessary bandwidth of any transmitted signal?
A. Emission Designators
B. Emission Zones
C. Band Plans
D. Modulation Indicators

A8B01
(A)
Page 8-2

A8B02
Which of the following describe the three most-used symbols of an ITU
emission designator?
A. Type of modulation, transmitted bandwidth and modulation code
   designator
B. Bandwidth of the modulating signal, nature of the modulating signal and
   transmission rate of signals
C. Type of modulation, nature of the modulating signal and type of
   information to be transmitted
D. Power of signal being transmitted, nature of multiplexing and
   transmission speed

A8B03
If the first symbol of an ITU emission designator is J, representing a single-
sideband, suppressed-carrier signal, what information about the emission is
described?
A. The nature of any signal multiplexing
B. The type of modulation of the main carrier
C. The maximum permissible bandwidth
D. The maximum signal level, in decibels

A8B04
If the first symbol of an ITU emission designator is G, representing a phase-
modulated signal, what information about the emission is described?
A. The nature of any signal multiplexing
B. The maximum permissible deviation
C. The nature of signals modulating the main carrier
D. The type of modulation of the main carrier

A8B05
If the first symbol of an ITU emission designator is P, representing a
sequence of unmodulated pulses, what information about the emission is
described?
A. The type of modulation of the main carrier
B. The maximum permissible pulse width
C. The nature of signals modulating the main carrier
D. The nature of any signal multiplexing

A8B06
If the second symbol of an ITU emission designator is 3, representing a
single channel containing analog information, what information about the
emission is described?
A. The nature of signals modulating the main carrier
B. The maximum permissible deviation
C. The maximum signal level, in decibels
D. The type of modulation of the main carrier

A8B07
If the second symbol of an ITU emission designator is 1, representing a single channel containing quantized, or digital information, what information about the emission is described?
A. The maximum transmission rate, in bauds
B. The maximum permissible deviation
C. The nature of signals modulating the main carrier
D. The type of information to be transmitted

A8B08
If the third symbol of an ITU emission designator is D, representing data transmission, telemetry or telecommand, what information about the emission is described?
A. The maximum transmission rate, in bauds
B. The maximum permissible deviation
C. The nature of signals modulating the main carrier
D. The type of information to be transmitted

A8B09
If the third symbol of an ITU emission designator is A, representing telegraphy for aural reception, what information about the emission is described?
A. The maximum transmission rate, in words per minute
B. The type of information to be transmitted
C. The nature of signals modulating the main carrier
D. The maximum number of different signal elements

A8B10
If the third symbol of an ITU emission designator is B, representing telegraphy for automatic reception, what information about the emission is described?
A. The maximum transmission rate, in bauds
B. The type of information to be transmitted
C. The type of modulation of the main carrier
D. The transmission code is Baudot

A8B11
If the third symbol of an ITU emission designator is F, representing television (video), what information about the emission is described?
A. The maximum frequency variation of the color-burst pulse
B. The picture scan rate is fast
C. The type of modulation of the main carrier
D. The type of information to be transmitted

A8B07
(C)
Page 8-2

A8B08
(D)
Page 8-2

A8B09
(B)
Page 8-2

A8B10
(B)
Page 8-2

A8B11
(D)
Page 8-2

A8C Modulation methods; modulation index; deviation ratio

A8C01
How can an FM-phone signal be produced?
A. By modulating the supply voltage to a Class-B amplifier
B. By modulating the supply voltage to a Class-C amplifier
C. By using a reactance modulator on an oscillator
D. By using a balanced modulator on an oscillator

A8C02
How can the unwanted sideband be removed from a double-sideband signal generated by a balanced modulator to produce a single-sideband phone signal?
A. By filtering
B. By heterodyning
C. By mixing
D. By neutralization

A8C03
What is meant by modulation index?
A. The processor index
B. The ratio between the deviation of a frequency modulated signal and the modulating frequency
C. The FM signal-to-noise ratio
D. The ratio of the maximum carrier frequency deviation to the highest audio modulating frequency

A8C04
In an FM-phone signal, what is the term for the ratio between the deviation of the frequency modulated signal and the modulating frequency?
A. FM compressibility
B. Quieting index
C. Percentage of modulation
D. Modulation index

A8C05
How does the modulation index of a phase-modulated emission vary with RF carrier frequency (the modulated frequency)?
A. It increases as the RF carrier frequency increases
B. It decreases as the RF carrier frequency increases
C. It varies with the square root of the RF carrier frequency
D. It does not depend on the RF carrier frequency

A8C06

In an FM-phone signal having a maximum frequency deviation of 3000 Hz either side of the carrier frequency, what is the modulation index when the modulating frequency is 1000 Hz?

A. 3
B. 0.3
C. 3000
D. 1000

A8C06
(A)
Page 8-8

A8C07

What is the modulation index of an FM-phone transmitter producing an instantaneous carrier deviation of 6 kHz when modulated with a 2-kHz modulating frequency?

A. 6000
B. 3
C. 2000
D. 1/3

A8C07
(B)
Page 8-8

A8C08

What is meant by deviation ratio?

A. The ratio of the audio modulating frequency to the center carrier frequency
B. The ratio of the maximum carrier frequency deviation to the highest audio modulating frequency
C. The ratio of the carrier center frequency to the audio modulating frequency
D. The ratio of the highest audio modulating frequency to the average audio modulating frequency

A8C08
(B)
Page 8-7

A8C09

In an FM-phone signal, what is the term for the maximum deviation from the carrier frequency divided by the maximum audio modulating frequency?

A. Deviation index
B. Modulation index
C. Deviation ratio
D. Modulation ratio

A8C09
(C)
Page 8-7

A8C10

What is the deviation ratio of an FM-phone signal having a maximum frequency swing of plus or minus 5 kHz and accepting a maximum modulation rate of 3 kHz?

A. 60
B. 0.16
C. 0.6
D. 1.66

A8C10
(D)
Page 8-8

A8C11
(A)
Page 8-8

A8C11
What is the deviation ratio of an FM-phone signal having a maximum frequency swing of plus or minus 7.5 kHz and accepting a maximum modulation rate of 3.5 kHz?
A. 2.14
B. 0.214
C. 0.47
D. 47

A8D Electromagnetic radiation; wave polarization; signal-to-noise (S/N) ratio

A8D01
(C)
Page 8-9

A8D01
What are electromagnetic waves?
A. Alternating currents in the core of an electromagnet
B. A wave consisting of two electric fields at right angles to each other
C. A wave consisting of an electric field and a magnetic field at right angles to each other
D. A wave consisting of two magnetic fields at right angles to each other

A8D02
(A)
Page 8-9

A8D02
At approximately what speed do electromagnetic waves travel in free space?
A. 300 million meters per second
B. 468 million meters per second
C. 186,300 feet per second
D. 300 million miles per second

A8D03
(C)
Page 8-10

A8D03
Why don't electromagnetic waves penetrate a good conductor for more than a fraction of a wavelength?
A. Electromagnetic waves are reflected by the surface of a good conductor
B. Oxide on the conductor surface acts as a magnetic shield
C. The electromagnetic waves are dissipated as eddy currents in the conductor surface
D. The resistance of the conductor surface dissipates the electromagnetic waves

A8D04
(D)
Page 8-9

A8D04
Which of the following best describes electromagnetic waves traveling in free space?
A. Electric and magnetic fields become aligned as they travel
B. The energy propagates through a medium with a high refractive index
C. The waves are reflected by the ionosphere and return to their source
D. Changing electric and magnetic fields propagate the energy across a vacuum

A8D05
What is meant by horizontally polarized electromagnetic waves?
A. Waves with an electric field parallel to the Earth
B. Waves with a magnetic field parallel to the Earth
C. Waves with both electric and magnetic fields parallel to the Earth
D. Waves with both electric and magnetic fields perpendicular to the Earth

A8D06
What is meant by circularly polarized electromagnetic waves?
A. Waves with an electric field bent into a circular shape
B. Waves with a rotating electric field
C. Waves that circle the Earth
D. Waves produced by a loop antenna

A8D07
What is the polarization of an electromagnetic wave if its electric field is perpendicular to the surface of the Earth?
A. Circular
B. Horizontal
C. Vertical
D. Elliptical

A8D08
What is the polarization of an electromagnetic wave if its magnetic field is parallel to the surface of the Earth?
A. Circular
B. Horizontal
C. Elliptical
D. Vertical

A8D09
What is the polarization of an electromagnetic wave if its magnetic field is perpendicular to the surface of the Earth?
A. Horizontal
B. Circular
C. Elliptical
D. Vertical

A8D10
What is the polarization of an electromagnetic wave if its electric field is parallel to the surface of the Earth?
A. Vertical
B. Horizontal
C. Circular
D. Elliptical

A8D05
(A)
Page 8-10

A8D06
(B)
Page 8-10

A8D07
(C)
Page 8-10

A8D08
(D)
Page 8-10

A8D09
(A)
Page 8-10

A8D10
(B)
Page 8-10

A8D11
(D)
Page 8-11

A8D11
What is the primary source of noise that can be heard in an HF-band receiver with an antenna connected?
A. Detector noise
B. Man-made noise
C. Receiver front-end noise
D. Atmospheric noise

A8D12
(A)
Page 8-11

A8D12
What is the primary source of noise that can be heard in a VHF/UHF-band receiver with an antenna connected?
A. Receiver front-end noise
B. Man-made noise
C. Atmospheric noise
D. Detector noise

A8E AC waveforms: Sine wave, square wave, sawtooth wave

A8E01
(B)
Page 8-12

A8E01
What is a sine wave?
A. A constant-voltage, varying-current wave
B. A wave whose amplitude at any given instant can be represented by a point on a wheel rotating at a uniform speed
C. A wave following the laws of the trigonometric tangent function
D. A wave whose polarity changes in a random manner

A8E02
(C)
Page 8-12

A8E02
Starting at a positive peak, how many times does a sine wave cross the zero axis in one complete cycle?
A. 180 times
B. 4 times
C. 2 times
D. 360 times

A8E03
(D)
Page 8-12

A8E03
How many degrees are there in one complete sine wave cycle?
A. 90 degrees
B. 270 degrees
C. 180 degrees
D. 360 degrees

A8E04
(A)
Page 8-13

A8E04
What is the period of a wave?
A. The time required to complete one cycle
B. The number of degrees in one cycle
C. The number of zero crossings in one cycle
D. The amplitude of the wave

A8E05
What is a square wave?
A. A wave with only 300 degrees in one cycle
B. A wave that abruptly changes back and forth between two voltage levels and remains an equal time at each level
C. A wave that makes four zero crossings per cycle
D. A wave in which the positive and negative excursions occupy unequal portions of the cycle time

A8E05
(B)
Page 8-13

A8E06
What is a wave called that abruptly changes back and forth between two voltage levels and remains an equal time at each level?
A. A sine wave
B. A cosine wave
C. A square wave
D. A sawtooth wave

A8E06
(C)
Page 8-13

A8E07
What sine waves added to a fundamental frequency make up a square wave?
A. A sine wave 0.707 times the fundamental frequency
B. All odd and even harmonics
C. All even harmonics
D. All odd harmonics

A8E07
(D)
Page 8-13

A8E08
What type of wave is made up of a sine wave of a fundamental frequency and all its odd harmonics?
A. A square wave
B. A sine wave
C. A cosine wave
D. A tangent wave

A8E08
(A)
Page 8-13

A8E09
What is a sawtooth wave?
A. A wave that alternates between two values and spends an equal time at each level
B. A wave with a straight line rise time faster than the fall time (or vice versa)
C. A wave that produces a phase angle tangent to the unit circle
D. A wave whose amplitude at any given instant can be represented by a point on a wheel rotating at a uniform speed

A8E09
(B)
Page 8-13

A8E10
(C)
Page 8-13

A8E10
What type of wave has a rise time significantly faster than the fall time (or vice versa)?
A. A cosine wave
B. A square wave
C. A sawtooth wave
D. A sine wave

A8E11
(A)
Page 8-13

A8E11
What type of wave is made up of sine waves of a fundamental frequency and all harmonics?
A. A sawtooth wave
B. A square wave
C. A sine wave
D. A cosine wave

A8F AC measurements: peak, peak-to-peak and root-mean-square (RMS) value; peak-envelope-power (PEP) relative to average

A8F01
(B)
Page 8-15

A8F01
What is the peak voltage at a common household electrical outlet?
A. 240 volts
B. 170 volts
C. 120 volts
D. 340 volts

A8F02
(C)
Page 8-15

A8F02
What is the peak-to-peak voltage at a common household electrical outlet?
A. 240 volts
B. 120 volts
C. 340 volts
D. 170 volts

A8F03
(A)
Page 8-15

A8F03
What is the RMS voltage at a common household electrical power outlet?
A. 120-V AC
B. 340-V AC
C. 85-V AC
D. 170-V AC

A8F04
(A)
Page 8-15

A8F04
What is the RMS value of a 340-volt peak-to-peak pure sine wave?
A. 120-V AC
B. 170-V AC
C. 240-V AC
D. 300-V AC

A8F05
What is the equivalent to the root-mean-square value of an AC voltage?
A. The AC voltage found by taking the square of the average value of the peak AC voltage
B. The DC voltage causing the same heating of a given resistor as the peak AC voltage
C. The AC voltage causing the same heating of a given resistor as a DC voltage of the same value
D. The AC voltage found by taking the square root of the average AC value

A8F06
What would be the most accurate way of determining the RMS voltage of a complex waveform?
A. By using a grid dip meter
B. By measuring the voltage with a D'Arsonval meter
C. By using an absorption wavemeter
D. By measuring the heating effect in a known resistor

A8F07
For many types of voices, what is the approximate ratio of PEP to average power during a modulation peak in a single-sideband phone signal?
A. 2.5 to 1
B. 25 to 1
C. 1 to 1
D. 100 to 1

A8F08
In a single-sideband phone signal, what determines the PEP-to-average power ratio?
A. The frequency of the modulating signal
B. The speech characteristics
C. The degree of carrier suppression
D. The amplifier power

A8F09
What is the approximate DC input power to a Class B RF power amplifier stage in an FM-phone transmitter when the PEP output power is 1500 watts?
A. 900 watts
B. 1765 watts
C. 2500 watts
D. 3000 watts

A8F05
(C)
Page 8-14

A8F06
(D)
Page 8-14

A8F07
(A)
Page 8-16

A8F08
(B)
Page 8-16

A8F09
(C)
Page 7-15

A8F10
(B)
Page 7-15

A8F10
What is the approximate DC input power to a Class C RF power amplifier
stage in a RTTY transmitter when the PEP output power is 1000 watts?
A. 850 watts
B. 1250 watts
C. 1667 watts
D. 2000 watts

A8F11
(D)
Page 7-15

A8F11
What is the approximate DC input power to a Class AB RF power amplifier
stage in an unmodulated carrier transmitter when the PEP output power is
500 watts?
A. 250 watts
B. 600 watts
C. 800 watts
D. 1000 watts

Subelement
A9

A9 — ANTENNAS AND FEED LINES

[5 Exam Questions — 5 Groups]

**A9A Basic antenna parameters: radiation resistance and reactance
(including wire dipole, folded dipole, gain, beamwidth, efficiency)**

A9A01
(C)
Page 9-2

A9A01
What is meant by the radiation resistance of an antenna?
A. The combined losses of the antenna elements and feed line
B. The specific impedance of the antenna
C. The equivalent resistance that would dissipate the same amount of
   power as that radiated from an antenna
D. The resistance in the atmosphere that an antenna must overcome to be
   able to radiate a signal

A9A02
(This question number was omitted when the VEC Question Pool
Committee released this question pool. There is no question numbered
A9A02.)

A9A03
(A)
Page 9-2

A9A03
Why would one need to know the radiation resistance of an antenna?
A. To match impedances for maximum power transfer
B. To measure the near-field radiation density from a transmitting antenna
C. To calculate the front-to-side ratio of the antenna
D. To calculate the front-to-back ratio of the antenna

A9A04
What factors determine the radiation resistance of an antenna?
A. Transmission-line length and antenna height
B. Antenna location with respect to nearby objects and the conductors'
   length/diameter ratio
C. It is a physical constant and is the same for all antennas
D. Sunspot activity and time of day

A9A05
What is the term for the ratio of the radiation resistance of an antenna to
the total resistance of the system?
A. Effective radiated power
B. Radiation conversion loss
C. Antenna efficiency
D. Beamwidth

A9A06
What is included in the total resistance of an antenna system?
A. Radiation resistance plus space impedance
B. Radiation resistance plus transmission resistance
C. Transmission-line resistance plus radiation resistance
D. Radiation resistance plus ohmic resistance

A9A07
What is a folded dipole antenna?
A. A dipole one-quarter wavelength long
B. A type of ground-plane antenna
C. A dipole whose ends are connected by a one-half wavelength piece of wire
D. A hypothetical antenna used in theoretical discussions to replace the
   radiation resistance

A9A08
How does the bandwidth of a folded dipole antenna compare with that of a
simple dipole antenna?
A. It is 0.707 times the bandwidth
B. It is essentially the same
C. It is less than 50%
D. It is greater

A9A09
What is meant by antenna gain?
A. The numerical ratio relating the radiated signal strength of an antenna to
   that of another antenna
B. The numerical ratio of the signal in the forward direction to the signal in
   the back direction
C. The numerical ratio of the amount of power radiated by an antenna
   compared to the transmitter output power
D. The final amplifier gain minus the transmission-line losses (including any
   phasing lines present)

A9A04
(B)
Page 9-2

A9A05
(C)
Page 9-3

A9A06
(D)
Page 9-3

A9A07
(C)
Page 9-4

A9A08
(D)
Page 9-4

A9A09
(A)
Page 9-11

A9A10
What is meant by antenna bandwidth?
A. Antenna length divided by the number of elements
B. The frequency range over which an antenna can be expected to perform well
C. The angle between the half-power radiation points
D. The angle formed between two imaginary lines drawn through the ends of the elements

A9A11
How can the approximate beamwidth of a beam antenna be determined?
A. Note the two points where the signal strength of the antenna is down 3 dB from the maximum signal point and compute the angular difference
B. Measure the ratio of the signal strengths of the radiated power lobes from the front and rear of the antenna
C. Draw two imaginary lines through the ends of the elements and measure the angle between the lines
D. Measure the ratio of the signal strengths of the radiated power lobes from the front and side of the antenna

A9A12
How is antenna efficiency calculated?
A. (radiation resistance / transmission resistance) x 100%
B. (radiation resistance / total resistance) x 100%
C. (total resistance / radiation resistance) x 100%
D. (effective radiated power / transmitter output) x 100%

A9A13
How can the efficiency of an HF grounded vertical antenna be made comparable to that of a half-wave dipole antenna?
A. By installing a good ground radial system
B. By isolating the coax shield from ground
C. By shortening the vertical
D. By lengthening the vertical

A9B Free-space antenna patterns: E and H plane patterns, (i.e., azimuth and elevation in free-space); gain as a function of pattern; antenna design (computer modeling of antennas)

A9B01
What determines the free-space polarization of an antenna?
A. The orientation of its magnetic field (H Field)
B. The orientation of its free-space characteristic impedance
C. The orientation of its electric field (E Field)
D. Its elevation pattern

A9B02
Which of the following describes the free-space radiation pattern shown in
Figure A9-1?
A. Elevation pattern
B. Azimuth pattern
C. Bode pattern
D. Bandwidth pattern

A9B02
(B)
Page 9-13

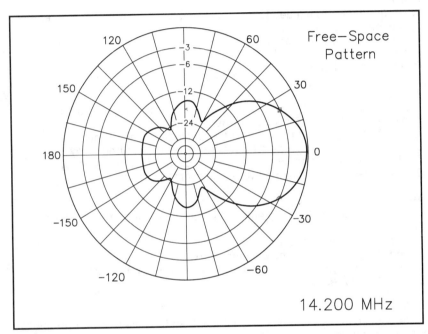

Figure A9-1 — Refer to questions A9B02, A9B03, A9B04 and A9B11.

A9B03
In the free-space H-Field radiation pattern shown in Figure A9-1, what is
the 3-dB beamwidth?
A. 75 degrees
B. 50 degrees
C. 25 degrees
D. 30 degrees

A9B03
(B)
Page 9-13

A9B04
(B)
Page 9-13

A9B04
In the free-space H-Field pattern shown in Figure A9-1, what is the front-to-back ratio?
A. 36 dB
B. 18 dB
C. 24 dB
D. 14 dB

A9B05
(D)
Page 9-15

A9B05
What information is needed to accurately evaluate the gain of an antenna?
A. Radiation resistance
B. E-Field and H-Field patterns
C. Loss resistance
D. All of these choices

A9B06
(D)
Page 9-15

A9B06
Which is NOT an important reason to evaluate a gain antenna across the whole frequency band for which it was designed?
A. The gain may fall off rapidly over the whole frequency band
B. The feedpoint impedance may change radically with frequency
C. The rearward pattern lobes may vary excessively with frequency
D. The dielectric constant may vary significantly

A9B07
(B)
Page 9-16

A9B07
What usually occurs if a Yagi antenna is designed solely for maximum forward gain?
A. The front-to-back ratio increases
B. The feedpoint impedance becomes very low
C. The frequency response is widened over the whole frequency band
D. The SWR is reduced

A9B08
(A)
Page 9-16

A9B08
If the boom of a Yagi antenna is lengthened and the elements are properly retuned, what usually occurs?
A. The gain increases
B. The SWR decreases
C. The front-to-back ratio increases
D. The gain bandwidth decreases rapidly

A9B09
(B)
Page 9-11

A9B09
What type of computer program is commonly used for modeling antennas?
A. Graphical analysis
B. Method of Moments
C. Mutual impedance analysis
D. Calculus differentiation with respect to physical properties

A9B10
What is the principle of a "Method of Moments" analysis?
A. A wire is modeled as a series of segments, each having a distinct value of current
B. A wire is modeled as a single sine-wave current generator
C. A wire is modeled as a series of points, each having a distinct location in space
D. A wire is modeled as a series of segments, each having a distinct value of voltage across it

A9B10
(A)
Page 9-11

A9B11
In the free-space H-field pattern shown in Figure A9-1, what is the front-to-side ratio?
A. 12 dB
B. 14 dB
C. 18 dB
D. 24 dB

A9B11
(B)
Page 9-13

A9C Antenna patterns: elevation above real ground; ground effects as related to polarization; take-off angles as a function of height above ground.

A9C01
What type of antenna pattern over real ground is shown in Figure A9-2?
A. Elevation pattern
B. Azimuth pattern
C. E-Plane pattern
D. Polarization pattern

A9C01
(A)
Page 9-14

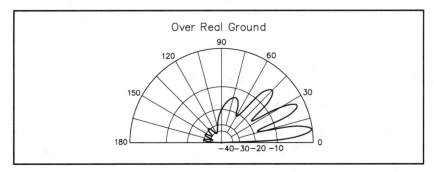

Figure A9-2 — Refer to questions A9C01, A9C09, A9C10 and A9C11.

Advanced Class (Element 4A) Question Pool—With Answers   **10-109**

A9C02
(B)
Page 9-12

A9C02
How would the electric field be oriented for a Yagi with three elements
mounted parallel to the ground?
A. Vertically
B. Horizontally
C. Right-hand elliptically
D. Left-hand elliptically

A9C03
(A)
Page 9-16

A9C03
What strongly affects the shape of the far-field, low-angle elevation pattern
of a vertically polarized antenna?
A. The conductivity and dielectric constant of the soil
B. The radiation resistance of the antenna
C. The SWR on the transmission line
D. The transmitter output power

A9C04
(D)
Page 9-16

A9C04
The far-field, low-angle radiation pattern of a vertically polarized antenna
can be significantly improved by what measures?
A. Watering the earth surrounding the base of the antenna
B. Lengthening the ground radials more than a quarter wavelength
C. Increasing the number of ground radials from 60 to 120
D. None of these choices

A9C05
(D)
Page 9-16

A9C05
How is the far-field elevation pattern of a vertically polarized antenna
affected by being mounted over seawater versus rocky ground?
A. The low-angle radiation decreases
B. The high-angle radiation increases
C. Both the high- and low-angle radiation decrease
D. The low-angle radiation increases

A9C06
(B)
Page 9-16

A9C06
How is the far-field elevation pattern of a horizontally polarized antenna
affected by being mounted one wavelength high over seawater versus
rocky ground?
A. The low-angle radiation greatly increases
B. The effect on the radiation pattern is minor
C. The high-angle radiation increases greatly
D. The nulls in the elevation pattern are filled in

A9C07
(B)
Page 9-16

A9C07
Why are elevated-radial counterpoises popular with vertically polarized
antennas?
A. They reduce the far-field ground losses
B. They reduce the near-field ground losses, compared to on-ground radial
systems using more radials
C. They reduce the radiation angle
D. None of these choices

A9C08
If only a modest on-ground radial system can be used with an eighth-wavelength-high, inductively loaded vertical antenna, what would be the best compromise to minimize near-field losses?
A. 4 radial wires, 1 wavelength long
B. 8 radial wires, a half-wavelength long
C. A wire-mesh screen at the antenna base, an eighth-wavelength square
D. 4 radial wires, 2 wavelengths long

A9C08
(C)
Page 9-16

A9C09
In the antenna radiation pattern shown in Figure A9-2, what is the elevation angle of the peak response?
A. 45 degrees
B. 75 degrees
C. 7.5 degrees
D. 25 degrees

A9C09
(C)
Page 9-15

A9C10
In the antenna radiation pattern shown in Figure A9-2, what is the front-to-back ratio?
A. 15 dB
B. 28 dB
C. 3 dB
D. 24 dB

A9C10
(B)
Page 9-15

A9C11
In the antenna radiation pattern shown in Figure A9-2, how many elevation lobes appear in the forward direction?
A. 4
B. 3
C. 1
D. 7

A9C11
(A)
Page 9-15

A9D Losses in real antennas and matching: resistivity losses, losses in resonating elements (loading coils, matching networks, etc. {i.e., mobile, trap}); SWR bandwidth; efficiency

A9D01
What is the approximate input terminal impedance at the center of a folded dipole antenna?
A. 300 ohms
B. 72 ohms
C. 50 ohms
D. 450 ohms

A9D01
(A)
Page 9-5

A9D02
(A)
Page 9-7

A9D02
For a shortened vertical antenna, where should a loading coil be placed to minimize losses and produce the most effective performance?
A. Near the center of the vertical radiator
B. As low as possible on the vertical radiator
C. As close to the transmitter as possible
D. At a voltage node

A9D03
(C)
Page 9-7

A9D03
Why should an HF mobile antenna loading coil have a high ratio of reactance to resistance?
A. To swamp out harmonics
B. To maximize losses
C. To minimize losses
D. To minimize the Q

A9D04
(D)
Page 9-7

A9D04
Why is a loading coil often used with an HF mobile antenna?
A. To improve reception
B. To lower the losses
C. To lower the Q
D. To tune out the capacitive reactance

A9D05
(A)
Page 9-7

A9D05
What is a disadvantage of using a trap antenna?
A. It will radiate harmonics
B. It can only be used for single-band operation
C. It is too sharply directional at lower frequencies
D. It must be neutralized

A9D06
(D)
Page 9-5

A9D06
What is an advantage of using a trap antenna?
A. It has high directivity in the higher-frequency bands
B. It has high gain
C. It minimizes harmonic radiation
D. It may be used for multiband operation

A9D07
(B)
Page 9-7

A9D07
What happens at the base feedpoint of a fixed length HF mobile antenna as the frequency of operation is lowered?
A. The resistance decreases and the capacitive reactance decreases
B. The resistance decreases and the capacitive reactance increases
C. The resistance increases and the capacitive reactance decreases
D. The resistance increases and the capacitive reactance increases

A9D08
What information is necessary to design an impedance matching system for an antenna?
A. Feedpoint radiation resistance and loss resistance
B. Feedpoint radiation reactance
C. Transmission-line characteristic impedance
D. All of these choices

A9D09
How must the driven element in a 3-element Yagi be tuned to use a "hairpin" matching system?
A. The driven element reactance is capacitive
B. The driven element reactance is inductive
C. The driven element resonance is higher than the operating frequency
D. The driven element radiation resistance is higher than the characteristic impedance of the transmission line

A9D10
What is the equivalent lumped-constant network for a "hairpin" matching system on a 3-element Yagi?
A. Pi network
B. Pi-L network
C. L network
D. Parallel-resonant tank

A9D11
What happens to the bandwidth of an antenna as it is shortened through the use of loading coils?
A. It is increased
B. It is decreased
C. No change occurs
D. It becomes flat

A9D12
What is an advantage of using top loading in a shortened HF vertical antenna?
A. Lower Q
B. Greater structural strength
C. Higher losses
D. Improved radiation efficiency

A9D08
(D)
Page 9-18

A9D09
(A)
Page 9-18

A9D10
(C)
Page 9-18

A9D11
(B)
Page 9-7

A9D12
(D)
Page 9-10

A9E Feed lines: coax vs. open-wire; velocity factor; electrical length; transformation characteristics of line terminated in impedance not equal to characteristic impedance.

A9E01
(D)
Page 9-20

A9E01
What is the velocity factor of a transmission line?
A. The ratio of the characteristic impedance of the line to the terminating impedance
B. The index of shielding for coaxial cable
C. The velocity of the wave on the transmission line multiplied by the velocity of light in a vacuum
D. The velocity of the wave on the transmission line divided by the velocity of light in a vacuum

A9E02
(A)
Page 9-20

A9E02
What is the term for the ratio of the actual velocity at which a signal travels through a transmission line to the speed of light in a vacuum?
A. Velocity factor
B. Characteristic impedance
C. Surge impedance
D. Standing wave ratio

A9E03
(B)
Page 9-21

A9E03
What is the typical velocity factor for a coaxial cable with polyethylene dielectric?
A. 2.70
B. 0.66
C. 0.30
D. 0.10

A9E04
(C)
Page 9-20

A9E04
What determines the velocity factor in a transmission line?
A. The termination impedance
B. The line length
C. Dielectrics in the line
D. The center conductor resistivity

A9E05
(D)
Page 9-20

A9E05
Why is the physical length of a coaxial cable transmission line shorter than its electrical length?
A. Skin effect is less pronounced in the coaxial cable
B. The characteristic impedance is higher in the parallel feed line
C. The surge impedance is higher in the parallel feed line
D. RF energy moves slower along the coaxial cable

A9E06
What would be the physical length of a typical coaxial transmission line that is electrically one-quarter wavelength long at 14.1 MHz? (Assume a velocity factor of 0.66.)

$\frac{20}{4} = 5$

A. 20 meters
B. 2.33 meters
C. 3.51 meters
D. 0.25 meters

A9E06
(C)
Page 9-21

A9E07
What would be the physical length of a typical coaxial transmission line that is electrically one-quarter wavelength long at 7.2 MHz? (Assume a velocity factor of 0.66.)

$\frac{\sqrt{40}}{4} < 10 \quad < 10$

A. 10.5 meters
B. 6.88 meters
C. 24 meters
D. 50 meters

A9E07
(B)
Page 9-21

A9E08
What is the physical length of a parallel conductor feed line that is electrically one-half wavelength long at 14.10 MHz? (Assume a velocity factor of 0.95.)
A. 15 meters
B. 20.2 meters
C. 10.1 meters
D. 70.8 meters

A9E08
(C)
Page 9-21

A9E09
What is the physical length of a twin lead transmission feed line at 3.65 MHz? (Assume a velocity factor of 0.8.)
A. Electrical length times 0.8
B. Electrical length divided by 0.8
C. 80 meters
D. 160 meters

A9E09
(A)
Page 9-21

A9E10
What parameter best describes the interactions at the load end of a mismatched transmission line?
A. Characteristic impedance
B. Reflection coefficient
C. Velocity factor
D. Dielectric Constant

A9E10
(B)
Page 9-23

A9E11
(D)
Page 9-23

A9E12
(A)
Page 9-23

A9E11
Which of the following measurements describes a mismatched transmission line?
A. An SWR less than 1:1
B. A reflection coefficient greater than 1
C. A dielectric constant greater than 1
D. An SWR greater than 1:1

A9E12
What characteristic will 450-ohm ladder line have at 50 MHz, as compared to 0.195-inch-diameter coaxial cable (such as RG-58)?
A. Lower loss in dB/100 feet
B. Higher SWR
C. Smaller reflection coefficient
D. Lower velocity factor

10-116   Chapter 10

APPENDIX A

US Customary—Metric Conversion Factors

International System of Units (SI)—Metric Units

Prefix	Symbol	Multiplication Factor		
exa	E	10^{18}	=	1 000 000 000 000 000 000
peta	P	10^{15}	=	1 000 000 000 000 000
tera	T	10^{12}	=	1 000 000 000 000
giga	G	10^{9}	=	1 000 000 000
mega	M	10^{6}	=	1 000 000
kilo	k	10^{3}	=	1 000
hecto	h	10^{2}	=	100
deca	da	10^{1}	=	10
(unit)		10^{0}	=	1
deci	d	10^{-1}	=	0.1
centi	c	10^{-2}	=	0.01
milli	m	10^{-3}	=	0.001
micro	μ	10^{-6}	=	0.000001
nano	n	10^{-9}	=	0.000000001
pico	p	10^{-12}	=	0.000000000001
femto	f	10^{-15}	=	0.000000000000001
atto	a	10^{-18}	=	0.000000000000000001

Linear
1 metre (m) = 100 centimetres (cm) = 1000 millimetres (mm)

Area
$1 \ m^2 = 1 \times 10^4 \ cm^2 = 1 \times 10^6 \ mm^2$

Volume
$1 \ m^3 = 1 \times 10^6 \ cm^3 = 1 \times 10^9 \ mm^3$
1 litre (l) = 1000 cm^3 = $1 \times 10^6 \ mm^3$

Mass
1 kilogram (kg) = 1 000 grams (g)
(Approximately the mass of 1 litre of water)
1 metric ton (or tonne) = 1 000 kg

US Customary Units

Linear Units
12 inches (in) = 1 foot (ft)
36 inches = 3 feet = 1 yard (yd)
1 rod = 5½ yards = 16½ feet
1 statute mile = 1 760 yards = 5 280 feet
1 nautical mile = 6 076.11549 feet

Area
$1 \ ft^2 = 144 \ in^2$
$1 \ yd^2 = 9 \ ft^2 = 1 \ 296 \ in^2$
$1 \ rod^2 = 30\frac{1}{4} \ yd^2$
$1 \ acre = 4840 \ yd^2 = 43 \ 560 \ ft^2$
$1 \ acre = 160 \ rod^2$
$1 \ mile^2 = 640 \ acres$

Volume
$1 \ ft^3 = 1 \ 728 \ in^3$
$1 \ yd^3 = 27 \ ft^3$

Liquid Volume Measure
1 fluid ounce (fl oz) = 8 fluidrams = 1.804 in^3
1 pint (pt) = 16 fl oz
1 quart (qt) = 2 pt = 32 fl oz = 57¾ in^3
1 gallon (gal) = 4 qt = 231 in^3
1 barrel = 31½ gal

Dry Volume Measure
1 quart (qt) = 2 pints (pt) = 67.2 in^3
1 peck = 8 qt
1 bushel = 4 pecks = 2 150.42 in^3

Avoirdupois Weight
1 dram (dr) = 27.343 grains (gr) or (gr a)
1 ounce (oz) = 437.5 gr
1 pound (lb) = 16 oz = 7 000 gr
1 short ton = 2 000 lb, 1 long ton = 2 240 lb

Troy Weight
1 grain troy (gr t) = 1 grain avoirdupois
1 pennyweight (dwt) or (pwt) = 24 gr t
1 ounce troy (oz t) = 480 grains
1 lb t = 12 oz t = 5 760 grains

Apothecaries' Weight
1 grain apothecaries' (gr ap) = 1 gr t = 1 gr a
1 dram ap (dr ap) = 60 gr
1 oz ap = 1 oz t = 8 dr ap = 480 fr
1 lb ap = 1 lb t = 12 oz ap = 5 760 gr

A-1

Multiply →
Metric Unit = Conversion Factor × US Customary Unit

← Divide
Metric Unit ÷ Conversion Factor = US Customary Unit

Metric Unit =	Conversion Factor ×	US Unit
(Length)		
mm	25.4	inch
cm	2.54	inch
cm	30.48	foot
m	0.3048	foot
m	0.9144	yard
km	1.609	mile
km	1.852	nautical mile
(Area)		
mm^2	645.16	$inch^2$
cm^2	6.4516	in^2
cm^2	929.03	ft^2
m^2	0.0929	ft^2
cm^2	8361.3	yd^2
m^2	0.83613	yd^2
m^2	4047	acre
km^2	2.59	mi^2
(Mass)	(Avoirdupois Weight)	
grams	0.0648	grains
g	28.349	oz
g	453.59	lb
kg	0.45359	lb
tonne	0.907	short ton
tonne	1.016	long ton

Metric Unit =	Conversion Factor ×	US Unit
(Volume)		
mm^3	16387.064	in^3
cm^3	16.387	in^3
m^3	0.028316	ft^3
m^3	0.764555	yd^3
ml	16.387	in^3
ml	29.57	fl oz
ml	473	pint
ml	946.333	quart
l	28.32	ft^3
l	0.9463	quart
l	3.785	gallon
l	1.101	dry quart
l	8.809	peck
l	35.238	bushel
(Mass)	(Troy Weight)	
g	31.103	oz t
g	373.248	lb t
(Mass)	(Apothecaries' Weight)	
g	3.387	dr ap
g	31.103	oz ap
g	373.248	lb ap

Standard Resistance Values

Numbers in **bold** type are ± 10% values. Others are 5% values.

Ohms

										Megohms				
1.0	3.6	**12**	43	**150**	510	**1800**	6200	**22000**	75000		0.24	0.62	1.6	4.3	11.0
1.1	**3.9**	13	**47**	160	**560**	2000	**6800**	24000	**82000**		0.27	**0.68**	1.8	**4.7**	**12.0**
1.2	4.3	**15**	51	**180**	620	**2200**	7500	**27000**	91000		0.30	0.75	2.0	5.1	**13.0**
1.3	**4.7**	16	**56**	200	**680**	2400	**8200**	30000	**100000**		**0.33**	**0.82**	**2.2**	**5.6**	**15.0**
1.5	5.1	**18**	62	**220**	750	**2700**	9100	**33000**	110000		0.36	0.91	2.4	6.2	16.0
1.6	**5.6**	20	**68**	240	**820**	3000	**10000**	36000	**120000**		**0.39**	**1.0**	2.7	**6.8**	**18.0**
1.8	6.2	**22**	75	**270**	910	**3300**	11000	**39000**	130000		0.43	1.1	3.0	7.5	20.0
2.0	**6.8**	24	**82**	300	**1000**	3600	**12000**	43000	**150000**		**0.47**	**1.2**	**3.3**	**8.2**	**22.0**
2.2	7.5	**27**	91	**330**	1100	**3900**	13000	**47000**	160000		0.51	1.3	3.6	9.1	
2.4	**8.2**	30	**100**	360	**1200**	4300	**15000**	51000	**180000**		**0.56**	**1.5**	**3.9**	**10.0**	
2.7	9.1	**33**	110	**390**	1300	**4700**	16000	**56000**	200000						
3.0	**10.0**	36	**120**	430	**1500**	5100	**18000**	62000	**220000**						
3.3	11.0	**39**	130	**470**	1600	**5600**	20000	**68000**							

Resistor Color Code

Color	Sig. Figure	Decimal Multiplier	Tolerance (%)	Color	Sig. Figure	Decimal Multiplier	Tolerance (%)
Black	0	1		Violet	7	10,000,000	
Brown	1	10		Gray	8	100,000,000	
Red	2	100		White	9	1,000,000,000	
Orange	3	1,000		Gold	—	0.1	5
Yellow	4	10,000		Silver	—	0.01	10
Green	5	100,000		No color	—		20
Blue	6	1,000,000					

Standard Values for 1000-V Disc-Ceramic Capacitors

pF	pF	pF	pF
3.3	39	250	1000
5	47	270	1200
6	50	300	1500
6.8	51	330	1800
8	56	360	2000
10	68	390	2500
12	75	400	2700
15	82	470	3000
18	100	500	3300
20	120	510	3900
22	130	560	4700
24	150	600	5000
25	180	680	5600
27	200	750	6800
30	220	820	8200
33	240	910	10000

Common Values for Small Electrolytic Capacitors

µF	V*	µF	V*
33	6.3	10	35
33	10	22	35
100	10	33	35
220	10	47	35
330	10	100	35
470	10	220	35
10	16	330	35
22	16	470	35
33	16	1000	35
47	16	1	50
100	16	2.2	50
220	16	3.3	50
470	16	4.7	50
1000	16	10	50
2200	16	33	50
4.7	25	47	50
22	25	100	50
33	25	220	50
47	25	330	50
100	25	470	50
220	25	10	63
330	25	22	63
470	25	47	63
1000	25	1	100
2200	25	10	100
4.7	35	33	100

*Working voltage

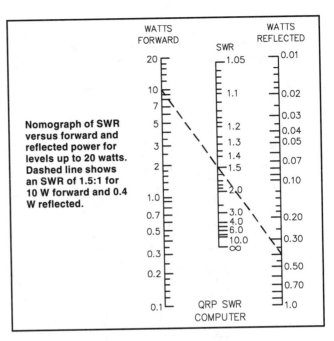

Nomograph of SWR versus forward and reflected power for levels up to 20 watts. Dashed line shows an SWR of 1.5:1 for 10 W forward and 0.4 W reflected.

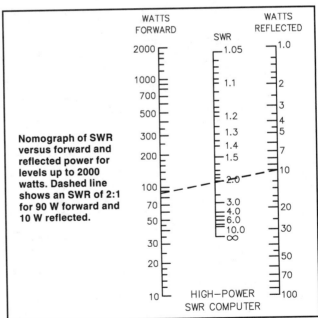

Nomograph of SWR versus forward and reflected power for levels up to 2000 watts. Dashed line shows an SWR of 2:1 for 90 W forward and 10 W reflected.

Fractions of an Inch with Metric Equivalents

Fractions Of An Inch		Decimals Of An Inch	Millimeters	Fractions Of An Inch		Decimals Of An Inch	Millimeters
	1/64	0.0156	0.397		33/64	0.5156	13.097
1/32		0.0313	0.794	17/32		0.5313	13.494
	3/64	0.0469	1.191		35/64	0.5469	13.891
1/16		0.0625	1.588	9/16		0.5625	14.288
	5/64	0.0781	1.984		37/64	0.5781	14.684
3/32		0.0938	2.381	19/32		0.5938	15.081
	7/64	0.1094	2.778		39/64	0.6094	15.478
1/8		0.1250	3.175	5/8		0.6250	15.875
	9/64	0.1406	3.572		41/64	0.6406	16.272
5/32		0.1563	3.969	21/32		0.6563	16.669
	11/64	0.1719	4.366		43/64	0.6719	17.066
3/16		0.1875	4.763	11/16		0.6875	17.463
	13/64	0.2031	5.159		45/64	0.7031	17.859
7/32		0.2188	5.556	23/32		0.7188	18.256
	15/64	0.2344	5.953		47/64	0.7344	18.653
1/4		0.2500	6.350	3/4		0.7500	19.050
	17/64	0.2656	6.747		49/64	0.7656	19.447
9/32		0.2813	7.144	25/32		0.7813	19.844
	19/64	0.2969	7.541		51/64	0.7969	20.241
5/16		0.3125	7.938	13/16		0.8125	20.638
	21/64	0.3281	8.334		53/64	0.8281	21.034
11/32		0.3438	8.731	27/32		0.8438	21.431
	23/64	0.3594	9.128		55/64	0.8594	21.828
3/8		0.3750	9.525	7/8		0.8750	22.225
	25/64	0.3906	9.922		57/64	0.8906	22.622
13/32		0.4063	10.319	29/32		0.9063	23.019
	27/64	0.4219	10.716		59/64	0.9219	23.416
7/16		0.4375	11.113	15/16		0.9375	23.813
	29/64	0.4531	11.509		61/64	0.9531	24.209
15/32		0.4688	11.906	31/32		0.9688	24.606
	31/64	0.4844	12.303		63/64	0.9844	25.003
1/2		0.5000	12.700	1		1.0000	25.400

A-4

Schematic Symbols Used in Circuit Diagrams

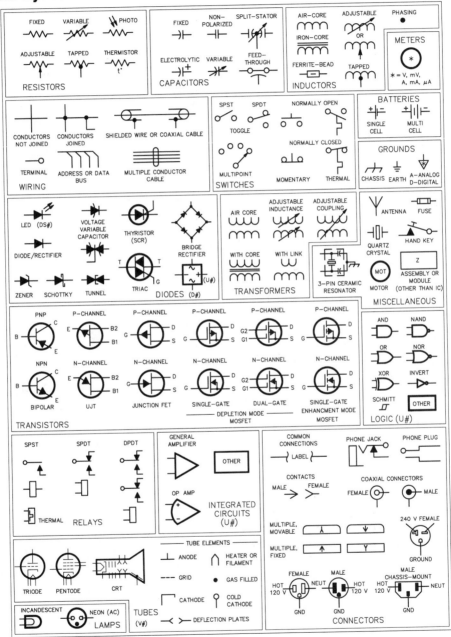

APPENDIX B

EQUATIONS USED IN THIS BOOK

$$dB = 10 \log \left(\frac{P_1}{P_2} \right)$$
(Equation 1-1)

$$D\,(\text{mi}) = 1.415 \times \sqrt{H\,(\text{ft})}$$
(Equation 3-1)

$$D\,(\text{km}) = 4.123 \times \sqrt{H\,(\text{m})}$$
(Equation 3-2)

$$\text{Error} = f(\text{Hz}) \times \frac{\text{counter error}}{1,000,000}$$
(Equation 4-1)

$$\frac{f_H}{f_V} = \frac{n_V}{n_H}$$
(Equation 4-2)

$$f_V = \frac{n_H}{n_V}\, f_H$$
(Equation 4-3)

$$\text{Noise Floor} = \text{Theoretical MDS} + \text{noise figure}$$
(Equation 4-4)

$$\text{Blocking Dynamic Range} = |\text{ Noise Floor } - \text{ Blocking Level }|$$

(Equation 4-5)

$$f_{IMD\,1} = 2f_1 + f_2$$

(Equation 4-6)

$$f_{IMD\,2} = 2f_1 - f_2$$

(Equation 4-7)

$$f_{IMD\,3} = 2f_2 + f_1$$

(Equation 4-8)

$$f_{IMD\,4} = 2f_2 - f_1$$

(Equation 4-9)

$$1 \text{ volt} = \frac{1 \text{ joule}}{1 \text{ coulomb}}$$

(Equation 5-1)

$$W = \frac{V^2 C}{2}$$

(Equation 5-2)

$$W = \frac{I^2 L}{2}$$

(Equation 5-3)

$$\tan \theta = \frac{\text{side opposite}}{\text{side adjacent}}$$

(Equation 5-4)

$$\sin \theta = \frac{\text{side opposite}}{\text{hypotenuse}}$$

(Equation 5-5)

$$C^2 = A^2 + B^2$$

(Equation 5-6)

$$C = \sqrt{A^2 + B^2}$$

(Equation 5-7)

$$I_R = \frac{E}{R}, \; I_L = \frac{E}{X_L} \text{ and } I_C = \frac{E}{X_C}$$

(Equation 5-8)

$$f_r = \frac{1}{2 \pi \sqrt{L\,C}}$$

(Equation 5-9)

$$L = \frac{1}{\left(2 \pi f_r\right)^2 C}$$

(Equation 5-10)

$$C = \frac{1}{\left(2 \pi f_r\right)^2 L}$$

(Equation 5-11)

$$Q = \frac{X}{R}$$

(Equation 5-12)

$$Q = \frac{R}{X}$$

(Equation 5-13)

$$\Delta f = \frac{f_r}{Q}$$

(Equation 5-14)

$$P = I\,E$$

(Equation 5-15)

$$P = I^2\,R$$

(Equation 5-16)

$$P = \frac{E^2}{R}$$

(Equation 5-17)

$$\text{Power factor} = \frac{P_{REAL}}{P_{APPARENT}}$$

(Equation 5-18)

$$P_{REAL} = P_{APPARENT} \times \text{Power factor}$$

(Equation 5-19)

$$\text{Power factor} = \cos \theta$$

(Equation 5-20)

$$dB = 10 \log \left(\frac{P_2}{P_1} \right) \qquad \text{(Equation 5-21)}$$

$$R_T = \frac{R_1 \times R_2}{R_1 + R_2} \qquad \text{(Equation 5-22)}$$

$$\beta = \frac{I_c}{I_b} \qquad \text{(Equation 6-1)}$$

$$\alpha = \frac{I_c}{I_e} \qquad \text{(Equation 6-2)}$$

$$L = \frac{A_L \times N^2}{10,000} \qquad \text{(Equation 6-3)}$$

$$N = 100 \sqrt{\frac{L}{A_L}} \qquad \text{(Equation 6-4)}$$

$$L = \frac{A_L \times N^2}{1,000,000} \qquad \text{(Equation 6-5)}$$

$$N = 1000 \sqrt{\frac{L}{A_L}} \qquad \text{(Equation 6-6)}$$

$$R_{e-b} = \frac{26}{I_e} \qquad \text{(Equation 7-1)}$$

$$A_V = \frac{R_L}{R_{e-b}} \qquad \text{(Equation 7-2)}$$

B-4

$$R_b = \beta R_{e-b} \qquad \text{(Equation 7-3)}$$

$$A_V = \frac{R_L}{R_E} \qquad \text{(Equation 7-4)}$$

$$P_{IN} = P_{OUT} + P_D \qquad \text{(Equation 7-5)}$$

$$\text{Efficiency} = \frac{P_{OUT}}{P_{IN}} \times 100\% = \frac{P_{OUT}}{P_{OUT} + P_D} \times 100\% \qquad \text{(Equation 7-6)}$$

$$\text{Efficiency} = \frac{\text{RF power out}}{\text{DC power in}} \times 100\% \qquad \text{(Equation 7-7)}$$

$$\text{THD} = \frac{V_H}{V_F} \times 100\% \qquad \text{(Equation 7-8)}$$

$$P = I \times E \qquad \text{(Equation 7-9)}$$

$$I = \frac{P}{E} \qquad \text{(Equation 7-10)}$$

$$Z = \frac{E}{I} \qquad \text{(Equation 7-11)}$$

$$F = N f_r \qquad \text{(Equation 7-12)}$$

$$\text{deviation ratio} = \frac{D_{max}}{M} \qquad \text{(Equation 8-1)}$$

$$\text{modulation index} = \frac{D_{max}}{m} \qquad \text{(Equation 8-2)}$$

$$f = \frac{1}{T} \qquad \text{(Equation 8-3)}$$

$$V_{peak} = V_{RMS} \times \sqrt{2} = V_{RMS} \times 1.414 \qquad \text{(Equation 8-4)}$$

$$V_{RMS} = \frac{V_{peak}}{\sqrt{2}} = V_{peak} \times 0.707 \qquad \text{(Equation 8-5)}$$

$$V_{avg} = V_{peak} \times 0.646 \qquad \text{(Equation 8-6)}$$

$$V_{avg} = V_{RMS} \times 0.899 \qquad \text{(Equation 8-7)}$$

$$V_{RMS} \times I_{RMS} = P_{avg} \qquad \text{(Equation 8-8)}$$

$$V_{peak} \times I_{peak} = P_{peak} = 2 \times P_{avg} \qquad \text{(Equation 8-9)}$$

$$\text{Efficiency} = \frac{R_R}{R_T} \times 100\% \qquad \text{(Equation 9-1)}$$

$$\lambda = \frac{v}{f} \qquad \text{(Equation 9-2)}$$

$$\lambda \text{ (meters)} = \frac{300}{f \text{ (MHz)}} \qquad \text{(Equation 9-3)}$$

$$\lambda \text{ (feet)} = \frac{984}{f \text{ (MHz)}} \qquad \text{(Equation 9-4)}$$

$$V = \frac{\text{speed of wave (in line)}}{\text{speed of light (in vacuum)}} \qquad \text{(Equation 9-5)}$$

$$V = \frac{1}{\sqrt{\varepsilon}}$$

(Equation 9-6)

$$\text{Length (meters)} = \frac{300}{f\ (\text{MHz})} \times V$$

(Equation 9-7)

$$\text{Length (feet)} = \frac{984}{f\ (\text{MHz})} \times V$$

(Equation 9-8)

GLOSSARY OF KEY WORDS

Absorption — The loss of energy from an electromagnetic wave as it travels through any material. The energy may be converted to heat or other forms. Absorption usually refers to energy lost as the wave travels through the ionosphere.

Absorption wavemeter — A device for measuring frequency or wavelength. It takes some power from the circuit under test when the meter is tuned to the same resonant frequency.

Alpha (α) — The ratio of transistor collector current to emitter current. It is between 0.92 and 0.98 for a junction transistor.

Alpha cutoff frequency — A term used to express the useful upper frequency limit of a transistor. The point at which the gain of a common-base amplifier is 0.707 times the gain at 1 kHz.

Amplifier transfer function — A graph or equation that relates the input and output of an amplifier under various conditions.

Amplitude modulation (AM) — A method of combining an information signal and a radio-frequency (RF) carrier signal. The amplitude of the RF signal is varied in a way that is controlled by the information signal. For fax (A3C) or SSTV (J3F or A3F) transmissions it refers to a method of superimposing picture information on the RF signal.

Amplitude modulation — A method of superimposing an information signal on an RF carrier wave in which the amplitude of the RF envelope (carrier and sidebands) is varied in relation to the information signal strength.

Anode — The terminal that connects to the positive supply lead for current to flow through a device.

Antenna — An electric circuit designed specifically to radiate the energy applied to it in the form of electromagnetic waves. An antenna is reciprocal; a wave moving past it will induce a current in the circuit also. Antennas are used to transmit and receive radio waves.

Antenna bandwidth — A range of frequencies over which the antenna SWR will be below some specified value.

Antenna efficiency — The ratio of the radiation resistance to the total resistance of an antenna system, including losses.

Apparent power — The product of the RMS current and voltage values in a circuit without consideration of the phase angle between them.

Aurora — A disturbance of the atmosphere around the magnetic poles of the Earth, caused by an interaction between electrically charged particles from the sun and the magnetic field of the Earth. Often a display of colored lights is produced, which is visible to those who are close enough to the magnetic-polar regions. Auroras can disrupt HF radio communication and enhance VHF communication. They are classified as visible auroras and radio auroras.

Automatic control — The operation of an amateur station without a control operator present at the control point. In §97.3 (a) (6), the FCC defines automatic control as "The use of devices and procedures for control of a station when it is transmitting so that compliance with the FCC Rules is achieved without the control operator being present at a control point."

Automatic gain control — An amplifier circuit designed to provide a relatively constant output amplitude over a wide range of input values.

Auxiliary station — An amateur station, other than in a message forwarding system, that is transmitting communications point-to-point within a system of co-operating amateur stations. {§97.3 (a) (7)}

Avalanche point — That point on a diode characteristic curve where the amount of reverse current increases greatly for small increases in reverse bias voltage.

Average power — The product of the RMS current and voltage values associated with a purely resistive circuit, equal to one half the peak power when the applied voltage is a sine wave.

Back EMF — An opposing electromotive force (voltage) produced by a changing current in a coil. It can be equal to the applied EMF under some conditions.

Balanced modulator — A circuit used in a single-sideband suppressed-carrier transmitter to combine a voice signal and an RF signal. The balanced modulator isolates the input signals from each other and the output, so that only the difference of the two input signals reaches the output.

Band-pass filter — A circuit that allows signals to go through it only if they are within a certain range of frequencies. It attenuates signals above and below this range.

Bandwidth — The frequency range (measured in hertz — Hz) over which a signal is stronger than some specified amount below the peak signal level. For example, if a certain signal is at least half as strong as the peak power level over a range of ±3 kHz, the signal has a 3-dB bandwidth of 6 kHz.

Base loading — The technique of inserting a coil at the bottom of an electrically short vertical antenna in order to cancel the capacitive reactance of the antenna, producing a resonant antenna system.

Baud — A unit of signaling speed equal to the number of discrete conditions or events per second. (For example, if the duration of a pulse is 3.33 milliseconds, the signaling rate is 300 bauds or the reciprocal of 0.00333 seconds.)

Beamwidth — As related to directive antennas, the width (measured in degrees) of a major lobe between the two directions at which the relative power is one half (– 3 dB) its value at the peak of the lobe.

Beta (β) — The ratio of transistor collector current to base current. Betas show wide variations, even between individual devices of the same type.

Beta cutoff frequency — The point at which the gain of a common-emitter amplifier is 0.707 times the gain at 1 kHz.

Bipolar junction transistor — A transistor made of two PN semiconductor junctions using two layers of similar-type material (N or P) with a third layer of the opposite type between them.

Butterworth filter — A filter whose passband frequency response is as flat as possible. The design is based on a Butterworth polynomial to calculate the input/output characteristics.

Capacitive coupling (of a dip meter) — A method of transferring energy from a dip-meter oscillator to a tuned circuit by means of an electric field.

Capture effect — An effect especially noticed with FM and PM systems whereby the strongest signal to reach the demodulator is the one to be received. You cannot tell whether weaker signals are present.

Cathode — The terminal that connects to the negative supply lead for current to flow through a device

Cathode-ray tube (CRT) — An electron-beam tube in which the beam can be focused on a luminescent screen. The spot position can be varied to produce a pattern on the screen. CRTs are used in oscilloscopes and as the "picture tube" in television receivers.

Cavity — A high-Q tuned circuit that passes energy at one frequency with little or no attenuation but presents a high impedance to another nearby frequency.

Center loading — A technique for adding a series inductor at or near the center of an antenna element in order to cancel the capacitive reactance of the antenna. This technique is usually used with elements that are less than $1/4$ wavelength.

Chebyshev filter — A filter whose passband and stopband frequency response has an equal-amplitude ripple, and a sharper transition to the stop band than does a Butterworth filter. The design is based on a Chebyshev polynomial to calculate the input/output characteristics.

Circular polarization — Describes an electromagnetic wave in which the electric and magnetic fields are rotating. If the electric field vector is rotating in a clockwise sense, then it is called right-hand polarization and if the electric field vector is rotating in a counter-clockwise sense, it is called left-hand polarization.

Circulator — A passive device with three or more ports or input/output terminals. It can be used to combine the output from several transmitters to one antenna. A circulator acts as a one-way valve to allow radio waves to travel in one direction (to the antenna) but not in another (to the receiver).

Control link — A device used by a control operator to manipulate the station adjustment controls from a location other than the station location. A control link provides the means of control between a control point and a remotely controlled station.

Cross modulation — A type of intermodulation caused by the carrier of a desired signal being modulated by an unwanted signal in a receiver.

Crystal-lattice filter — A filter that employs piezoelectric crystals (usually quartz) as the reactive elements. They are most often used in the IF stages of a receiver or transmitter.

D'Arsonval meter movement — A type of meter movement in which a coil is suspended between the poles of a permanent magnet. DC flowing through the coil causes it to rotate an amount proportional to the current. A pointer attached to the coil indicates the amount of deflection on a scale.

Decibel (dB) — One tenth of a bel, denoting a logarithm of the ratio of two power levels — dB = 10 log (P2/P1). Power gains and losses are expressed in decibels.

Depletion region — An area around the semiconductor junction where the charge density is very small. This creates a potential barrier for current to flow across the junction. In general, the region is thin when the junction is forward biased, and becomes thicker under reverse-bias conditions. Also called the **transition region**.

Desensitization — A reduction in receiver sensitivity caused by the receiver front end being overloaded by noise or RF from a local transmitter.

Detector — A circuit used in a receiver to recover the modulation (voice or other information) signal from the RF signal.

Deviation — The peak difference between an instantaneous frequency of the modulated wave and the unmodulated-carrier frequency in an FM system.

Deviation ratio — The ratio of the maximum frequency deviation to the maximum modulating frequency in an FM system.

Dielectric — An insulating material. A dielectric is a medium in which it is possible to maintain an electric field with little or no additional direct-current energy supplied after the field has been established.

Dielectric constant (ε) — Relative figure of merit for an insulating material. This is the property that determines how much electric energy can be stored in a unit volume of the material per volt of applied potential.

Dielectric materials — Materials in which it is possible to maintain an electric field with little or no additional energy being supplied. Insulating materials or nonconductors.

Dip meter — A tunable RF oscillator that supplies energy to another circuit resonant at the frequency that the oscillator is tuned to. A meter indicates when the most energy is being coupled out of the circuit by showing a dip in indicated current.

Dipole — An antenna with two elements in a straight line that are fed in the center; literally, two poles. For amateur work, dipoles are usually operated near half-wave resonance.

Direct sequence (DS) spread spectrum — A **spread-spectrum** technique in which a very fast binary bit stream is used to shift the phase of an RF carrier.

Doping — The addition of impurities to a semiconductor material, with the intent to provide either excess electrons or positive charge carriers (holes) in the material.

Double-balanced mixer (DBM) — A mixer circuit that is balanced for both inputs, so that only the sum and the difference frequencies, but neither of the input frequencies, appear at the output. There will be no output unless both input signals are present.

Duplexer — A device, usually employing cavities, to allow a transmitter and receiver to be connected simultaneously to one antenna. Most often, as in the case of a repeater, the transmitter and receiver operate at the same time on different frequencies.

Dynamic range — The ability of a receiver to tolerate strong signals outside the band-pass range. Blocking dynamic range and intermodulation distortion (IMD) dynamic range are the two most common dynamic range measurements used to predict receiver performance.

Effective radiated power (ERP) — The relative amount of power radiated in a specific direction from an antenna, taking system gains and losses into account.

Electric field — A region through which an electric force will act on an electrically charged object.

Electric force — A push or pull exerted through space by one electrically charged object on another.

Electromagnetic radiation — Another term for electromagnetic waves, consisting of an electric field and a magnetic field that are at right angles to each other.

Electromagnetic waves — A disturbance moving through space or materials in the form of changing electric and magnetic fields.

Elliptical filter — A filter with equal-amplitude passband ripple and points of infinite attenuation in the stop band. The design is based on an elliptical function to calculate the input/output characteristics.

Emission designators — A method of identifying the characteristics of a signal from a radio transmitter using a series of three characters following the ITU system.

Emission types — A method of identifying the signals from a radio transmitter using a "plain English" format that simplifies the ITU **emission designators**.

Emissions — Any signals produced by a transmitter that reach the antenna connector to be radiated.

Equinoxes — One of two spots on the orbital path of the Earth around the sun, at which it crosses a horizontal plane extending through the center of the sun. The *vernal equinox* marks the beginning of spring and the *autumnal equinox* marks the beginning of autumn .

Facsimile (fax) — The process of scanning pictures or images and converting the information into signals that can be used to form a likeness of the copy in another location. The pictures are often printed on paper for permanent display.

Fast-scan television (ATV) — A television system used by amateurs that employs the same video-signal standards as commercial TV.

Field — The region of space through which any of the invisible forces in nature, such as gravity, electric force or magnetic forces, act.

Folded dipole — An antenna consisting of two (or more) parallel, closely spaced halfwave wires connected at their ends. One of the wires is fed at its center.

Forward bias — A voltage applied across a semiconductor junction so that it will tend to produce current.

Frequency, f — A property of an electromagnetic wave that refers to the number of complete alternations (or oscillations) made in one second.

Frequency counter — A digital-electronic device that counts the cycles of an electromagnetic wave for a certain amount of time and gives a digital readout of the frequency.

Frequency discriminator — A circuit used to recover the audio from an FM signal. The output amplitude depends on the deviation of the received signal from a center (carrier) frequency.

Frequency hopping (FH) spread spectrum — A **spread-spectrum** technique in which the transmitter frequency is changed rapidly according to a pseudo-random list of channels.

Frequency modulation (FM) — A method of combining an information signal and a radio-frequency (RF) carrier signal. The instantaneous frequency of the RF signal is varied by an amount that depends on the frequency of the information signal at that instant. For fax (F3C) or SSTV (J3F) transmissions it refers to a method of superimposing picture information on the radio-frequency carrier. G3C and G3F refer to a method of varying the phase of the carrier wave. There is no practical difference in the way you receive phase-modulated signals from frequency-modulated ones.

Frequency standard — A circuit or device used to produce a highly accurate reference frequency. The frequency standard may be a crystal oscillator in a marker generator or a radio broadcast, such as from WWV, with a carefully controlled transmit frequency.

Gain — An increase in the effective power radiated by an antenna in a certain desired direction. This is at the expense of power radiated in other directions.

Gray scale — A photographic term that defines a series of neutral densities (based on the percentage of incident light that is reflected from a surface), ranging from white to black.

Ground-wave propagation — An effect usually observed on the amateur 160 and 80-meter bands, in which the radio waves are bent slightly (diffracted) by the rounded edge of the surface of the Earth. Daytime propagation out to 120 miles or more is possible with ground-wave propagation.

Half-power points — Those points on the response curve of a resonant circuit where the power is one half its value at resonance.

Half section — A basic L-section building block of image-parameter filters.

High-pass filter — A filter that allows signals above the cutoff frequency to pass through. It attenuates signals below the cutoff frequency.

Horizontal synchronization pulse — Part of a TV signal used by the receiver to keep the **cathode-ray tube (CRT)** electron-beam scan in step with the camera scanning beam. This pulse is transmitted at the beginning of each horizontal scan line.

Hot-carrier diode — A type of diode in which a small metal dot is placed on a single semiconductor layer. It is superior to a point-contact diode in most respects.

Image rejection — The ability of a receiver (or receiver stage) to prevent unwanted signals from mixing with the local oscillator signal and producing a signal at the **intermediate frequency (IF)**.

Image signal — An unwanted signal that mixes with a receiver local oscillator to produce a signal at the desired intermediate frequency.

Inductive coupling (of a dip meter) — A method of transferring energy from a dip-meter oscillator to a tuned circuit by means of a magnetic field between two coils.

Intermediate frequency (IF) — The output frequency of a mixing stage in a superheterodyne receiver. Signal processing (filter and amplification) stages tuned to this frequency allow efficient signal processing.

Intermodulation distortion (IMD) — A type of interference that results from the unwanted mixing of two strong signals, producing a signal on an unintended frequency. The resulting mixing products can interfere with desired signals on those frequencies. "Intermod" usually occurs in a nonlinear stage or device.

International Telecommunication Union (ITU) — The international organization with responsibility for dividing the range of communications frequencies between the various services for the entire world.

Isolator — A passive attenuator in which the loss in one direction is much greater than the loss in the other.

Joule — The unit of energy in the metric system of measure.

K index — A geomagnetic-field measurement that is updated every three hours at Boulder, Colorado. Changes in the K index can be used to indicate HF propagation conditions. Rising values generally indicate disturbed conditions while falling values indicate improving conditions.

L network — A combination of a capacitor and an inductor, one of which is connected in series with the signal lead while the other is shunted to ground.

Light-emitting diode — A device that uses a semiconductor junction to produce light when current flows through it.

Line A — A line roughly parallel to, and south of, the US - Canadian border. {See §97.3 (a) (27).}

Linear electronic voltage regulator — A type of voltage-regulator circuit that varies either the current through a fixed dropping resistor or the resistance of the dropping element itself. The conduction of the control element varies in direct proportion to the line voltage or load current.

Linear polarization — Describes the orientation of the electric-field component of an electromagnetic wave. The electric field can be vertical or horizontal with respect to the earth's surface, resulting in either a vertically or a horizontally polarized wave. (Also called **plane polarization**.)

Lissajous figure — An oscilloscope pattern obtained by connecting one sine wave to the vertical amplifier and another sine wave to the horizontal amplifier. The two signals must be harmonically related to produce a stable pattern.

Loading coil — An inductor that is inserted in an antenna element or transmission line for the purpose of producing a resonant system at a specific frequency.

Low-pass filter — A filter that allows signals below the cutoff frequency to pass through. It attenuates signals above the cutoff frequency.

Magnetic field — A region through which a magnetic force will act on a magnetic object.

Magnetic force — A push or pull exerted through space by one magnetically charged object on another.

Major lobe of radiation — A three-dimensional area that contains the maximum radiation peak in the space around an antenna. The field strength decreases from the peak level, until a point is reached where it starts to increase again. The area described by the radiation maximum is known as the major lobe.

Marker generator — An RF signal generator that produces signals at known frequency intervals. A marker generator is often a crystal oscillator that is rich in harmonics, usually for the purpose of calibrating a receiver dial.

Maximum average forward current — The highest average current that can flow through the diode in the forward direction for a specified junction temperature.

Minimum discernible signal (MDS) — The smallest input signal level that can just be detected above the receiver internal noise. Also called **noise floor**.

Minor lobe of radiation — Those areas of an antenna pattern where there is some increase in radiation, but not as much as in the major lobe. Minor lobes normally appear at the back and sides of the antenna.

Mixer — A circuit that takes two or more input signals, and produces an output that includes the sum and difference of those signal frequencies.

Modulation index — The ratio of the maximum frequency deviation of the modulated wave to the instantaneous frequency of the modulating signal.

Modulator — A circuit designed to superimpose an information signal on an RF carrier wave.

Monolithic microwave integrated circuit (MMIC) — A small pill-sized amplifying device that simplifies amplifier designs for microwave-frequency circuits. An MMIC has an input lead, an output lead and two ground leads.

Multipath — A fading effect caused by the transmitted signal traveling to the receiving station over more than one path.

N-type material — Semiconductor material that has been treated with impurities to give it an excess of electrons. We call this a "donor material."

National Radio Quiet Zone — The area in Maryland, Virginia and West Virginia bounded by 39° 15' N on the north, 78° 30' W on the east, 37° 30' N on the south and 80° 30' W on the west {§97.3 (a) (30)}.

Neon lamp — A cold-cathode (no heater or filament), gas-filled tube used to give a visual indication of voltage in a circuit, or of an RF field.

Neutralization — Feeding part of the output signal from an amplifier back to the input so it arrives out of phase with the input signal. This negative feedback neutralizes the effect of positive feedback caused by coupling between the input and output circuits in the amplifier. The negative-feedback signal is usually supplied by connecting a capacitor from the output to the input circuit.

Noise figure — A ratio of the noise output power to the noise input power when the input termination is at a standard temperature of 290 K. It is a measure of the noise generated in the receiver circuitry.

Noise floor — The smallest input signal level that can just be detected above the receiver internal noise. Also called **minimum discernible signal (MDS)**.

Oscillator — A circuit built by adding positive feedback to an amplifier. It produces an alternating current signal with no input signal except the dc operating voltages.

Oscilloscope — A device using a cathode-ray tube to display the waveform of an electric signal with respect to time or as compared with another signal.

P-type material — A semiconductor material that has been treated with impurities to give it an electron shortage. This creates excess positive charge carriers, or "holes," so it becomes an "acceptor material."

Parallel-resonant circuit — A circuit including a capacitor, an inductor and sometimes a resistor, connected in parallel, and in which the inductive and capacitive reactances are equal at the applied-signal frequency. The circuit impedance is a maximum, and the current through the circuit is a minimum at the resonant frequency.

Parasitics — Undesired oscillations or other responses in an amplifier.

Peak envelope power (PEP) — An expression used to indicate the power level in a signal. It is found by squaring the RMS voltage and dividing by the load resistance. The average power of the RF envelope during a modulation peak. (Used for modulated RF signals.)

Peak envelope voltage (PEV) — The maximum peak voltage occurring in a complex waveform.

Peak inverse voltage (PIV) — The maximum instantaneous anode-to-cathode reverse voltage that is to be applied to a diode.

Peak power — The product of peak voltage and peak current in a resistive circuit. (Used with sine-wave signals.)

Peak voltage — A measure of voltage on an ac waveform taken from the centerline (0 V) and the maximum positive or negative level.

Peak-to-peak (P-P) voltage — A measure of the voltage taken between the negative and positive peaks on a cycle.

Pedersen ray — A high-angle radio wave that penetrates deeper into the F region of the ionosphere, so the wave is bent less than a lower-angle wave, and thus travels for some distance through the F region, returning to Earth at a distance farther than normally expected for single-hop propagation.

Period, T — The time it takes to complete one cycle of an ac waveform.

Phase — A representation of the relative time or space between two points on a waveform, or between related points on different waveforms. Also the time interval between two events in a regularly recurring cycle.

Phase angle — If one complete cycle of a waveform is divided into 360 equal parts, then the phase relationship between two points or two waves can be expressed as an angle.

Phase modulation — A method of superimposing an information signal on an RF carrier wave in which the phase of an RF carrier wave is varied in relation to the information signal strength.

Phase modulator — A device capable of modulating an ac signal by varying the reactance of an amplifier circuit in response to the modulating signal. (The modulating signal may be voice, data, video or some other kind.) The circuit capacitance or inductance changes in response to an audio input signal. Used in PM (or FM) systems, this circuit acts as a variable reactance in an amplifier tank circuit.

Phase noise — Undesired variations in the phase of an oscillator signal. Phase noise is usually associated with phase-locked loop (PLL) oscillators.

Photocell — A solid-state device in which the voltage and current-conducting characteristics change as the amount of light striking the cell changes.

Photodetector — A device that produces an amplified signal that changes with the amount of light striking a light-sensitive surface.

Phototransistor — A bipolar transistor constructed so the base-emitter junction is exposed to incident light. When light strikes this surface, current is generated at the junction, and this current is then amplified by transistor action.

Pi network output-coupling circuits — A combination of two like reactances (coil or capacitor) and one of the opposite type. The single component is connected in series with the signal lead and the two others are shunted to ground, one on either side of the series element.

Piezoelectric effect — The physical deformation of a crystal when a voltage is applied across the crystal surfaces.

PIN diode — A diode consisting of a relatively thick layer of nearly pure semiconductor material (intrinsic semiconductor) with a layer of P-type material on one side and a layer of N-type material on the other.

Plane polarization — Describes the orientation of the electric-field component of an electromagnetic wave. The electric field can be vertical or horizontal with respect to the earth's surface, resulting in either a vertically or a horizontally polarized wave. (Also called **linear polarization**.)

PN junction — The contact area between two layers of opposite-type semiconductor material.

Point-contact diode — A diode that is made by a pressure contact between a semiconductor material and a metal point.

Polarization — A property of an electromagnetic wave that tells whether the electric field of the wave is oriented vertically or horizontally. The polarization sense can change from vertical to horizontal under some conditions, and can even be gradually rotating either in a clockwise (right-hand-circular polarization) or a counterclockwise (left-hand-circular polarization) direction.

Potential energy — Stored energy. This stored energy can do some work when it is "released." For example, electrical energy can be stored as an electric field in a capacitor or as a magnetic field in an inductor. This stored energy can produce a current in a circuit when it is released.

Power — The time rate of transferring or transforming energy, or the rate at which work is done. In an electric circuit, power is calculated by multiplying the voltage applied to the circuit by the current through the circuit.

Power factor — The ratio of real power to apparent power in a circuit. Also calculated as the cosine of the phase angle between current and voltage in a circuit.

Product detector — A detector circuit whose output is equal to the product of a beat-frequency oscillator (BFO) and the modulated RF signal applied to it.

Pseudonoise (PN) — A signal that *appears* to be noise because of its random properties. A pseudonoise signal is used to produce (or receive) spread-spectrum communication.

Q — A quality factor describing how closely a practical coil or capacitor approaches the characteristics of an ideal component

Radiation resistance — The equivalent resistance that would dissipate the same amount of power as is radiated from an antenna. It is calculated by dividing the radiated power by the square of the RMS antenna current.

Radio horizon — The position at which a direct wave radiated from an antenna becomes tangent to the surface of the Earth. Note that as the wave continues past the horizon, the wave gets higher and higher above the surface.

Ratio detector — A circuit used to demodulate FM signals. The output is the ratio of voltages from either side of a discriminator-transformer secondary.

Reactance modulator — A device capable of modulating an ac signal by varying the reactance of an oscillator circuit in response to the modulating signal. (The modulating signal may be voice, data, video or some other kind.) The circuit capacitance or inductance changes in response to an audio input signal. Used in FM systems, this circuit acts as a variable reactance in an oscillator tank circuit.

Reactive power — The apparent power in an inductor or capacitor. The product of RMS current through a reactive component and the RMS voltage across it. Also called wattless power

Real power — The actual power dissipated in a circuit, calculated to be the product of the apparent power times the phase angle between the voltage and current.

Rectangular coordinates — A graphical system used to represent length and direction of physical quantities for the purpose of finding other unknown quantities.

Reflection coefficient (ρ) — The ratio of the reflected voltage at a given point on a transmission line to the incident voltage at the same point. The reflection coefficient is also equal to the ratio of reflected and incident currents.

Remote control — The operation of an Amateur Radio station using a **control link** to manipulate the station operating adjustments from somewhere other than the station location.

Repeater operation — Operation of an amateur station that simultaneously retransmits the transmission of another amateur station on a different channel or channels.

Resonant frequency — That frequency at which a circuit including capacitors and inductors presents a purely resistive impedance. The inductive reactance in the circuit is equal to the capacitive reactance.

Reverse bias — A voltage applied across a semiconductor junction so that it will tend to prevent current.

Root-mean-square (RMS) voltage — A measure of the effective value of an ac voltage.

Sawtooth wave — A waveform consisting of a linear ramp and then a return to the original value. It is made up of sine waves at a fundamental frequency and all harmonics.

Selective fading — A variation of radio-wave field intensity that is different over small frequency changes. It may be caused by changes in the material that the wave is traveling through or changes in transmission path, among other things.

Semiconductor material — A material with resistivity between that of metals and insulators. Pure semiconductor materials are usually doped with impurities to control the electrical properties.

Sensitivity — A measure of the minimum input signal level that will produce a certain audio output from a receiver.

Series-resonant circuit — A circuit including a capacitor, an inductor and sometimes a resistor, connected in series, and in which the inductive and capacitive reactances are equal at the applied-signal frequency. The circuit impedance is at a minimum, and the current is a maximum at the resonant frequency.

Signal-to-noise ratio — Signal input power divided by noise input power or signal output power divided by noise output power.

Silicon-controlled rectifier (SCR) — A bistable semiconductor device that can be switched between the off and on states by a control voltage.

Sine wave — A single-frequency waveform that can be expressed in terms of the mathematical sine function.

Single-sideband, suppressed-carrier signal — A radio signal in which only one of the two sidebands generated by amplitude modulation is transmitted. The other sideband and the RF carrier wave are removed before the signal is transmitted.

Skin effect — A condition in which ac flows in the outer portions of a conductor. The higher the signal frequency, the less the electric and magnetic fields penetrate the conductor and the smaller the effective area of a given wire for carrying the electrons.

Slope detection — A method for using an AM receiver to demodulate an FM signal. The signal is tuned to be part way down the slope of the receiver IF filter curve.

Slow-scan television (SSTV) — A television system used by Amateurs to transmit pictures within a signal bandwidth allowed on the HF bands by the FCC. It takes approximately 8 seconds to send a single black and white SSTV frame, and between 12 seconds and 4½ minutes for the various color systems currently in use.

Solar wind — Electrically charged particles emitted by the sun, and traveling through space. The wind strength depends on how severe the disturbance on the sun was. These charged particles may have a sudden impact on radio communications when they arrive at the atmosphere of the Earth.

Sporadic-E propagation — A type of radio-wave propagation that occurs when *dense* patches of ionization form in the E layer of the ionosphere. These "clouds" reflect radio waves, extending the possible VHF communications range.

Spread-spectrum (SS) communication — A communications method in which the RF bandwidth of the transmitted signal is much larger than that needed for traditional modulation schemes, and in which the RF bandwidth is independent of the modulation content. The frequency or phase of the RF carrier changes very rapidly according to a particular pseudorandom sequence. SS systems are resistant to interference because signals not using the same spreading sequence code are suppressed in the receiver. The FCC refers to this as a form of bandwidth-expansion modulation.

Spurious emissions — Any **emission** that is not part of the desired signal. The FCC defines this term as "an emission, on frequencies outside the necessary bandwidth of a transmission, the level of which may be reduced without affecting the information being transmitted."

Square wave — A periodic waveform that alternates between two values, and spends an equal time at each level. It is made up of sine waves at a fundamental frequency and all odd harmonics.

SSTV scan converter — A device that uses digital signal-processing techniques to change the output from a normal TV camera into an SSTV signal or to change a received SSTV signal to one that can be displayed on a normal TV.

Summer solstice — One of two spots on the orbital path of the Earth around the sun at which it reaches a point farthest from a horizontal plane extending through the center of the sun. With the north pole inclined toward the sun, it marks the beginning of summer in the northern hemisphere.

Surface-mount package — An electronic component without wire leads, designed to be soldered directly to copper-foil pads on a circuit board.

Switching regulator — A voltage-regulator circuit in which the output voltage is controlled by turning the pass element on and off at a high rate, often several kilohertz. The control-element duty cycle is proportional to the line or load conditions.

Sync — Having two or more signals in step with each other, or occurring at the same time. A pulse on a TV or fax signal that ensures the transmitted and received images start at the same point.

Telecommand — The FCC defines telecommand as a one-way transmission to initiate, modify, or terminate functions of a device at a distance. {§97.3 (a) (41)} A telecommand station is used to remotely control an amateur station through a radio link.

Thevenin's Theorem — Any combination of voltage sources and impedances, no matter how complex, can be replaced by a single voltage source and a single impedance that will present the same voltage and current to a load circuit.

Thyristor — Another name for a **silicon-controlled rectifier (SCR)**.

Top loading — The addition of inductive reactance (a coil) or capacitive reactance (a capacitance hat) at the end of a driven element opposite the feed point. It is intended to increase the electrical length of the radiator.

Toroid — A coil wound on a donut-shaped ferrite or powdered-iron form.

Transition region — An area around the semiconductor junction where the charge density is very small. This creates a potential barrier for current to flow across the junction. In general, the region is thin when the junction is forward biased, and becomes thicker under reverse-bias conditions. Also called the **depletion region**.

Traps — Parallel LC networks inserted in an antenna element to provide multiband operation.

Triac — A bidirectional SCR, primarily used to control ac voltages.

Tropospheric ducting — A type of radio-wave propagation whereby the VHF communications range is greatly extended. Certain weather conditions cause portions of the troposphere to act like a duct or waveguide for the radio signals.

Tunnel diode — A diode with an especially thin depletion region, so that it exhibits a negative resistance characteristic.

Unijunction transistor (UJT) — A three-terminal, single-junction device that exhibits negative resistance and switching characteristics unlike bipolar transistors.

Varactor diode — A component whose capacitance varies as the reverse-bias voltage is changed. This diode has a voltage-variable capacitance.

Velocity factor — An expression of how fast a radio wave will travel through a material. It is usually stated as a fraction of the speed the wave would have in free space (where the wave would have its maximum velocity). Velocity factor is also sometimes specified as a percentage of the speed of a radio wave in free space.

Vertical synchronization pulse — Part of a TV signal used by the receiver to keep the CRT electron-beam scan in step with the camera scanning beam. This pulse returns the beam to the top edge of the screen at the proper time.

Wavelength — The distance between two points that describe one complete cycle of a wave.

White noise — A random noise that covers a wide frequency range across the RF spectrum. It is characterized by a hissing sound in your receiver speaker.

Winter solstice — One of two spots on the orbital path of the Earth around the sun at which it reaches a point farthest from a horizontal plane extending through the center of the sun. With the north pole inclined away from the sun, it marks the beginning of winter in the northern hemisphere.

Zener diode — A diode that is designed to be operated in the reverse-breakdown region of its characteristic curve.

Zener voltage — A reverse-bias voltage that produces a sudden change in apparent resistance across the diode junction, from a large value to a small value.

Zero beat — The condition that occurs when two signals are at exactly the same frequency. The beat frequency between the two signals is zero. When two operators in a QSO are transmitting on the same frequency the stations are zero beat.

ABOUT THE AMERICAN RADIO RELAY LEAGUE

The seed for Amateur Radio was planted in the 1890s, when Guglielmo Marconi began his experiments in wireless telegraphy. Soon he was joined by dozens, then hundreds, of others who were enthusiastic about sending and receiving messages through the air—some with a commercial interest, but others solely out of a love for this new communications medium. The United States government began licensing Amateur Radio operators in 1912.

By 1914, there were thousands of Amateur Radio operators—hams—in the United States. Hiram Percy Maxim, a leading Hartford, Connecticut, inventor and industrialist saw the need for an organization to band together this fledgling group of radio experimenters. In May 1914 he founded the American Radio Relay League (ARRL) to meet that need.

Today ARRL, with more than 170,000 members, is the largest organization of radio amateurs in the United States. The League is a not-for-profit organization that:

- promotes interest in Amateur Radio communications and experimentation
- represents US radio amateurs in legislative matters, and
- maintains fraternalism and a high standard of conduct among Amateur Radio operators.

At League headquarters in the Hartford suburb of Newington, the staff helps serve the needs of members. ARRL is also International Secretariat for the International Amateur Radio Union, which is made up of similar societies in more than 100 countries around the world.

ARRL publishes the monthly journal *QST*, as well as newsletters and many publications covering all aspects of Amateur Radio. Its headquarters station, W1AW, transmits bulletins of interest to radio amateurs and Morse code practice sessions. The League also coordinates an extensive field organization, which includes volunteers who provide technical information for radio amateurs and public-service activities. ARRL also represents US amateurs with the Federal Communications Commission and other government agencies in the US and abroad.

Membership in ARRL means much more than receiving *QST* each month. In addition to the services already described, ARRL offers membership services on a personal level, such as the ARRL Volunteer Examiner Coordinator Program and a QSL bureau.

Full ARRL membership (available only to licensed radio amateurs) gives you a voice in how the affairs of the organization are governed. League policy is set by a Board of Directors (one from each of 15 Divisions). Each year, half of the ARRL Board of Directors stands for election by the full members they represent. The day-to-day operation of ARRL HQ is managed by an Executive Vice President and a Chief Financial Officer.

No matter what aspect of Amateur Radio attracts you, ARRL membership is relevant and important. There would be no Amateur Radio as we know it today were it not for the ARRL. We would be happy to welcome you as a member! (An Amateur Radio license is not required for Associate Membership.) For more information about ARRL and answers to any questions you may have about Amateur Radio, write or call:

ARRL Educational Activities Dept
225 Main Street
Newington CT 06111-1494
(203) 666-1541

Prospective new amateurs call:
800-32-NEW HAM (800-326-3942)

INDEX

T

U

V

NOTES

NOTES

NOTES

NOTES

NOTES

NOTES

FEEDBACK

Please use this form to give us your comments on this book and what you'd like to see in future editions.

Where did you purchase this book?

☐ From ARRL directly ☐ From an ARRL dealer

Is there a dealer who carries ARRL publications within:

☐ 5 miles ☐ 15 miles ☐ 30 miles of your location? ☐ Not sure.

License class:

☐ Novice ☐ Technician ☐ Technician with HF privileges
☐ General ☐ Advanced ☐ Extra

Name _____ ARRL member? ☐ Yes ☐ No

_____ Call sign _____

Daytime Phone ( ) _____ Age _____

Address _____

City, State/Province, ZIP/Postal Code_____

If licensed, how long? _____

Other hobbies _____

_____

Occupation _____

For ARRL use only	ACLM
Edition	4 5 6 7 8 9 10 11 12
Printing	1 2 3 4 5 6 7 8 9 10 11 12

From _____

_____

_____

EDITOR, ADVANCED CLASS LICENSE MANUAL
AMERICAN RADIO RELAY LEAGUE
225 MAIN ST
NEWINGTON CT 06111-1494

·· please fold and tape ···